C000292266

Birds of the Cotswolds

A new breeding atlas

Birds of the Cotswolds

A new breeding atlas

Iain Main and Dave Pearce *(principal editors)*
Tim Hutton *(species accounts)*
Peter Dymott
Martin Wright
Ian Ralphs

North Cotswold
Ornithological Society

Liverpool University Press

First published 2009 by
Liverpool University Press
4 Cambridge Street
Liverpool L69 7ZU

British Library Cataloguing-in-Publication data
A British Library CIP record is available

ISBN 978-1-84631-210-6 cased

Edited, designed and typeset by BBR (www.bbr-online.com)

Figure 1 and distribution map relief image
© Maps in Minutes™/Collins Bartholomew 2007.
All other cartography by Iain Main and BBR

Cover photographs:
front: Redstart by Andrew Carey (andrewcareyphotography.co.uk)
rear: Buzzard by Rob Brookes

Printed and bound by Gutenberg Press, Malta
Gutenberg Press prints for BirdLife Malta

The paper used for this book is FSC-certified and totally
chlorine-free. FSC (the Forest Stewardship Council) is
an international network to promote responsible
management of the world's forests.

Contents

Maps and diagrams

Tables

Preface

The North Cotswold Ornithological Society (NCOS) was formed in early 1983, on the initiative of Martin Wright following his move to the Cotswolds from neighbouring Oxfordshire, where he had been an active member of the Banbury Ornithological Society (BOS). After attempting with limited success to organize a Gloucestershire-wide Lapwing survey, he decided to form a birdwatching society with five other local birdwatchers, covering the more restricted area of the Cotswolds. Based largely on the model of the BOS, whose encouragement in the early days is gratefully acknowledged, the NCOS recording area was chosen to consist of twelve 10 km squares lying immediately to the southwest of the BOS area. With the later expansion into a thirteenth square, since 1991 the NCOS area of activity has covered the area bounded by Cirencester Park and Fairford to the south, and Chipping Campden and Shipston-on-Stour to the north. This includes the high ground of the Cotswolds, the escarpment, parts of the Severn Vale and the Vale of Evesham, and the upper reaches of the Thames Valley.

The founding objective of the society was to participate in fieldwork to increase the ornithological knowledge of the North Cotswolds in particular and, in general, to cooperate with other organizations in the collection of ornithological information. From the outset it has been a committed fieldwork organization, dedicated to the recording and surveying of birds in its area. These studies have been initiated both locally and in cooperation with national bodies such as the British Trust for Ornithology (BTO) and the Royal Society for the Protection of Birds (RSPB).

The fledgling society quickly decided to conduct a breeding atlas survey, in order to document the bird-life and habitats of its designated area. The survey lasted five years (1983–87) during which NCOS membership increased eightfold, and culminated in the publication of *An Atlas of Cotswold Breeding Birds* in 1990 (Wright *et al* 1990). In its second year the society participated in a survey of the important ornithological areas of the River Windrush and its tributaries. It also introduced a programme of 'Quarterly Walks' in which members walked the same route four times a year, recording all the birds they saw. In addition, in most years a species or habitat of particular interest was surveyed. In 1995 a Winter Random Square Survey was established (modelled on one designed by the BOS) in which members make two-hour counts in November and February, in randomly selected 1 km squares. That survey has helped to expose, for example, the steady and rapid increase of both Buzzard and Woodpigeon populations. In 1995 an informal Abundant Breeding Bird Survey was begun, using the methods of the first breeding atlas survey of Britain and Ireland (1968–72). Coverage was rather uneven, but the data hinted at distribution changes occurring since that survey and led to a decision to embark on another complete atlas survey. Exactly 20 years after the first Atlas, this book presents the results of that new survey.

Amongst national schemes, NCOS members have contributed to *The Atlas of Wintering Birds in Britain and Ireland*; *The New Atlas of Breeding Birds in Britain and Ireland 1988–1991*; a Corn Bunting survey (1992); a Lapwing survey (1994); a Rookery survey (1996); 'Winter Farmland Walks' (1999–2002) and the RSPB surveys of the Cotswold Hills Environmentally Sensitive Area of 1997 and 2002. At the time of writing, NCOS members were involved in survey work for the ongoing BTO/JNCC/RSPB Breeding Bird Survey as well as the new Gloucestershire county atlas and the BTO/BWI/SOC atlas for Britain and Ireland (both running from 2007 to 2011).

NCOS is a registered charity (number 1127045) governed by a written constitution. One of its stated objects is to publish, for public benefit, the results of its ornithological studies, in the form of an annual report and occasional special reports, of which this book is one.

NCOS is always pleased to welcome new members who are interested in recording. If you fall into that category, then contact the Membership Secretary for details at North Cotswold Ornithological Society, 60 King William Drive, Cheltenham, Gloucestershire, GL53 7RP.

Fieldworkers

The following is a list of NCOS members and other helpers who provided breeding records during the survey.

Gordon Avery
Geoffrey Bailey
Arthur Ball
Mabs Barratt
Terry Barratt*
Roger Batham
Bob Beale
Nigel Birch*
Jonathon Bowley
Chris Britton
Rob Brookes*
Beverley Bulford
Ian Cox
Chris Digger*
Duncan Dine*
Rebecca Dine
Peter Dymott*
Mark Farmer
Terry Fenton
John Fleming
Anne Francis
Caroline Gibbs
Robert Gibbs
Ray Goodwin
Mark Grieve
Fraser Hart
Ken Heron
Tim Hutton*
Derek Jackson
Phil Janka
Jenny Jones
Jo Jones
Mick Jones
Don Jowett*
Gordon Kirk
Roy Lester
Andy Lewis*
Chris McLaren
Iain Main‡
Jill Main
Tony Marx
Chris Oldershaw
Peter Ormerod
Steve Owen*
Dave Pearce*
Pam Perry
Tony Perry*
Don Potter
Ian Ralphs*
Edward Rice
John Sanders
Beryl Smith
Peter Smith
Sue Smith
Michael Stainton
Richard Tyler
Jeremy Voaden*
Andy Warren*
Tim Westlake
Janet Weyers
Keith White
Simon Willes
Lesley Withall
Martin Wright†*

* Acted as a 10 km square steward for all or part of the survey.
† Fieldwork coordinator
‡ Data coordinator

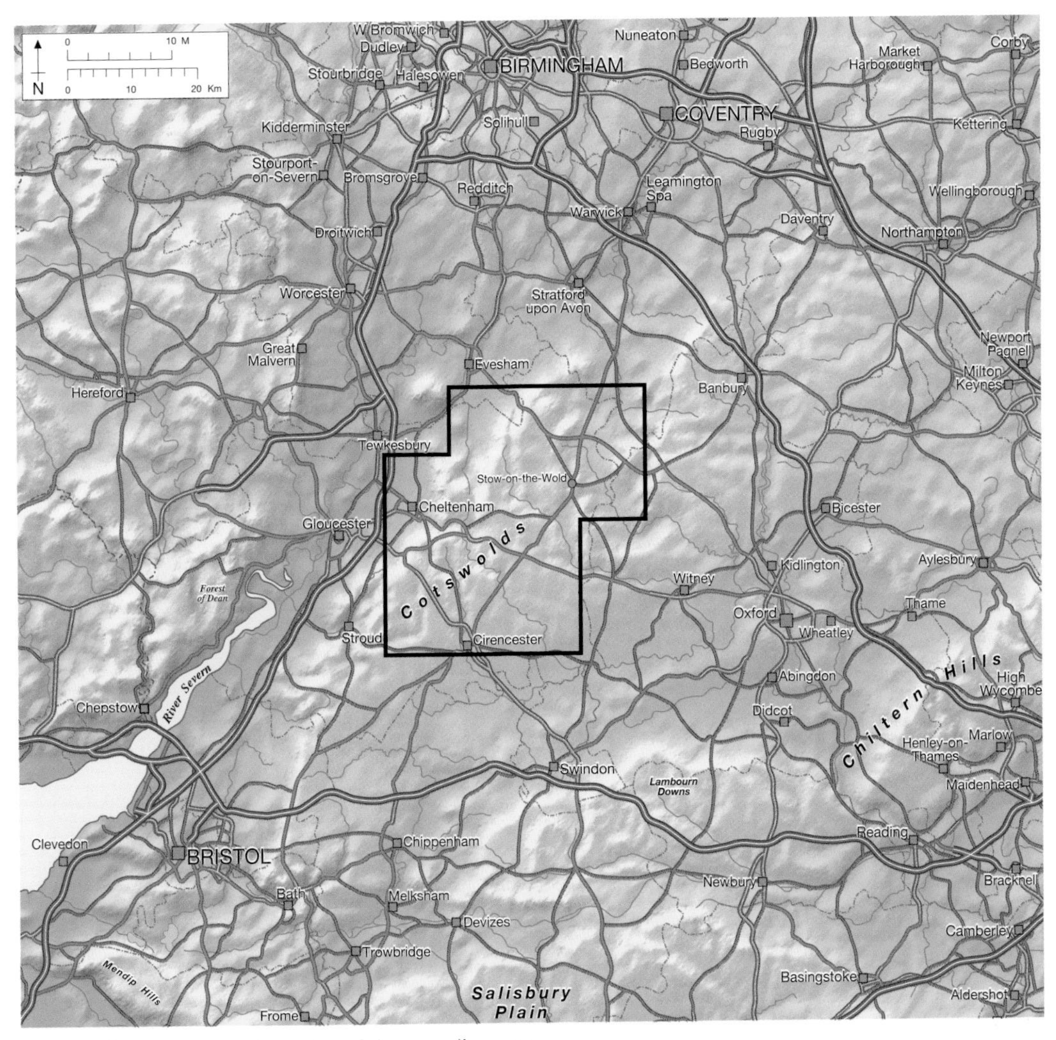

Figure 1. The geographical location of the recording area.

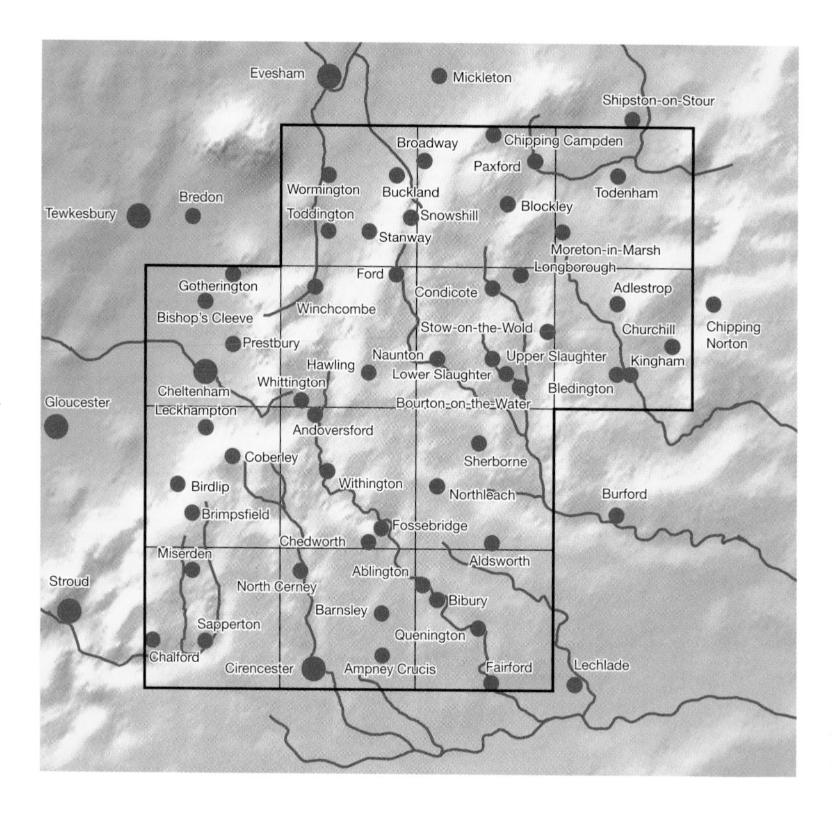

Figure 2. Towns and villages in and around the recording area.

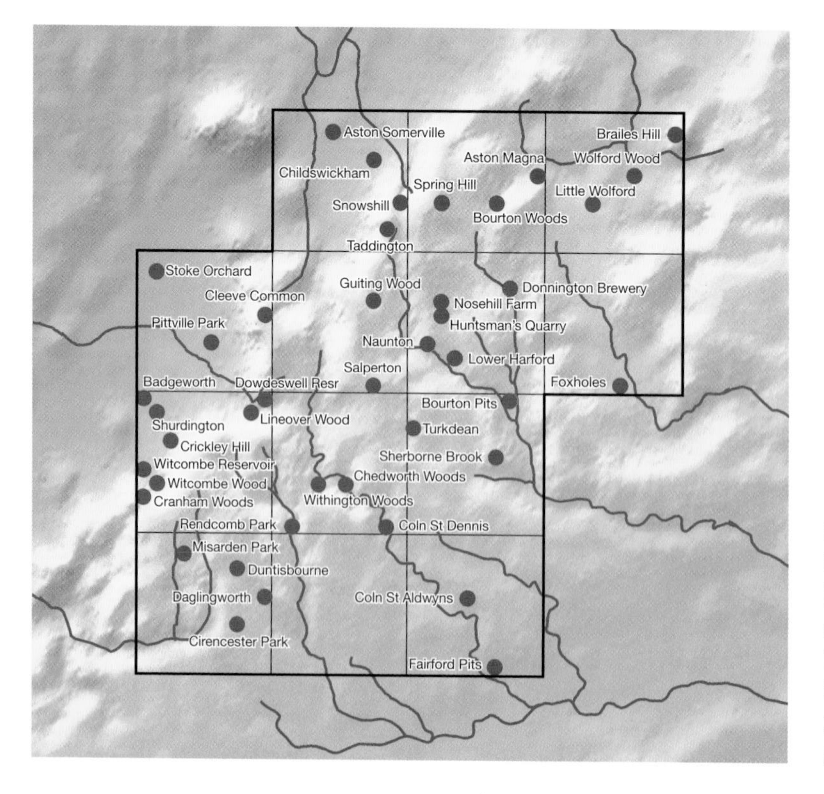

Figure 3. Sites mentioned in the species accounts (excluding places shown in Figure 2). The coordinates of all the places in Figures 2 and 3 are listed in Appendix B (page 222).

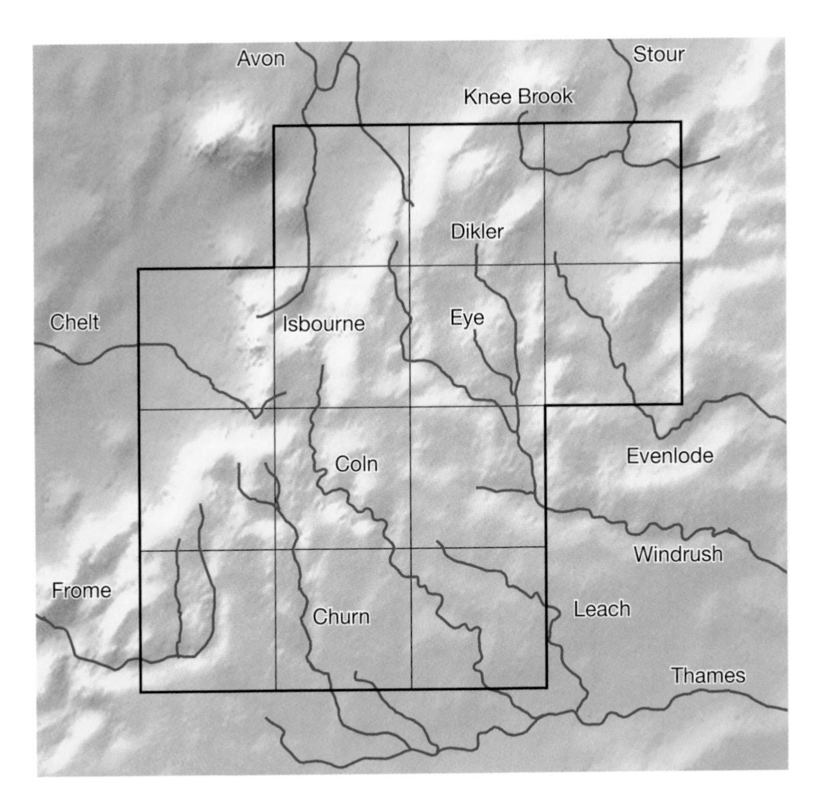

Figure 4. River systems in the recording area.

1. Introduction

The varied habitats in the Cotswolds support a good variety of inland breeding bird species. Shortly after its formation in 1983, North Cotswold Ornithological Society (NCOS) embarked on a major project to survey the distribution of breeding birds in its area. Fieldwork was carried out during the five breeding seasons 1983 to 1987, and the results were published in 1990 in the form of a distribution atlas (Wright *et al* 1990). That publication has served as a valuable starting point for monitoring subsequent changes.

This book presents the results of a new survey, carried out from 2003 to 2007, with the object of revealing changes that had occurred in the two decades since the previous Atlas. The start of the new survey coincided with coordinated national and local initiatives aimed at halting, and ultimately reversing, the continuing decline of many plant and animal species. Locally, Gloucestershire Biodiversity Partnership, of which NCOS is a member, had recently been launched and a scheme of conservation strategies—the Gloucestershire Biodiversity Action Plan (BAP)—was drawn up, defining a list of selected 'priority species' (Miller 2000). Ten of the priority bird species (Grey Partridge, Lapwing, Turtle Dove, Skylark, Song Thrush, Tree Sparrow, Linnet, Bullfinch, Reed Bunting and Corn Bunting) are farmland birds breeding in the Cotswolds and to this list may be added Spotted Flycatcher, one of the BAP woodland species. These 11 species had all declined nationally by at least 50% during the previous quarter-century (Baillie *et al* 2007). In an attempt to halt the decline, this group has been targeted by means of action plans for the various relevant habitats. It is hoped that this book will provide a new benchmark for monitoring the success of ongoing BAP activities.

2. The landscape and bird habitats of the Cotswolds

Sandwiched between Lias Clay below and Oxford Clay above, the dominant bedrock of the Cotswolds is Oolitic Limestone. As the strata dip towards the southeast, erosion has cut across them obliquely, wearing away the clays to a greater extent than the harder limestone. The result is an eroded limestone escarpment facing northwestwards, looking out over the Lias Clay of the Severn Vale, and dipping gently southeastwards—the dip slope—to give way eventually to Oxford Clay. At its most impressive the scarp edge rises from 60 m in the Vale to 330 m (or just over 1,000 feet) on Cleeve Common.

These diverse features of the Cotswolds combine to bring what is in many respects an upland landscape to southern England. Contrary to a common perception, the area is not a rural idyll caught in a time warp, but as much a working farmed landscape as anywhere else in southern England. Historically, much store has been set by its openness and unique nature, which culminated in designation of the Cotswolds Area of Outstanding Natural Beauty (AONB) in 1966.

The avifauna contains no single species typical of the Cotswolds landscape as a whole, but rather a suite of species loosely defined as farmland birds, which are widely (although often patchily) distributed over the area. Farmland birds are generally recognized as one of the fastest declining and most threatened bird communities in Britain.

In describing the various habitats within the recording area it is convenient to consider eight separate areas with their own characteristic landscape features as shown schematically in Figure 5. It should be noted that the boundaries between these areas are by no means as obvious or definite as they are on the diagram.

The scarp

The land use of the scarp is dominated by a mixture of permanent pasture (land that has been under grass for at least five years and has not been ploughed for other crops in that time) and woodland, much of the latter being valuable and diverse ancient semi-natural woodland. The survival here of so much woodland is due simply to the difficulty of intensively managing such a steeply sloping land form. Naturally exposed rock outcrops are largely absent from the scarp, with any exposed rock, where it does occur, being the result of former quarrying.

Ancient woodland on the scarp can be split into two distinct types: the beech-dominated stands southwest of Cheltenham, and the more mixed semi-natural stands with predominantly ash, oak, hazel (and occasionally lime) from Cheltenham northeastwards. Both retain moderate amounts of dead-wood habitats and hold a good variety of woodland bird species. The top of the scarp is a favoured place for singing Tree Pipits, and Ravens can often be seen using the wind to good effect.

Although springs and small streams arise frequently on the escarpment, the scarp is significantly breached by only three watercourses. In the north, the River Isbourne rises on Cleeve Common just above the small dammed pool commonly known as the Wash Pool, and flows almost directly north to enter the River Avon at Evesham. The Chelt has its source near Whittington and flows west through Cheltenham, eventually joining the Severn. The third breach is by the River Frome, which rises near Brimpsfield and flows south through a valley, before turning west near Sapperton.

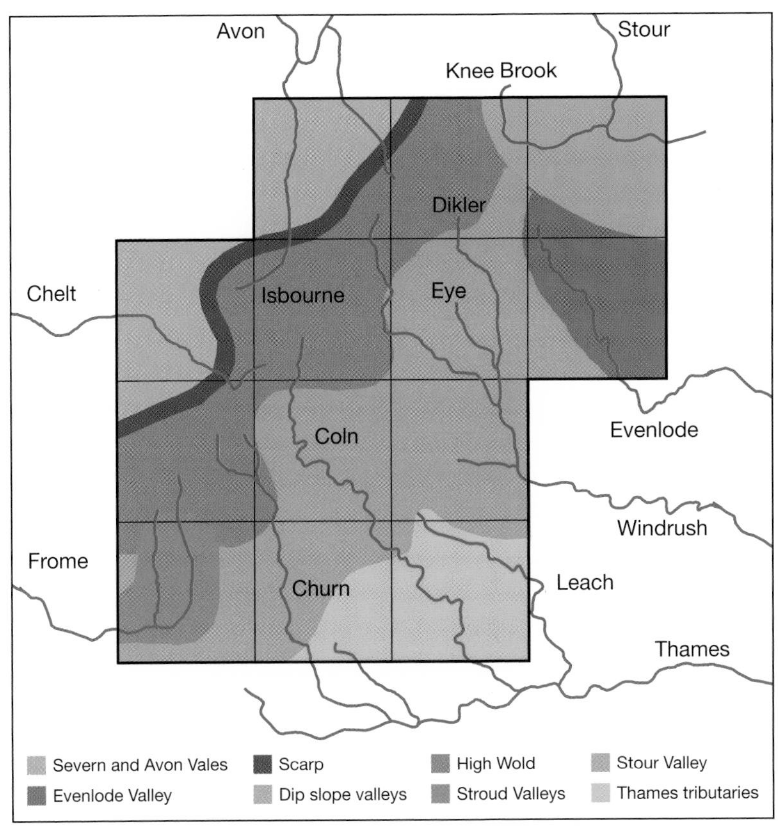

Figure 5. Schematic map of eight distinct landscapes within the recording area, showing river systems. The boundary between the High Wold and the dip slope valleys has been arbitrarily set at an altitude of approximately 250 m, and that between the dip slope valleys and the Thames tributaries at approximately 150 m.

ANDY LEWIS

The scarp and Lineover Wood.

DAVE PEARCE

The Wash Pool, Cleeve Common.

ANDY LEWIS

The High Wold near Condicote.

The High Wold

The High Wold is the highest part of the dip slope, its large open fields enclosed by low dry stone walls (and more recently wire fences) forming an open landscape. With its relatively well drained, light soils, the dominant land use is large-scale arable farming, a practice somewhat softened by the constancy of the two great classics of Cotswolds rural architecture: its old dry stone walls and derelict or converted barns. Within this open landscape there are blocks of plantation and semi-natural woodland such as Bourton and Guiting Woods with Withington and Chedworth Woods further down the dip slope.

The single notable expanse of permanent pasture still retained on the High Wold is the Cleeve Hill complex, near Cheltenham. Its greatest area (just over 400 ha) comprises Cleeve Common, a Site of Special Scientific Interest (SSSI) containing wide expanses of unimproved limestone grassland. The Common gives a glimpse of what much of the High Wold would have looked like 200 years ago, with large expanses of short-cropped grassland with scattered scrub. The Common today is still grazed by sheep and cattle, but less heavily than before and is therefore dominated by taller (and even semi-improved) grass swards, with dense and occasionally continuous areas of scrub which have to be mown in rotation to keep them in check. Meadow Pipits are found here at high densities, while Stonechats and Grasshopper Warblers are also present on the gorse-covered slopes.

The dip slope valleys

Slightly less exposed than the High Wold are the river valleys of the dip slope, where the limestone falls gently down to the southeast and gradually disappears below the tertiary sand and gravel deposits of the upper Thames Valley. Apart from the three northward- and westward-flowing Rivers Isbourne, Chelt and Frome mentioned

ANDY LEWIS

Chedworth Woods.

above, the dip slope is dissected by a series of streams and small rivers draining southeastwards into the Thames. These are the Churn, the Coln, the smaller Leach, and the Windrush with its tributaries the Dikler and the Eye. They tend to occupy shallow, broad valleys, but frequently these are quite steeply incised and meandering, especially in their upper (and occasionally dry or winterbourne) reaches. The waters here are fast-flowing, clear and clean. Dippers and Grey Wagtails are locally common on these reaches, and Little Egrets are increasingly frequent visitors to the southeast corner of the Coln Valley.

The dominant land use in the valley floors and sides is permanent pasture (mostly improved) with arable only on more gently sloping ground on the margins of the larger valleys. The valley sides are moderately or patchily wooded, but in some areas larger blocks are found, of which the extensive woodland of Cirencester Park is the most notable. Although often dominated by plantation forestry, these woods retain some semi-natural content and character.

The Thames tributaries

In the southeast of the recording area, the dip slope becomes open and mainly arable, with large fields, clipped hedgerows and wide horizons. The Cotswolds character is maintained by its limestone cottages and clear, fast-flowing streams, all tributaries of the Thames. The open farmland is interspersed with blocks of both broadleaved and coniferous woodland and the arable landscape is broken by river valleys, often steep sided, with permanent pasture beside the southeasterly flowing streams. These meadows are now almost entirely improved in character and only patchily grazed by cattle. Lapwing breed on the higher ground of this area and the flatter ground elsewhere. The area also holds good numbers of Barn Owls which are absent from the hilly areas to the immediate west.

Since the 1983–87 Atlas survey the area of open water in the Cotswold Water Park (eastern section) has increased and matured, with 13 lakes now lying wholly or partially within the recording area and occupying well over 1 km^2 in all. This area is referred to as Fairford Pits. Access to these waters can be difficult, as various plans for the commercial usage of the sites come and go. Great Crested Grebes, Little Grebes and Tufted Ducks are

IAN RALPHS

The River Churn above North Cerney.

IAN RALPHS

Barnsley Warren.

regular at these sites, but waders that relied on drained and predominantly bare ground protected from disturbance are now largely absent.

The Stroud Valleys

Occupying the southwest of the recording area, the northern Stroud Valleys are deep and often wooded, drained by the River Frome and its tributaries. Similar to the scarp in character, the hillsides and valley bottoms contain ancient broad-leaved woodland, with smaller areas of flower-rich hay meadows. With little or no arable land in the valley bottoms, however, many typical farmland species such as Yellowhammer and Skylark are absent. The Frome Valley is almost continuously wooded along its length by some of the most diverse and high-quality ancient semi-natural woodland anywhere in Britain. Arable farming prevails on the open and exposed tops, and the fast-flowing rivers host Dippers and Grey Wagtails.

The Evenlode Valley

Located to the north and east of Stow-on-the-Wold, the Evenlode Valley lies on Lias clay dividing the limestone of the Cotswolds from the ironstone to the east, and so begins to lose the typical Cotswolds character of clear, fast-flowing streams and honey-coloured stone cottages. The River Evenlode has its source near Moreton-in-Marsh, and flows gently southeastwards into Oxfordshire to join the Thames at Oxford. Within the recording area it is slow-flowing and meandering, and by the time it leaves our area at Kingham it is a substantial stream three to four metres wide. The valley itself is wide and open, and prone to flooding in winter. For this reason it is mainly a mixture of permanent pasture and arable, with small woods at a higher level. There are two significant areas of broad-leaved woodland: Wolford Wood northeast of Moreton-in-Marsh, and Foxholes Nature Reserve south of Kingham, both at one time the haunts of Willow Tits and Woodcocks. The permanent pasture of the valley bottom holds Reed Buntings, some Yellow Wagtails and, in recent years, a few pairs of Curlew.

The Severn and Avon Vales ('The Vale')

The geology of the Severn and Avon Vales is completely different from that of the escarpment, being sticky Lias clay covered in parts by sandy drift. Settlements are often placed on these areas of drift, and they are associated with market gardens to the west of Cheltenham and in the Avon Vale (the Vale of Evesham). Away from the scarp and its debris the land is generally flat, lying between 20 and 60 m above sea level. In both areas there are few copses or woods, and no sizeable bodies of open water apart from parks in Cheltenham and at Wormington.

Owing to the general lack of cover the bird-life of the pastures is undistinguished. The only relief is the presence of two landfill sites near Bishop's Cleeve which attract vast numbers of gulls throughout the year and have contained breeding Shelducks, Tufted Ducks and Little Ringed Plovers.

The Stour Valley

Dominated by Brailes Hill, the landscape in the northeast of the recording area becomes more rolling and open, with a 'Midlands' feel. These are the clay pastures, typified by arable fields bordered by mixed-species hedgerows, horse paddocks and large shooting estates. The main river is the Stour, a modest watercourse which, along with its tributary the Knee Brook, flows northwards to join the Avon near Stratford. Typical farmland birds abound: there are good populations of Lesser Whitethroat and Tree Sparrow, as well as several pairs of Turtle Dove and Barn Owl.

Other lakes and reservoirs

Standing water is on the whole scarce in the Cotswolds, with only the dip slope valleys retaining water where they have been dammed artificially. Probably the best example of these small valley lakes is at Donnington Brewery, but there are others associated with the fine old houses and estates. Lakes have also been created at Bourton-on-the-Water as gravel has been dug from the broader valley floors of the Dikler and Windrush, leaving a series of eight lakes with a variety of small pits. There are only two substantial reservoirs in the recording area, both at the foot of the scarp near Cheltenham. Although both are now decommissioned, Witcombe Reservoir is being managed as a sport fishery and Dowdeswell Reservoir as a nature reserve.

These lakes and reservoirs hold the expected waterfowl, and a few feeding Grey Herons and Cormorants. Several pairs of Great Crested and Little Grebe breed, and the exotic Red-crested Pochard is increasingly found at Bourton-on-the-Water.

Habitat changes since the mid-1980s

Breeding birds in the Cotswolds, as elsewhere, live in a habitat heavily influenced by human activity. In farming, the most obvious recent change in land use has been the 'set-aside' policy which was introduced in 1988 and in some cases impacted on the appearance of the landscape. A few species prospered: Lapwings found undisturbed areas to breed in, and Barn Owls may have benefited from the increased vole population. Set-aside payments were suspended in 2007.

Recent years have seen the introduction of a series of environmental stewardship schemes for farmers in England. These replaced simple crop subsidies with annual support payments tailored to individual farmers, reflecting the environmental impact of their activities as well as food production. Initiatives such as leaving wider arable field margins, delaying the ploughing-in of stubble, and the provision of 'beetle banks', were incorporated in a strategy for more sustainable farming practice. It is too early to tell how the new support arrangements will affect farmland birds, but at the time of writing (November 2008) impressions appear to be generally favourable. Populations of seed-eating birds such as Linnets seem to be on the rise after years of decline.

The ratio of pasture to arable land does not seem to have altered much over the period and the crops grown are much the same. There appear to be fewer cattle grazing in pastures although

ANDY LEWIS

The Evenlode Valley.

ANDY LEWIS

Witcombe Reservoir.

ANDY LEWIS

Stanway.

ROB BROOKES

Longborough.

there are probably more sheep now, and certainly more horses. Silage is increasingly replacing hay as fodder, possibly having an effect on ground-nesting birds. Sheep have an important part to play in maintaining grazed woodlands, a favourite habitat of Redstarts. Other changes have included the replacement of some hedgerows with barbed wire, and the drainage of wet areas (the latter being partially offset by the creation of many farm ponds).

In recent years government schemes have supported both the replacement of conifers by broad-leaved trees and the creation of new plantations. Trees have been planted as shelter-belts or in corners of fields inaccessible to modern farm machinery. All this means more woodland now and for the future but, on the other hand, steep valley slopes have also been planted, sometimes at the expense of flower-rich banks and habitat available for birds such as Tree Pipits. Another changing aspect of woodland habitat has been the growth of Grey Squirrel and deer populations which continue to be of concern to estate owners and foresters.

The Cotswolds are renowned for their stone houses, churches and traditional farm buildings. This has influenced the local bird-life, providing nesting sites for Swifts and House Sparrows in stone tiled roofs, and Swallows, Pied Wagtails and Barn Owls in farm buildings. In the past quarter-century a lot has changed: many barns and farm complexes have been converted to expensive housing, and villages and small towns have become smarter and tidier. Whilst there are very few houses in the Cotswolds nowadays which have not been modernized, their traditional exteriors have usually been maintained, leaving the picture-postcard image of the area unscathed. The rise in horse riding has also had an impact, with many villages having adjacent horse paddocks providing short grass for thrushes and Starlings, and stables providing nest sites for Swallows displaced by barn conversions. There has also been an increase in the feeding of garden birds, which has greatly helped some species: tits, finches and the Great Spotted Woodpecker in particular.

Overall, changes have been effected but they are generally subtle in nature so that the visual landscape of the Cotswolds and its habitats have not altered markedly over the last quarter-century.

3. Schematic bird habitat maps

Chapter 2 has described the general features of the landscape and habitats in the recording area. It is useful to consider how a given habitat is distributed across the area, using the survey mapping unit of a tetrad (a 2 km by 2 km square of the National Grid). By inspecting the Ordnance Survey (OS) 1:25,000 scale maps of the area, an estimate was made of the areas of each tetrad occupied by each of three habitat types: woodland, buildings and lakes. Also relevant are the maximum altitude, and perhaps the average steepness of the terrain. These five features are displayed as dot maps, similar to the breeding distribution dot maps discussed in Chapter 5, but using five or six symbol sizes instead of three. The size of the dot in a given tetrad indicates the area occupied by the habitat type within that tetrad (or the maximum altitude or steepness in the tetrad). The choice of quantity ranges represented by the different dot sizes aims to give an informative visual representation of the variation across the area. In each diagram the quantities are normalized to make the largest dot fill the tetrad. These maps have been compared with the species dot maps and any correlation between the presence of a scarce species and a particular habitat is discussed in the species accounts.

Woodland

Almost every tetrad in the recording area contains some woods or copses (Figure 6). At least two-thirds of tetrads have woods covering at least 5% of the total area of the tetrad, and in one-third of tetrads the coverage is at least 10%.

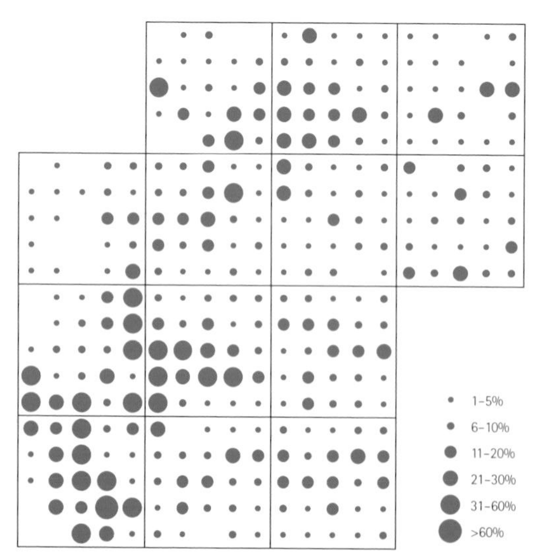

Figure 6. Habitat map: woodland. The symbol sizes indicate the percentage of each tetrad occupied by woods or copses. The most heavily wooded tetrad at 90% is in Cirencester Park.

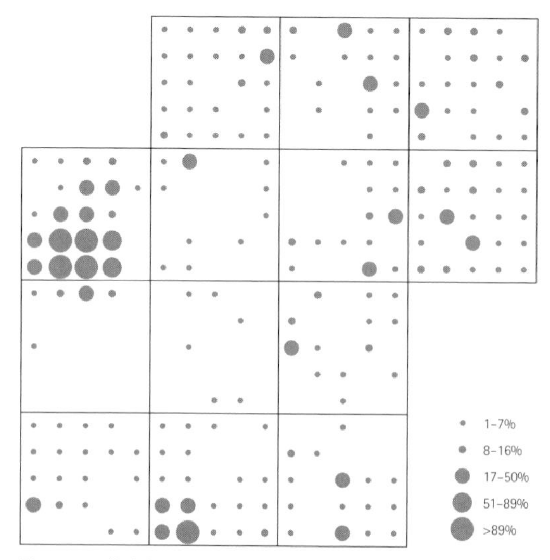

Figure 7. Habitat map: buildings. The symbol sizes indicate the percentage of each tetrad which may be classified as built-up. Four of the Cheltenham tetrads have built-up areas exceeding 90%.

Buildings

Outside Cheltenham and Cirencester, the area is largely rural, with few towns and villages (Figure 7). More than one-third of the tetrads contain no villages and another third contain a built-up area no greater than 5%. Excluding Cheltenham and Cirencester, only 10% of tetrads have a built-up area of 10% or more.

Lakes and other standing waterbodies

Perhaps surprisingly, half of the tetrads in the recording area have enough water (standing water or streams) to be shown on the 1:25,000 scale OS map. There are several major reservoirs and gravel-pit workings providing sizeable areas of water which are important for breeding swans, ducks, geese and grebes. One-fifth of the tetrads have some ponds or lakes larger than about 1 ha, *ie* 0.1 km by 0.1 km. These are shown on the habitat map (Figure 8) but there are also a number of smaller lakes, all of which provide still bodies of water.

Rivers and streams

Running water may attract different species compared with still water–in particular Dippers and Kingfishers. Rivers and named streams have been shown in Figure 4, but deciding the length of a small stream and any associated ditches in a given tetrad is difficult. Streams often split into several parallel paths, or meander, and the point upstream when it can no longer be considered as a separate habitat is hard to judge. Therefore no habitat map is shown, but a rough estimate is that one-quarter of tetrads have a stream which is at least the distance across a tetrad, *ie* 2 km, and more than one-third have some running water.

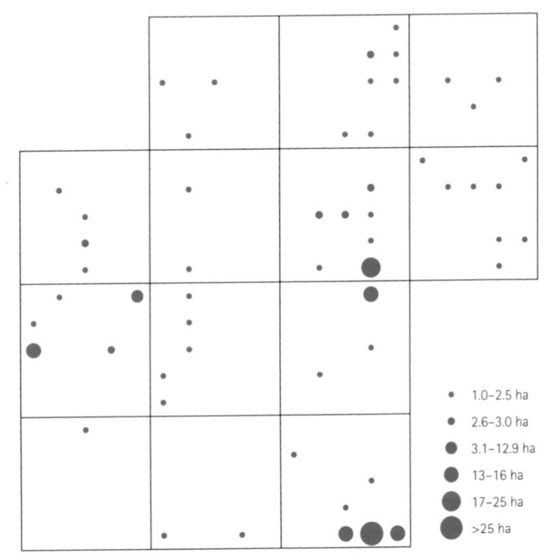

Figure 8. Habitat map: lakes and other standing waterbodies. The symbol sizes indicate the area in hectares of lakes or other waterbodies in each tetrad. (The area of a tetrad is 400 ha.) The largest symbol represents lakes near Fairford which occupy nearly 20% of the tetrad.

Countryside

To a first approximation, the area of a tetrad which is not taken up by woods, buildings or lakes is open countryside, mainly fields in a range of sizes from small paddocks to large fields of up to 25 ha (0.5 km by 0.5 km) in the southeast of the area. On this basis, four-fifths of the 13-square recording area is open countryside. This is reflected in the habitat maps, with most tetrads containing some woodland and half containing some still or running water. The only exceptions are first, one tetrad which is almost entirely woodland and secondly, five tetrads which are essentially built-up (although containing some parks, *etc*). The Atlas observer will have tried to visit the range of habitats in the tetrad, and so it is not surprising that, at least for relatively common birds, most species do not show a strong correlation with the habitat maps. Exceptions to this are mentioned in the species accounts.

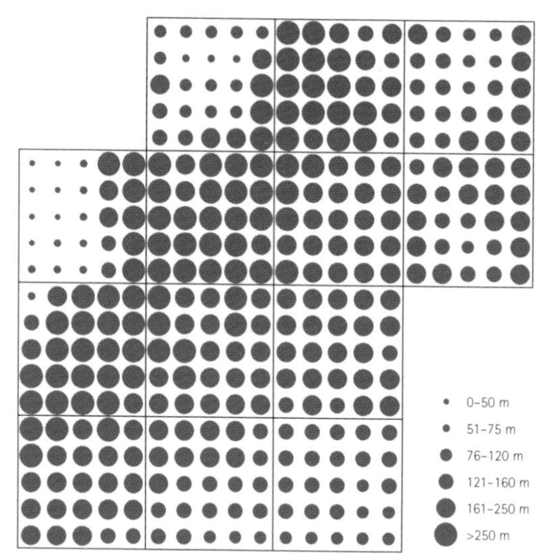

Figure 9. Habitat map: altitude. The symbol sizes indicate the maximum altitude in each tetrad. The highest area is Cleeve Common with an altitude of 330 m.

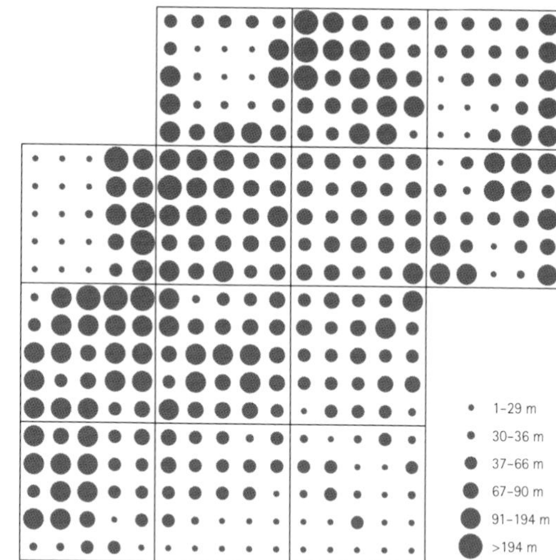

Figure 10. Habitat map: steepness. The symbol sizes indicate the relative differences between the maximum and minimum altitudes within each tetrad. As expected, the steepest tetrads are along the scarp, the maximum change in height over a tetrad being 245 m.

Terrain: altitude and steepness

The height above sea level has a significant influence on the distribution of breeding species. The altitude habitat map (Figure 9) shows the Vale, the scarp and the gentle reduction in height towards the southeast (the dip slope). Some species such as Stonechats, Redstarts, pipits and Grasshopper Warbler prefer to breed on the higher parts of the Cotswolds, but of course other aspects of the habitat must be just right as well. One species, the Tree Sparrow, is now distinctly uncommon away from the dip slope.

Some species appear to prefer hill *slopes*. A habitat map providing an impression of steepness (Figure 10) was obtained by plotting the average slope across a tetrad (maximum altitude minus minimum altitude within the tetrad). The scarp shows up well on such a map: this is generally an area of very rough pasture and, while avoided by Skylarks, is attractive to Tree Pipits.

4. Subjective impressions of bird population changes in the Cotswolds since the 1983–87 survey

The survey measured distributions, not population densities, of birds that breed in the Cotswolds. The two are of course interrelated, and it is useful to precede the quantitative results of the new survey with a brief qualitative description of changes in the numbers of some species since the end of the 1983–87 survey, as judged by members who took part in both surveys. These notes are written in autumn 2008, *ie* one year after the end of the new survey.

The species that has increased most obviously is the Buzzard. In the early 1980s it was a scarce bird, usually requiring a visit to Wales to be sure of spotting one. They did occur in the Atlas recording area, but only in a few scattered locations and, perhaps because of their scarcity, they were rarely seen in the territorial displays that are now familiar. For many years in the 1980s we waited for Buzzards to take up residence in the hills and valleys of the Cotswolds, just as we anticipate the arrival of Red Kites today. Buzzards now appear to be as common in the east of our area as in the west, and to breed throughout the area.

Following the Buzzard in becoming a familiar, if less common, breeding bird is the Raven. Having not been observed at all in the earlier survey, gradually Ravens in flight (regarded as non-breeders) were recorded in the Stroud Valleys, and now most 10 km squares probably hold several breeding pairs.

There are few other species which have increased in numbers, but four are worth mentioning, namely Woodpigeon, Blackcap, Greenfinch and Goldfinch. Woodpigeons, while numerous in the 1980s, have become so common and familiar in every habitat (particularly urban) that they could give the impression of being the area's most abundant bird. They certainly must have increased markedly over the years and, because of their size and year-round song, feature very prominently in any survey.

Blackcaps also appear to be doing well, even if heard more often than seen. They seem to be far more common in their prime habitat of deciduous woodland, and are moving into tall hedgerows and gardens. Winter visitors from central Europe have also become familiar garden birds, so that this species is now commonly seen throughout the year.

Greenfinches appeared to be almost entirely urban in the 1980s. Now they can be encountered in hedgerows and conifer plantations far from human habitation, a sign of increased numbers. By contrast, although Goldfinches have also become more common, it is in suburban gardens that the biggest expansion has been seen, possibly aided by the increasing pastime of feeding garden birds.

Unfortunately, many more species appear to have declined than increased, some quite dramatically. Those that have suffered major declines include Lapwing, Grey Partridge, Corn Bunting, Willow Tit, Lesser Spotted Woodpecker, Nightingale, Hawfinch, Turtle Dove, Tree Pipit, Cuckoo, Starling, House Sparrow and Tree Sparrow.

Lapwings still breed in some open arable areas, perhaps benefiting from set-aside practices, but 20 years ago they were widespread, nesting regularly in spring-sown cereals and other crops. Two other species of open country in serious trouble are Grey Partridge and Corn Bunting. Calling Grey Partridges were once a familiar Cotswolds sound, but now, despite apparently having large areas of suitable habitat, they are rarely encountered. Corn Buntings, although still fairly common in the high ground west of Stow-on-the-Wold, have declined dramatically outside this core area.

It seems curious that several woodland species also come into the declining category. Although many woods appear to be in the same condition as they were in the 1980s, and new woodland has been created, they appear to support fewer birds. Spotting a Willow Tit has become likely in only a few well-watched areas, whereas previously most broad-leaved woods would support a pair. A similar situation applies to the Lesser Spotted Woodpecker, which was once found in a variety of woodland habitats, but has now become one of our scarcest breeding birds.

Perhaps our biggest loss, though, is the Nightingale. In the 1980s several could be heard singing in Cirencester Park, and woodland north of Moreton-in-Marsh. By the 1990s these had disappeared, but we still had a healthy population in the Vale south of Broadway. Sadly these may have now gone, and the Nightingale, along with the Hawfinch, may well have the distinction of being one of only two species to have become extinct as breeding birds in the North Cotswolds since the 1980s.

Several other species appear to have fared badly. Turtle Doves were not uncommon in the 1980s, often occupying the same habitat as Tree Pipits—uncultivated slopes with some trees and new plantations. The Turtle Dove has almost completely disappeared while the Tree Pipit is mainly now found along the high and hilly reaches of the scarp.

The escarpment around Cleeve Common has probably achieved the distinction of being one of the few areas Cuckoos are regularly found. It is sad that a bird so familiar to all now requires a special journey to hear it.

In the 1980s the Starling and House Sparrow could be described as abundant. They are now decidedly scarce in many rural areas. Starlings have abandoned most woodland sites and there are now villages without breeding birds. Tree Sparrows have suffered more serious losses and now have a very localized distribution in the Cotswolds.

Many of these changes could not have been forecast, and we hope that this Atlas will serve as a new benchmark for the distribution of the breeding birds in our region. With a changing climate, and the series of harsh snowy winters during the early 1980s now only a distant memory, the future could bring even greater changes.

5. The survey in outline

The survey methods are fully described in Chapter 8, but to enable the species accounts to be read easily the basic details are given here.

In order to make the comparison as reliable as possible, the survey methods were based closely on those of 1983–87 in terms of mapping unit, categories of breeding and criteria to indicate breeding status. The mapping unit adopted was a 2 km by 2 km square of the National Grid known as a tetrad. However, the recording area was larger, consisting of thirteen 10 km grid squares (Figure 11) rather than the 12 squares used previously. It should be noted that whenever quantitative comparisons are made between the two surveys, the records in the thirteenth square are ignored.

Evidence of breeding was divided into the same three categories as the previous survey and given the shorthand descriptions *possible* (category 1), *probable* (category 2) and *confirmed* (category 3) breeding; and with one minor exception (described in Chapter 8) the same criteria were used to indicate breeding status for the new survey. The detailed criteria for each breeding category are shown in Figure 12. Possible breeding was indicated by the presence of the species in breeding habitat in the breeding season, but with no other signs of breeding activity. Observers were reminded that some species such as Buzzard, Swift, Swallow

	SP03	SP13	SP23
SO92	SP02	SP12	SP22
SO91	SP01	SP11	
SO90	SP00	SP10	

Figure 11. The Atlas recording area. Square SO92 was not included in the 1983–87 survey.

Possible Breeding (category 1)

H	Species observed in breeding season in possible nesting **HABITAT**

Probable Breeding (category 2)

S	**SINGING** male(s) present (or breeding calls heard) in the breeding season
P	**PAIR** observed in suitable nesting habitat in the breeding season
T	Permanent **TERRITORY** presumed through registration of territorial behaviour (song, *etc*) on at least two different days a week or more apart at the same place
D	**DISPLAY** and courtship
N	Visiting probable **NEST-SITE**
A	**AGITATED** behaviour or **ANXIETY** calls from adult
I	Brood patch on adult examined in the hand, indicating probably **INCUBATING**
B	**BUILDING** nest or excavating nest hole

Confirmed Breeding (category 3)

DD	**DISTRACTION DISPLAY** or injury feigning
UN	**USED NEST** or egg-shells found (occupied or laid within period of survey)
FL	Recently **FLEDGED** young (nidicolous species) or downy young (nidifugous species)
ON	Adults entering or leaving nest-site in circumstances indicating **OCCUPIED NEST** (including high nests or nest holes, the contents of which cannot be seen) or adult seen incubating
FY	Adult carrying **FOOD** for **YOUNG** or faecal sac
NE	**NEST** containing **EGGS**
NY	**NEST** with **YOUNG** seen or heard

Figure 12. Criteria for breeding records.

NCOS breeding atlas | Recording form issue 3.2

10 km square and tetrad	SP02	J	Village or landmark in tetrad	Lone Farm
Year		2006	Observer	A.N. Other
Early (E) or whole season (S)		S	('Early' means before the end of May)	

Species	Criterion	Category	Species	Criterion	Category
Little Grebe			Nightingale		
Great Crested Grebe			Redstart		
Grey Heron			Stonechat		
Mute Swan			Wheatear		
Canada Goose			Blackbird	ON	3
Mallard	FL	3	Song Thrush	FL	3
Tufted Duck			Mistle Thrush	T	2
Sparrowhawk	B	2	Grasshopper Warbler		
Buzzard	D	2	Sedge Warbler		
Kestrel	P	2	Reed Warbler		
Hobby			Lesser Whitethroat	H	1
Red-legged Partridge	P	2	Whitethroat	T	2
Grey Partridge			Garden Warbler		
Quail			Blackcap	A	2
Pheasant	P	2	Wood Warbler		
Moorhen	P	2	Chiffchaff	P	2
Coot	P	2	Willow Warbler	P	2
Little Ringed Plover			Goldcrest		
Lapwing			Spotted Flycatcher	T	2
Snipe			Long-tailed Tit	FL	3
Woodcock			Marsh Tit	T	2
Curlew			Willow Tit		
Redshank			Coal Tit	FL	3
Lesser Black-backed Gull			Blue Tit	NE	3
Herring Gull			Great Tit	DD	3
Feral Pigeon	FL	3	Nuthatch		
Stock Dove	P	2	Treecreeper		
Woodpigeon	FL	3	Jay	T	2
Collared Dove	FL	3	Magpie	UN	3
Turtle Dove			Jackdaw	NY	3
Cuckoo	T	2	Rook	UN	3
Barn Owl			Carrion Crow	P	2
Little Owl	ON	3	Raven		
Tawny Owl	S	2	Starling	FY	3
Swift	ON	3	House Sparrow	UN	3
Kingfisher	UN	3	Tree Sparrow		
Green Woodpecker	T	2	Chaffinch	P	2
Great Spotted Woodpecker	T	2	Greenfinch	P	2
Lesser Spotted Woodpecker	H	1	Goldfinch	P	2
Skylark	S	2	Linnet	B	2
Swallow	ON	3	Crossbill		
House Martin	ON	3	Bullfinch	P	2
Tree Pipit			Hawfinch		
Meadow Pipit			Yellowhammer	N	2
Yellow Wagtail			Reed Bunting		
Grey Wagtail	UN	3	Corn Bunting		
Pied Wagtail	FY	3	*Other species:*		
Dipper	ON	3	**Mandarin Duck**	P	2
Wren	P	2			
Dunnock	FL	3			
Robin	ON	3			

Enter the **criteria**. The **categories** are entered automatically.

Figure 13. A completed recording form. All tinted cells are locked.

and crows are very wide-ranging, and should be recorded in this category only in the immediate vicinity of a potential nest site.

Probable breeding required observation of one of a range of different criteria such as territorial behaviour (singing, displaying, *etc*), carrying nesting material or visiting a probable nest site. This category was also used for scarce species which clearly attempted to breed but subsequently were known to be unsuccessful for some reason.

To be confident of confirmed breeding, additional evidence was necessary, such as a nest with eggs, birds carrying food for their young or, ideally, young begging for food from their parents. A caution to observers was given in respect of species such as Mistle Thrush, which form family parties and fly with their young for considerable distances, so that the breeding might not have occurred in the tetrad where the young were observed.

Within the probable and confirmed categories, the criteria increase in confidence as one moves down the appropriate list. Thus seeing an adult carrying food for young is a surer indication of confirmed breeding than observing a bird feigning injury. However, observers were asked not to spend undue time trying to upgrade the criteria within a category, but to concentrate on upgrading the species from probable to confirmed breeding. The target status for any breeding species is of course for breeding to be confirmed but such evidence is more difficult to obtain.

Criteria S (singing), T (territorial behaviour) and B (building a nest) are discussed in more detail in Chapter 8.

The observer entered the highest criteria for the species recorded in each tetrad into recording forms (one for each tetrad). The recording form was primarily a computer spreadsheet (Figure 13) but a paper version was provided for those who preferred it. If repeat visits were made to a tetrad the observer either added any new records, or upgraded existing records (to a higher category or to a criterion giving greater confidence within the existing category). A separate, simpler, form was provided for the submission of casual records from observers merely passing through tetrads. At the end of each season, records were sent to the Atlas data coordinator who merged all the records for each tetrad and, when the survey was complete, produced a distribution map for each species.

General results of the survey

Table 1 shows the numbers of records obtained for the species in each breeding category. The survey produced a total of 16,030 breeding records of 112 species in the 325 tetrads of the recording area. Probable or confirmed breeding records were obtained for all but five of these species (shown in grey). The records were distributed between the three categories in the following way:

Category 1 (possible):	1,687 records (11%)
Category 2 (probable):	7,736 records (48%)
Category 3 (confirmed):	6,607 records (41%)

For a league table of the most widespread species see Appendix A (page 221). For the distribution of records by tetrad see Appendix C (page 224).

The species accounts and their interpretation

Two classes of species accounts are presented. There are main accounts (pages 23–205) for 91 species each occupying a double-page spread of standardized form, and most including 'old', 'new' and 'change' maps. These accounts are presented in the most recent taxonomic order proposed by the British Ornithologists' Union (BOU) in *The British List* (Dudley *et al* 2006). There are brief accounts (pages 206–11) for 25 additional species, also in *British List* order.

The scientific names are also those of *The British List*, while the English vernacular names are those which were familiar and still in common use, being the names used by the BTO when this book went to press. Where these differ from the vernacular names proposed by the BOU, the latter are added in parentheses. For most of the species discussed in this book the differences are slight, use of the BOU names being preferable only in an international context.

In a main species account, the principal distribution map presents the results for the new survey, with dots of three sizes used to represent the three breeding categories. A dot of given size plotted in a

Table 1. Numbers of tetrads in which each species was recorded, in each breeding category. For the species in grey only possible breeding records were obtained.

Species	Possible	Probable	Confirmed	Total
Mute Swan	8	14	32	54
Greylag Goose	1	3	5	9
Canada Goose	8	30	48	86
Shelduck	0	1	1	2
Mandarin Duck	5	20	4	29
Gadwall	0	0	1	1
Mallard	25	75	124	224
Red-crested Pochard	0	0	2	2
Tufted Duck	13	36	24	73
Ruddy Duck	0	0	2	2
Red-legged Partridge	33	185	24	242
Grey Partridge	17	43	5	65
Quail	1	21	1	23
Pheasant	13	235	69	317
Little Grebe	11	23	31	65
Great Crested Grebe	4	1	7	12
Cormorant	7	0	0	7
Little Egret	2	0	0	2
Grey Heron	62	7	7	76
Red Kite	7	1	0	8
Goshawk	1	1	1	3
Sparrowhawk	119	45	36	200
Buzzard	43	166	101	310
Kestrel	149	87	53	289
Hobby	25	4	9	38
Peregrine	3	0	2	5
Water Rail	2	0	0	2
Moorhen	37	45	124	206
Coot	8	19	77	104

Species	Possible	Probable	Confirmed	Total
Oystercatcher	0	1	0	1
Little Ringed Plover	0	0	1	1
Lapwing	19	59	34	112
Snipe	6	0	0	6
Woodcock	0	1	0	1
Curlew	3	7	2	12
Lesser Black-backed Gull	1	0	2	3
Herring Gull	0	0	2	2
Feral Pigeon	34	44	21	99
Stock Dove	21	232	43	296
Woodpigeon	2	171	152	325
Collared Dove	19	191	58	268
Turtle Dove	2	23	0	25
Cuckoo	5	114	0	119
Barn Owl	27	16	22	65
Little Owl	40	58	28	126
Tawny Owl	17	81	33	131
Long-eared Owl	0	0	1	1
Swift	44	66	77	187
Kingfisher	19	15	15	49
Green Woodpecker	28	171	73	272
Great Spotted Woodpecker	49	133	116	298
Lesser Spotted Woodpecker	3	6	3	12
Woodlark	0	1	0	1
Skylark	0	233	76	309
Sand Martin	0	1	1	2
House Martin	19	20	215	254
Swallow	13	43	250	306
Tree Pipit	1	31	7	39

given tetrad indicates the highest category recorded for the species in that tetrad during the five years of the survey: small dots for possible, medium dots for probable and large dots for confirmed breeding.

Below the main map there are two subsidiary dot maps. One of these is simply a reproduction of the distribution map from the 1983–87 survey; the other is a 'change map' intended to provide a visual presentation of distribution changes between the two surveys. The method adopted to illustrate changes is the same as that used (with a 10 km by 10 km mapping unit) to compare the results of 1968–72 and 1988–91 breeding bird surveys of Britain and Ireland (Sharrock 1976, Gibbons *et al* 1993). No symbol is plotted in the tetrad if, during both surveys, a species was either present or absent, in any breeding category. A symbol is plotted only if there has been an 'appearance' or a 'disappearance'. In addition, no distinction is made between breeding categories 2 and 3. A green dot indicates the occurrence of breeding in a tetrad where it had not been recorded in the previous survey, either in the possible breeding category (small dot) or in either of the two higher categories (large dot). A red dot indicates the reverse situation.

Species	Possible	Probable	Confirmed	Total
Meadow Pipit	7	8	3	18
Yellow Wagtail	18	18	10	46
Grey Wagtail	35	26	50	111
Pied Wagtail	46	93	142	281
Dipper	3	4	11	18
Wren	1	165	159	325
Dunnock	3	192	130	325
Robin	0	120	205	325
Nightingale	0	7	0	7
Redstart	4	40	14	58
Stonechat	2	3	2	7
Wheatear	11	0	0	11
Blackbird	0	70	255	325
Song Thrush	2	185	131	318
Mistle Thrush	30	142	124	296
Cetti's Warbler	0	1	0	1
Grasshopper Warbler	0	14	0	14
Sedge Warbler	2	14	3	19
Reed Warbler	0	12	2	14
Blackcap	0	255	60	315
Garden Warbler	3	124	8	135
Lesser Whitethroat	6	128	21	155
Whitethroat	3	207	87	297
Chiffchaff	5	250	56	311
Willow Warbler	7	224	24	255
Goldcrest	14	223	53	290
Spotted Flycatcher	31	62	55	148

Species	Possible	Probable	Confirmed	Total
Long-tailed Tit	36	76	172	284
Blue Tit	3	37	285	325
Great Tit	2	70	251	323
Coal Tit	15	142	94	251
Willow Tit	2	9	7	18
Marsh Tit	38	93	65	196
Nuthatch	14	123	58	195
Treecreeper	49	103	47	199
Jay	91	84	27	202
Magpie	39	146	116	301
Jackdaw	13	70	240	323
Rook	26	16	230	272
Carrion Crow	16	86	217	319
Raven	11	34	24	69
Starling	23	30	211	264
House Sparrow	5	64	206	275
Tree Sparrow	10	8	25	43
Chaffinch	1	122	202	325
Greenfinch	2	193	125	320
Goldfinch	10	187	122	319
Siskin	0	1	0	1
Linnet	15	212	70	297
Crossbill	3	3	0	6
Bullfinch	41	169	53	263
Yellowhammer	5	183	124	312
Reed Bunting	5	62	4	71
Corn Bunting	3	47	4	54
Totals	**1687**	**7736**	**6607**	**16030**
Species	**92**	**99**	**96**	**112**

Comparison with the 1983–87 survey

It is obvious that caution must be exercised when comparing the results of the two surveys, since there were inevitable differences in the details of how they were conducted. In particular, more birdwatchers were available for the new survey, which meant not only that more observer-hours could be contributed–the *survey effort*–but also that the uniformity of coverage was possibly different. These questions are discussed in detail in Chapter 9, where it is estimated that the survey effort for the new survey was about 70% greater than for the previous Atlas. It is argued that, for most species, a recorded increase in the number of occupied tetrads greater than about 50% is in most cases likely to reflect, at least partly, a real increase. If the 50% threshold is not met, such a conclusion is not justified. Because of the greater survey effort, any significant *decreases* recorded in the new survey can usually be assumed to be real.

The following tables therefore only give an indication of the species whose distributions may have increased or decreased in the two decades between the surveys. The individual species

Table 2. Species for which the number of occupied tetrads (in the 300-tetrad recording area) exceeds the corresponding number in the previous survey by at least 50%. Species with fewer than 10 records in the 1983–87 survey are excluded. For reasons explained in the species account (page 48) the figure of 74 for Grey Heron in the new survey is inflated, so the species should probably be excluded from this table also.

Species	83-87	03-07	% incr
Canada Goose	34	84	147
Tufted Duck	39	71	82
Grey Heron	44	74	68
Sparrowhawk	119	183	54
Buzzard	78	289	271
Kestrel	158	270	71
Barn Owl	14	64	357
Kingfisher	29	47	62
Green Woodpecker	152	250	64
Great Spotted Woodpecker	171	277	62
Grey Wagtail	45	103	129
Pied Wagtail	110	260	136
Whitethroat	158	273	73
Goldcrest	140	270	93
Long-tailed Tit	154	263	71
Coal Tit	129	236	83
Nuthatch	104	187	80
Treecreeper	111	188	69
Goldfinch	184	296	61
Linnet	153	280	83
Reed Bunting	37	67	81

Table 3. Species for which the number of occupied tetrads (in the 300-tetrad recording area) was lower than the corresponding number in the previous survey. Species with fewer than 10 records in the 1983–87 survey are excluded.

Species	83-87	03-07	% incr
Grey Partridge	150	64	-57
Lapwing	144	108	-25
Woodcock	13	1	-92
Turtle Dove	92	24	-74
Cuckoo	198	111	-44
Lesser Spotted Woodpecker	27	11	-59
Tree Pipit	87	35	-60
Meadow Pipit	23	13	-43
Dipper	35	17	-51
Nightingale	28	7	-75
Redstart	62	52	-16
Grasshopper Warbler	19	11	-42
Garden Warbler	168	125	-26
Wood Warbler	14	0	-100
Willow Warbler	278	240	-14
Spotted Flycatcher	152	142	-7
Willow Tit	48	17	-65
Starling	270	241	-11
Tree Sparrow	59	43	-27
Corn Bunting	80	53	-34

accounts give a more considered view of their fortunes.

Table 2 lists the 21 species for which the increase was found to meet the 50% threshold (omitting scarce species with fewer than 10 records). To those 21 species should be added Mandarin Duck and Raven, for which there were no breeding records in the previous Atlas whereas the new survey obtained 29 and 69 records respectively. These increases should be set against the list of species showing a *decrease* (Table 3). In compiling this table no threshold has been set since, statistical significance apart, such decreases in the face of a 70% increase in survey effort are very likely to be real.

Other factors may have affected the two surveys in different ways. It is at least possible that bird identification skills had become more widespread in the period, not to mention improvement in optical aids or in countryside mapping and access. These factors are harder to quantify, but are not thought to be significant in comparison with the differences in survey effort.

6. Main species accounts

In the main species accounts the following information is provided for each species:

The species' breeding status, *ie* 'resident breeder', 'migrant breeder' or 'introduced breeder'.

An indication if the species is one of those selected for the Gloucestershire Biodiversity Action Plan (BAP).

The UK conservation status of the species, *ie* **RED**, **AMBER**, or **GREEN** (Gregory *et al* 2002).

The UK national population trends for the species (Baillie *et al* 2007, summarized by Eaton *et al* 2008; see also Robinson 2005).

A table showing the numbers of records obtained for the species in each breeding category, for:

- the 2003–07 survey (all thirteen 10 km squares; 325 tetrads)
- the 2003–07 survey (omitting square SO92, which was not surveyed in 1983–87)
- the 1983–87 survey (twelve 10 km squares; 300 tetrads)

The second and third columns allow meaningful comparisons between the surveys.

ern Lapwing)

pwings g land in the g-sown icularly -facing tacular activity reeding is not found. l arable les and side. In a week robable as only e areas, ithwest

Migrant/resident breeder

Gloucestershire BAP species

UK conservation status: AMBER

Long-term UK trend (1970–2006): -47%

Recent population trend (1994–2007): -18%

Numbers of tetrads with breeding records:

	2003–07		1983–87
	13 squares	12 squares	12 squares
Possible	**19**	19	30
Probable	**59**	56	69
Confirmed	**34**	33	45
Totals	**112**	**108**	**144**

study improved coverage notwithstanding it is

The recent population trends are mostly derived from the BTO/JNCC/RSPB Breeding Bird Survey (BBS) which began in 1994 (Field & Gregory 1999). For most long-term trends, data from the BBS (where available) and its predecessor the Common Birds Census (CBC) (Marchant *et al* 1990) were combined. Exceptions include a few riverine species, for which the BTO Waterways Bird Survey (Marchant *et al* 1990) was a more reliable source; and Grey Heron, which has been the subject of annual surveys since 1928 (Marchant *et al* 2004).

Mute Swan

Cygnus olor

The Mute Swan is the largest and heaviest breeding bird in the region. Its huge, unmistakable nest, usually located near the bank of a river, pond or lake, makes breeding activity very easy to detect, and a successful outcome was determined by observation of family groups with cygnets. Breeding in the area appears to follow the norm for Britain, with nest occupation and egg-laying from early or mid-April.

In the survey area, breeding is concentrated in the valleys of the larger streams, particularly the Rivers Evenlode, Windrush and Coln, as well as lakes and ponds. In areas where breeding habitats are more thinly distributed, individual pairs were noted moving from one site to another over the time-span of the survey. Therefore, the distribution maps may slightly exaggerate the overall spread of the species. For example, it is believed that there was only a single pair in the Avon Vale for most of the survey, and that the birds moved from one site to another between years and sometimes within the same breeding season.

Resident breeder

UK conservation status: AMBER

Long-term UK trend (1970–2006): +151%

Recent population trend (1994–2007): 0%

Numbers of tetrads with breeding records:

	2003–07		1983–87
	13 squares	12 squares	12 squares
Possible	**8**	8	5
Probable	**14**	13	12
Confirmed	**32**	31	26
Totals	**54**	**52**	**43**

Compared with the survey in the 1980s, it seems that there has been little change in the total number of areas occupied, but the change map clearly suggests a slight overall northwards realignment in the species' distribution.

DAVE PEARCE

Species sponsored by Nigel Birch

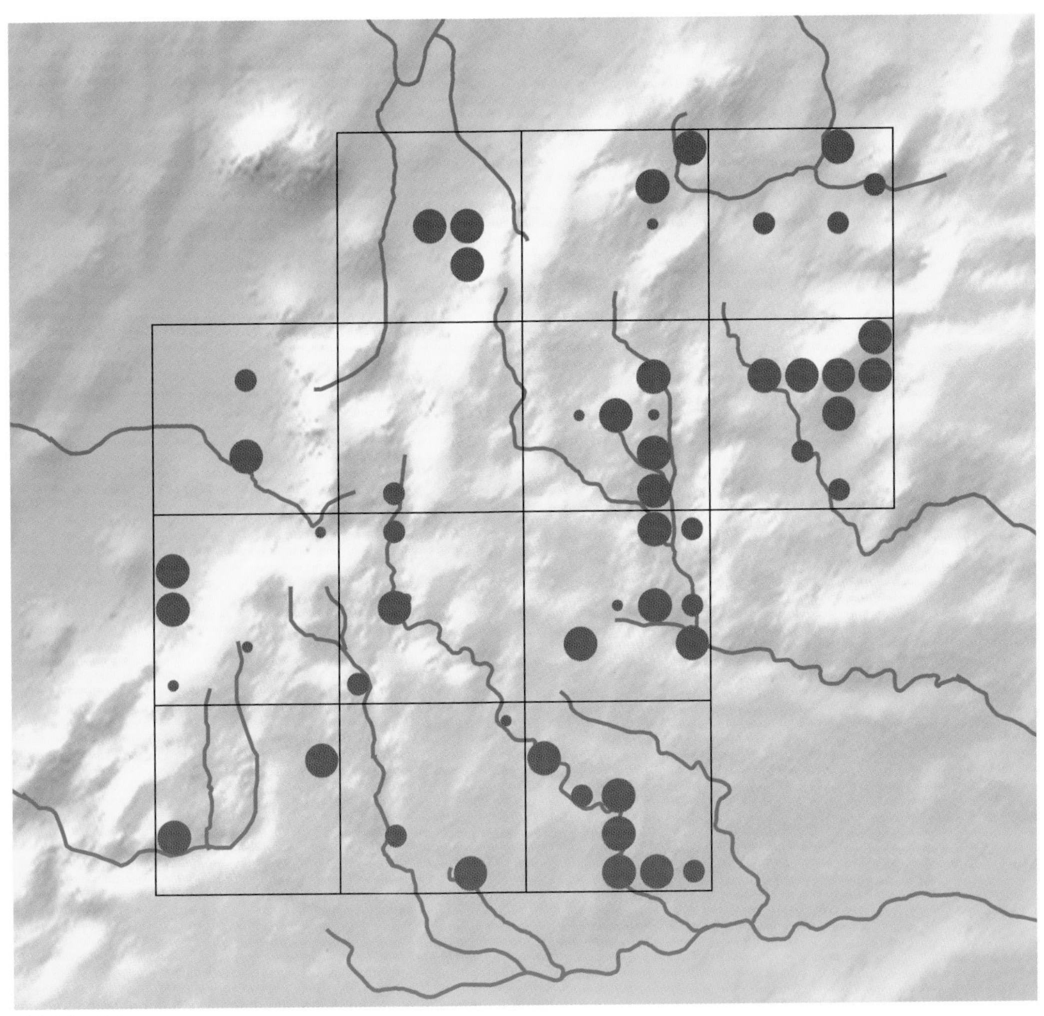
2003–07 survey

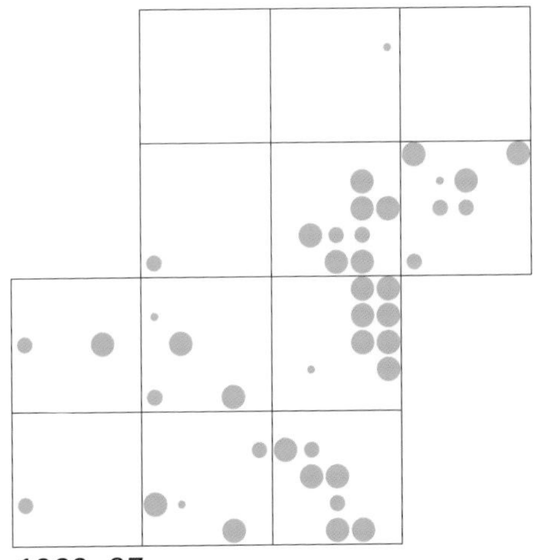
1983–87 survey

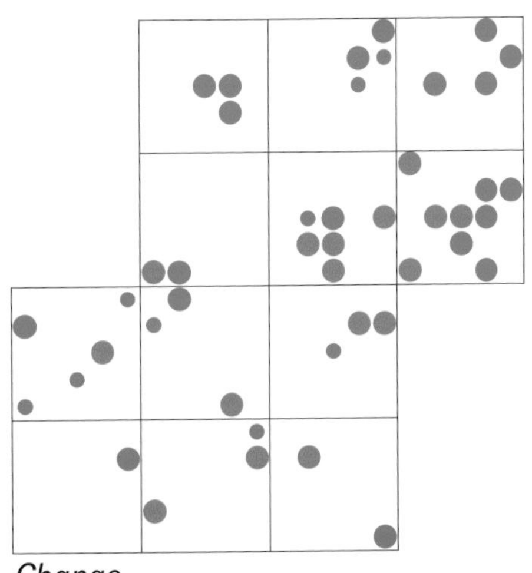
Change

Greylag Goose
Anser anser

Thanks to their size and loud voice, Greylags are easy to locate, and confirmation of breeding can readily be achieved by observation of family parties.

They are found in areas where there are large expanses of water. In this survey, breeding was confirmed in five tetrads in three specific areas: Bourton Pits, Fairford Pits and Witcombe Reservoir. Greylag Geese are fairly recent newcomers to the Bourton area, but have been established in the Fairford and Witcombe areas for some time. Data gathered by the NCOS would suggest that there has been an expansion in the population since the 1990s.

Greylags are a feral reintroduction to southern Britain, and the Cotswolds birds probably derive from a sizeable population in the Severn Vale area, particularly the Wildfowl and Wetlands Trust reserve at Slimbridge. There were no reports of Greylag Goose in the first Atlas. It is not clear whether this reflects a complete absence of the species during the 1983–87 survey or whether any present were considered to be merely escapes that had not formed a self-supporting population.

Introduced breeder

UK conservation status: Unlisted

Long-term UK trend: Not available

Recent population trend (1994–2007): +220%

Numbers of tetrads with breeding records:

	2003–07		1983–87
	13 squares	12 squares	12 squares
Possible	**1**	1	
Probable	**3**	3	
Confirmed	**5**	5	
Totals	**9**	**9**	

Although there were breeding records in the mid-1980s, the Greylag Goose was not accepted as a breeding resident in Gloucestershire until the mid-1990s.

RICHARD TYLER

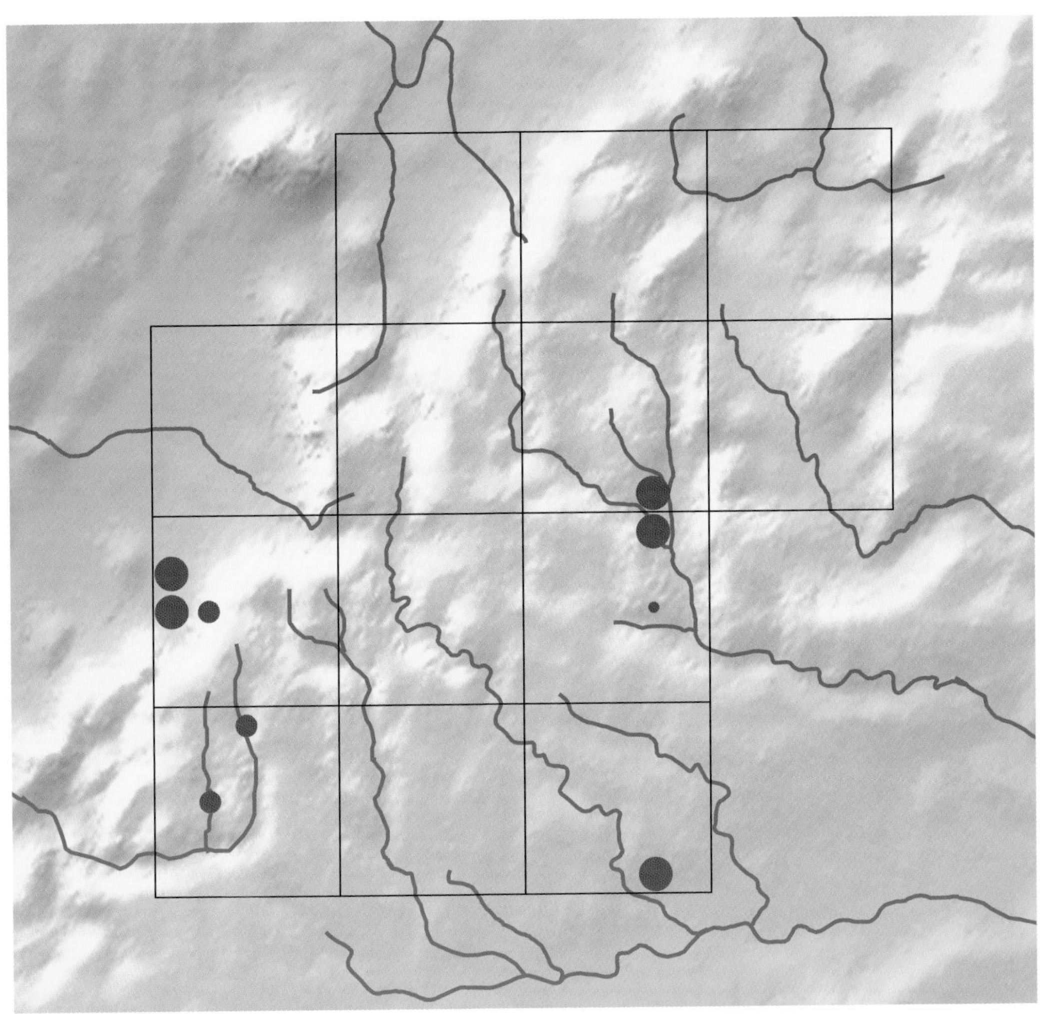

2003–07 survey

Canada Goose (Greater Canada Goose)

Branta canadensis

Although this large and noisy goose is easy to locate when in feeding flocks, it can be quite secretive when on the nest (which is often sheltered by vegetation and is frequently on an island) and when the young have first hatched. Despite this secretive nature, the data from this survey suggest that it is fairly easy to confirm breeding, and over half the records were in this category.

Breeding is usually in areas associated with still water–lakes, ponds and wet river valleys. The change map suggests a significant expansion (between two- and threefold) in the breeding distribution of the species. Most of this spread seems to be in the north and east of the region–particularly the Avon Vale, the Evenlode Valley and the Stour Valley area. Canada Geese are still absent from the Stroud Valleys, despite the existence of apparently suitable habitat.

The first Atlas commented: 'Breeding birds and even large goslings are very secretive ... It is ... quite possible that its numbers are significantly greater than the survey suggests.' Even taking this into account, and the greater effort during the later survey, it seems likely that the observed distribution expansion is the result of a population increase over the intervening 20 years. This is despite the possibility that a drop in water levels before and during the survey made some marginal sites (*eg* small ponds and generally low-lying wet areas) unsuitable. In the less marginal sites, Canada Geese tend to be colonial if sufficient suitable habitat is available, and some sites may hold several breeding pairs.

Introduced breeder

UK conservation status: Unlisted

Long-term UK trend: Not available

Recent population trend (1994–2007): +149%

Numbers of tetrads with breeding records:

	2003–07		1983–87
	13 squares	12 squares	12 squares
Possible	**8**	8	8
Probable	**30**	29	11
Confirmed	**48**	47	15
Totals	**86**	**84**	**34**

RICHARD TYLER

Species sponsored by Gordon and Jenny Kirk

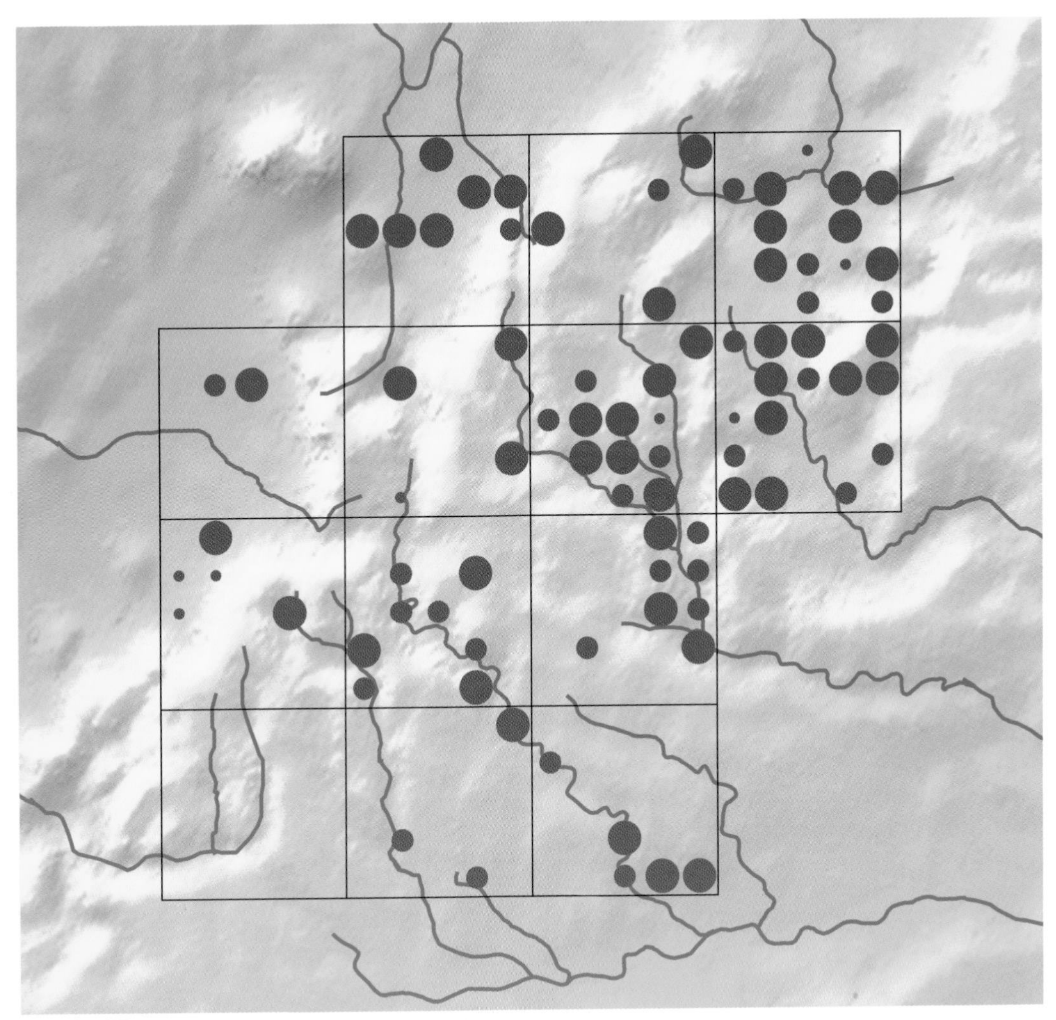

2003–07 survey

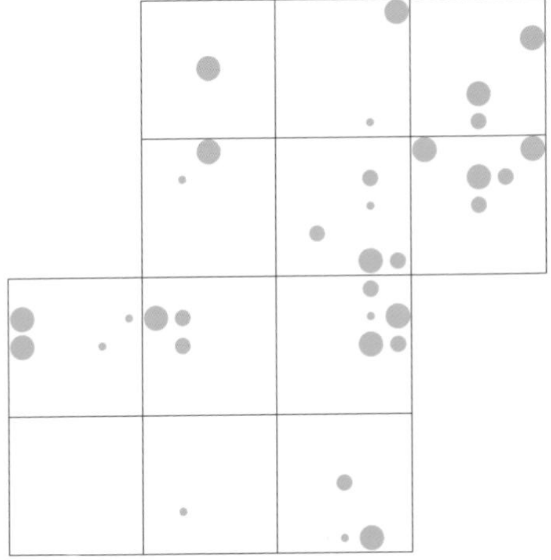

1983–87 survey

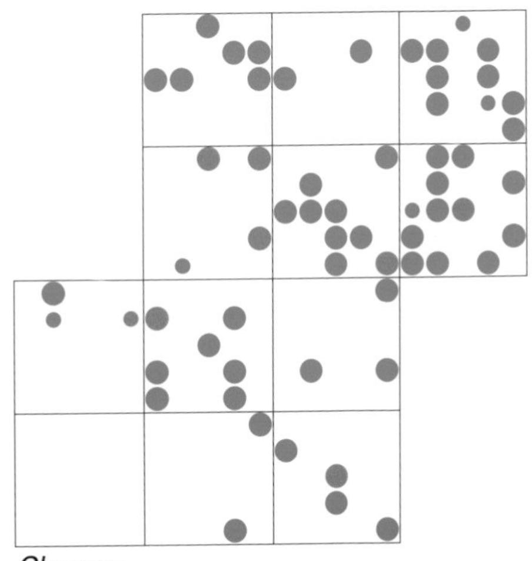

Change

Mandarin Duck

Aix galericulata

For most waterfowl, breeding can be confirmed by observing adults sitting on nests, or groups of newly hatched young. Mandarins are less easy to monitor, however, as they are secretive and nest in holes in trees. Young birds were only rarely seen during this survey. As a result, most records were of pairs of birds seen in suitable breeding habitat.

Although there were spring and summer records of Mandarins during the first survey in the 1980s, they were believed to have been of free-flying escapes or deliberately released birds, rather than the offspring of an established feral population. However, even then it was suggested that the species might have been breeding in the Cotswolds. The distribution map for the species in this survey suggests that it has become an established, if localized, breeder. Whether this is as a result of the major population in the Forest of Dean spreading, or of local releases, is not clear. Mandarins were found primarily associated with the Rivers Coln, Churn, Frome, Stour and Isbourne, and part of the Windrush complex.

Introduced breeder

UK conservation status: Unlisted

No population trends available

Numbers of tetrads with breeding records:

	2003–07		1983–87
	13 squares	12 squares	12 squares
Possible	5	5	
Probable	19	19	
Confirmed	5	5	
Totals	29	29	

Although breeding was confirmed in only five tetrads during the survey period, they are known to have bred in several of the other occupied tetrads in the years leading up to the survey, and so the map is likely to be a fair reflection of the species' distribution.

ROB BROOKES

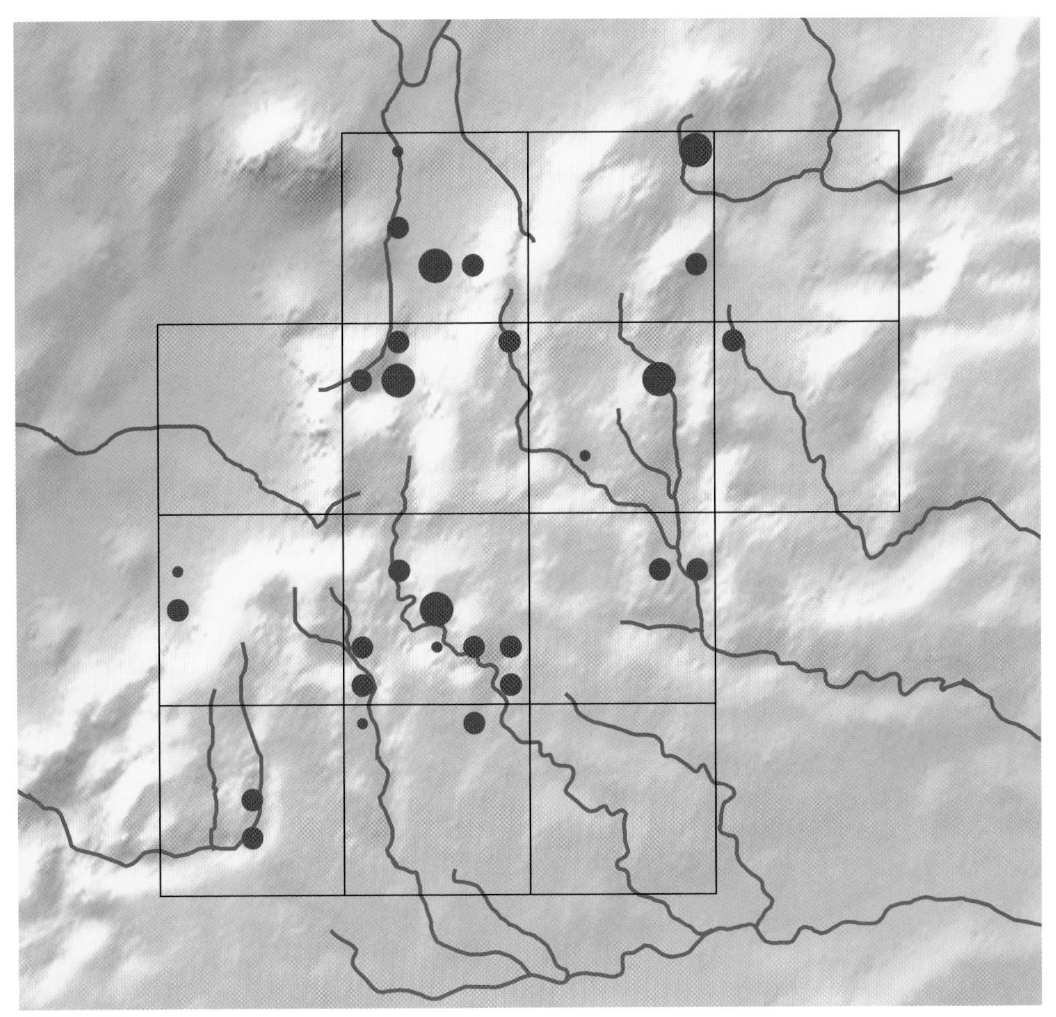

2003–07 survey

Mallard

Anas platyrhynchos

ROB BROOKES

Resident breeder

UK conservation status: GREEN

Long-term UK trend (1970–2006): +98%

Recent population trend (1994–2007): +27%

Numbers of tetrads with breeding records:

	2003–07		1983–87
	13 squares	12 squares	12 squares
Possible	**25**	23	17
Probable	**75**	71	73
Confirmed	**124**	113	77
Totals	**224**	**207**	**167**

There were two significant risks in surveying Mallards: erroneously recording partially domesticated individuals and assuming wrongly that a group of males and females on a pond include one or more pairs, as unpaired birds tend to congregate on any available body of water. In practice, identification of breeding pairs proved to be fairly easy owing to their behaviour of keeping close together and distinctly separate from other nearby ducks. Males could also sometimes be seen leading females to potential nest sites, and standing guard near nests. Although the Mallard's nest is difficult to find, being built on the ground, often in thick undergrowth, the sight of up to a dozen newly hatched ducklings pattering across ponds was a familiar occurrence and meant that there was a high percentage of confirmed breeding records. Mallards also have an extended breeding season–newly hatched young were seen from the end of March and through until August–and this helped in locating territories.

By far the most numerous waterfowl species, the Mallard is also the most widely distributed in the survey area, being recorded in about two-thirds of tetrads. As well as the major lakes and river systems, which hold a significant number of pairs, it was found on the smallest of ponds and in the most shallow of streams, including field drainage ditches, throughout the area.

The change map does not suggest any major change in geographical distribution of the species, the gains and losses being fairly evenly distributed across the survey area; the moderate increase in total number of occupied tetrads (about 18%) almost certainly reflects the greater survey effort.

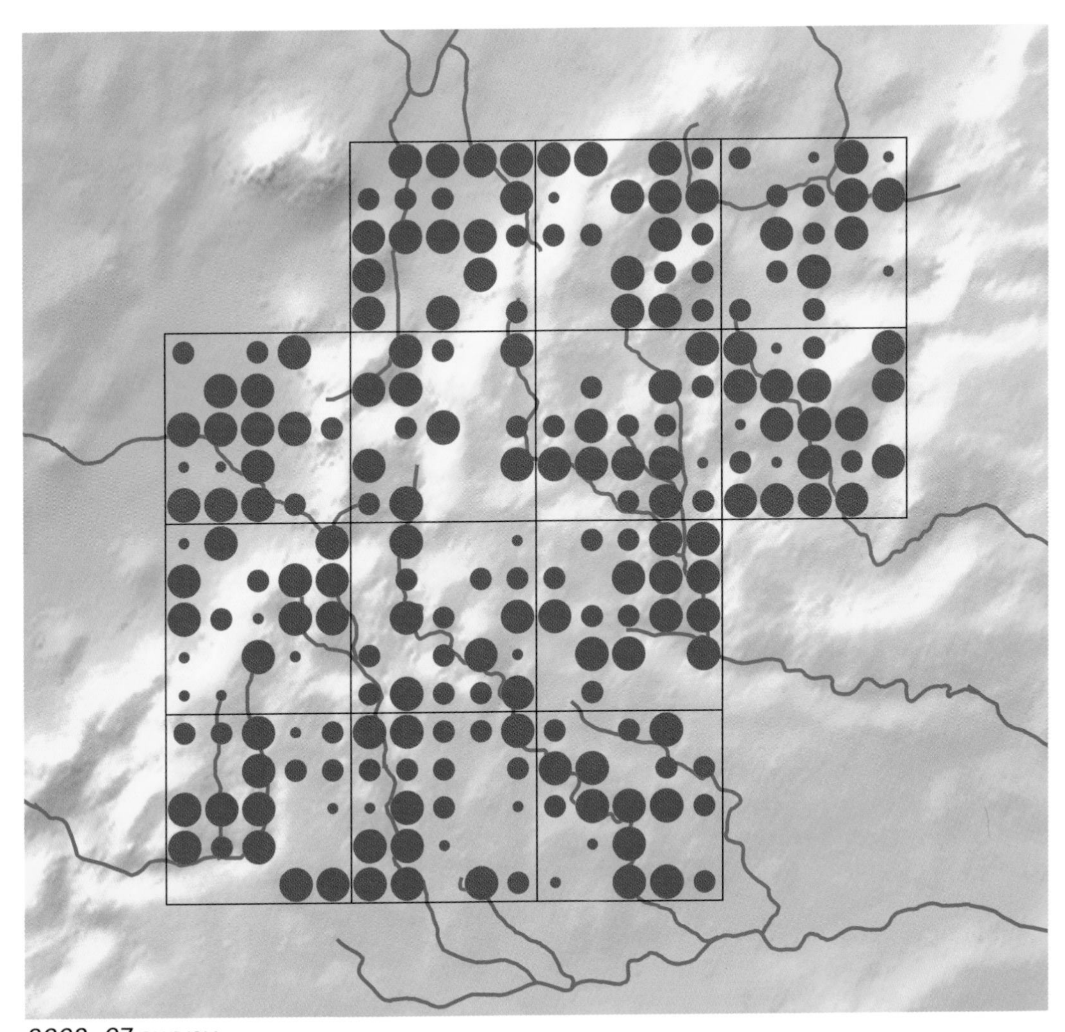

2003–07 survey

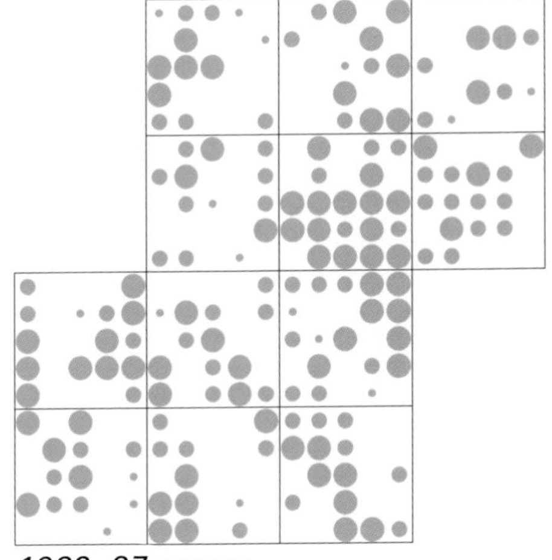

1983–87 survey

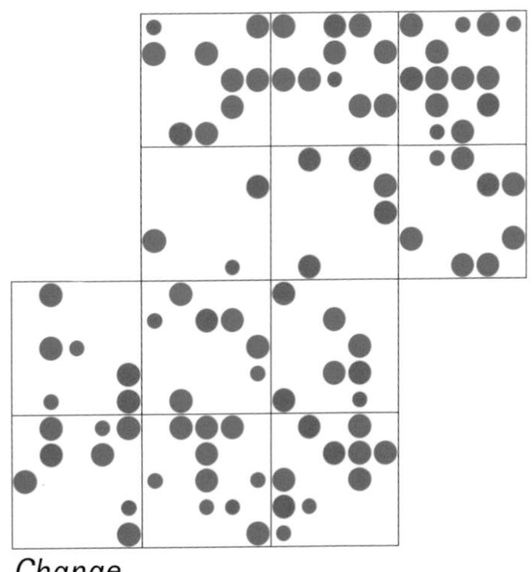

Change

Tufted Duck

Aythya fuligula

The Tufted Duck favours large bodies of water and especially those with islands, such as reservoirs, lakes and flooded gravel-pits. However, they also exploit slow-flowing rivers with bank-side vegetation in which to conceal nests, such as the Windrush southeast of Bourton-on-the-Water, and the Coln near Fairford.

It is a relatively late breeding bird with egg-laying starting in general in mid-May. During the survey it was noted that pairs of birds often seemed to sit around, apparently inactive, at known breeding sites well into June, before evidence of nesting was seen. As with most waterfowl, observation of family groups with newly hatched young made confirmation of breeding easy. However, as in the 1980s survey, breeding activity was also seen in many of the tetrads in which actual breeding was not subsequently confirmed.

The total number of occupied tetrads increased from 39 in the 1980s survey to 70 in this survey (in the same twelve 10 km squares), and the number of confirmed breeding records trebled from eight to 24. Although the greater survey effort might account for some of this increase, especially given the bias towards breeding later in the summer, it is large enough to suggest that there has probably been a significant expansion in the bird's distribution throughout the survey area. This agrees with the nationwide BTO/JNCC/RSPB Breeding Bird Survey data, which suggest a two-thirds increase in numbers since 1994.

Resident breeder

UK conservation status: **GREEN**

Long-term UK trend (1975–2006): +15%

Recent population trend (1994–2007): +67%

Numbers of tetrads with breeding records:

	2003–07		1983–87
	13 squares	12 squares	12 squares
Possible	**13**	13	6
Probable	**36**	33	25
Confirmed	**24**	24	8
Totals	**73**	**70**	**39**

ROB BROOKES

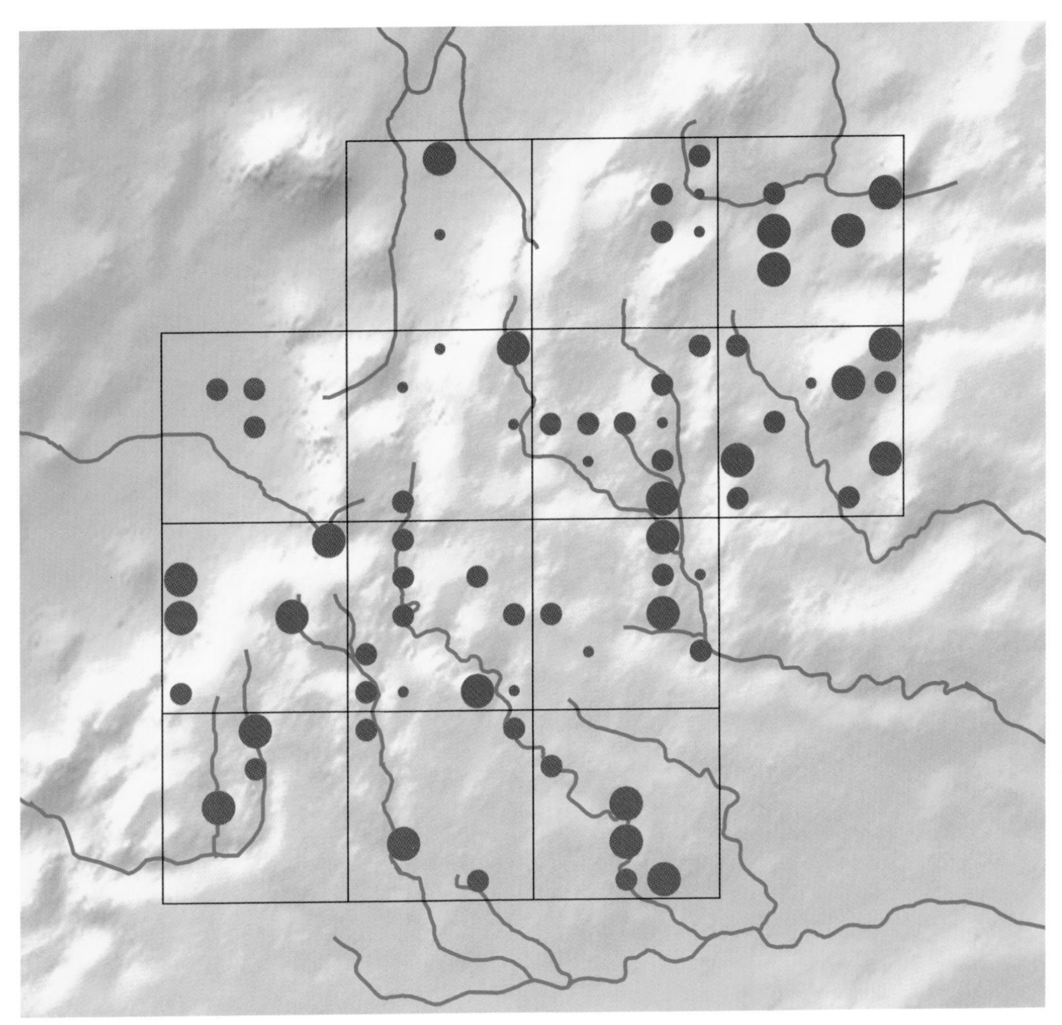

2003–07 survey

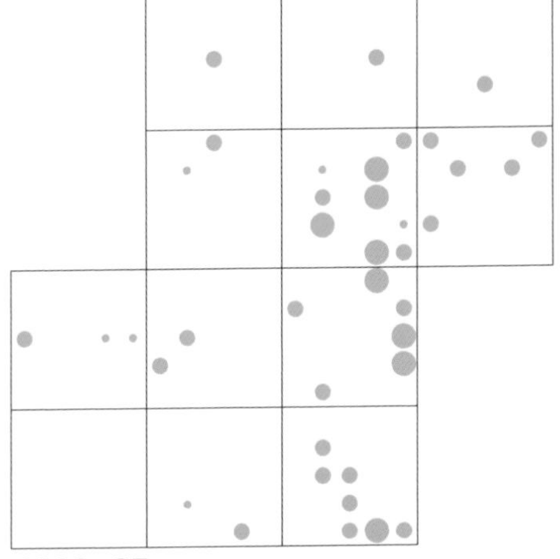

1983–87 survey

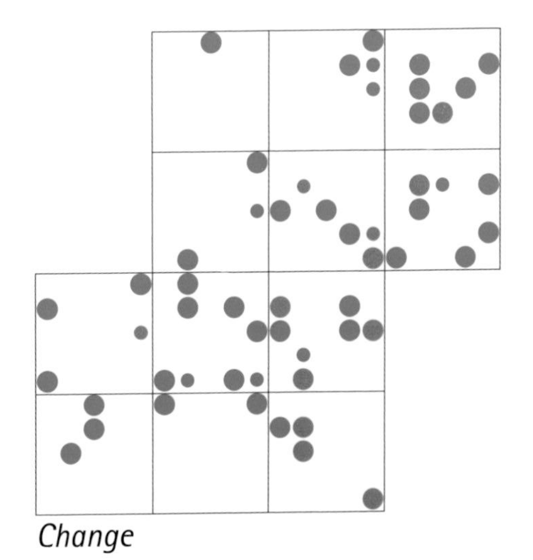

Change

Red-legged Partridge

Alectoris rufa

ROB BROOKES

Introduced breeder

UK conservation status: Unlisted

Long-term UK trend (1970–2006): -9%

Recent population trend (1994–2007): +43%

Numbers of tetrads with breeding records:

	2003–07		1983–87
	13 squares	12 squares	12 squares
Possible	**33**	30	16
Probable	**185**	175	118
Confirmed	**24**	23	20
Totals	**242**	**228**	**154**

Although not a native of the UK, the Red-legged ('French') Partridge is a fairly common bird over much of the survey area, especially the areas of arable land where they are bred and released. They are primarily birds of open arable country and are absent from pasture-dominated valley bottoms, deep uncultivated slopes and woodland.

Red-legged Partridges were regularly seen in pairs during spring and summer. Their breeding call is distinctive, making surveying quite straightforward. However, as was found in the 1983–87 survey, despite its relatively widespread occurrence, confirmation of breeding was difficult to achieve, despite the fact that two broods are often raised. The nests are usually hidden in deep vegetation and family parties are secretive in their habits, usually scuttling around unseen in fields of cereals. It is believed, however, that the breeding success rate of released birds is low, and that the population might not be sustained without repeated releases.

The change map hints at an expansion in distribution of birds attempting to breed, especially in the west of the area. This is probably of minor significance, as their distribution is influenced by the locations of their release sites. Like the Pheasant, it is predominantly a seed-eater and can be fairly easily lured back to feeding stations, making it a convenient game bird to breed and release. The increased level of surveying may have had an effect on the findings, although probably not enough to explain all of the 48% increase in total records.

Species sponsored by Chris Oldershaw

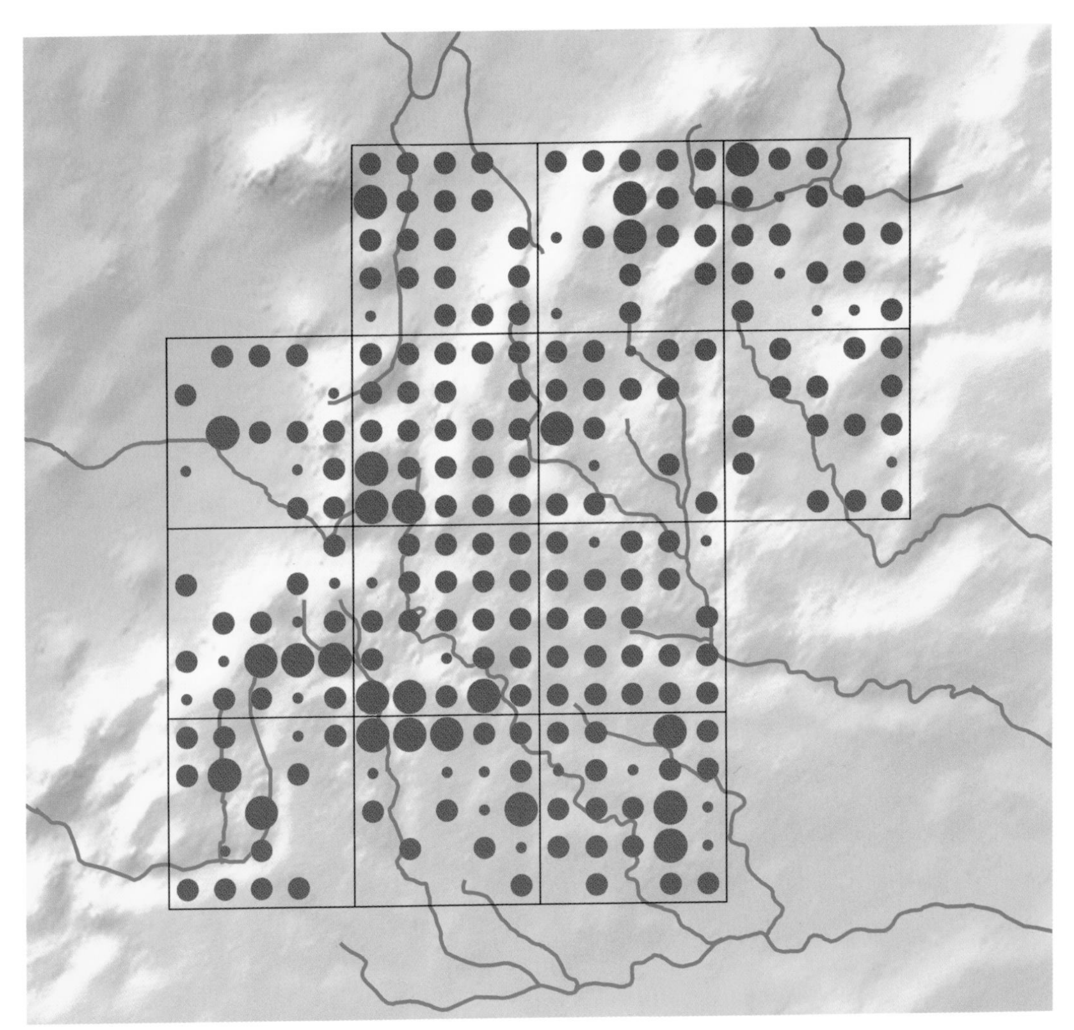

2003–07 survey

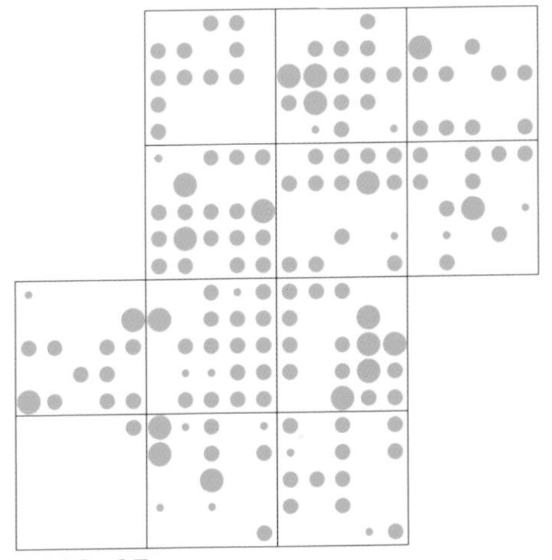

1983–87 survey

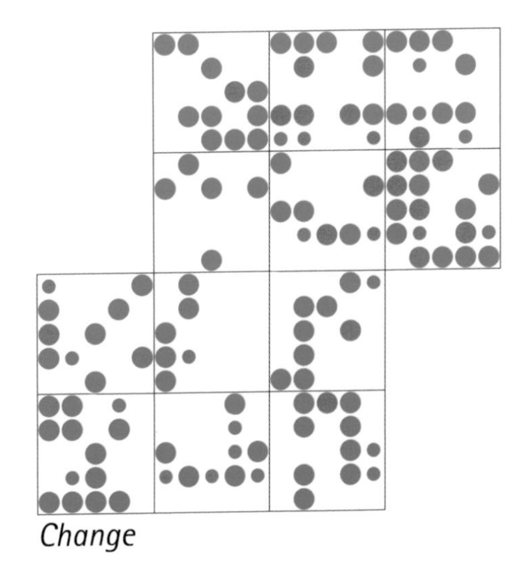

Change

Grey Partridge

Perdix perdix

Grey Partridges are secretive birds, tending to crouch low when danger is perceived. This makes locating them somewhat difficult, although they do take flight more readily than Red-legged Partridges, and the 'rusty gate' call made by both sexes can indicate their presence, especially in the breeding season. Confirmation of breeding also proved to be difficult, although this is probably as much an indication of the thinness of their distribution as their secretiveness.

In the survey in the 1980s, it was noted that the distributions of Grey and Red-legged Partridges were similar, although numbers of the former were believed to be somewhat lower. It was further commented that, despite a national decline, the native species was believed to be holding its own in the Atlas recording area. This is clearly no longer the case. The number of tetrads in which Grey Partridges were found was 60% lower than in the 1980s survey, which is consistent with the national population decline. Allowing for the greater survey effort, a 75% reduction in number of occupied tetrads is a realistic estimate.

Resident breeder

Gloucestershire BAP species

UK conservation status: RED

Long-term UK trend (1970–2006): -88%

Recent population trend (1994–2007): -39%

Numbers of tetrads with breeding records:

	2003–07		1983–87
	13 squares	12 squares	12 squares
Possible	17	17	14
Probable	43	43	119
Confirmed	5	4	17
Totals	65	64	150

Grey Partridges are released much less widely than either Red-legged Partridges or Pheasants. Efforts made on some farms to improve the habitat for Grey Partridges appear to have been unsuccessful. This may be because the majority of Grey Partridges now in the Cotswolds are derived from released stock selected by breeding in captivity and so their breeding success in the wild is poor. It seems that the medium-term future of the species as a regular Cotswolds breeding bird is under serious threat. This is particularly poignant since the previous Atlas depicts the Grey Partridge on the cover as a representative breeding species of the Cotswolds.

ROB BROOKES

Species sponsored by Ian Boyd

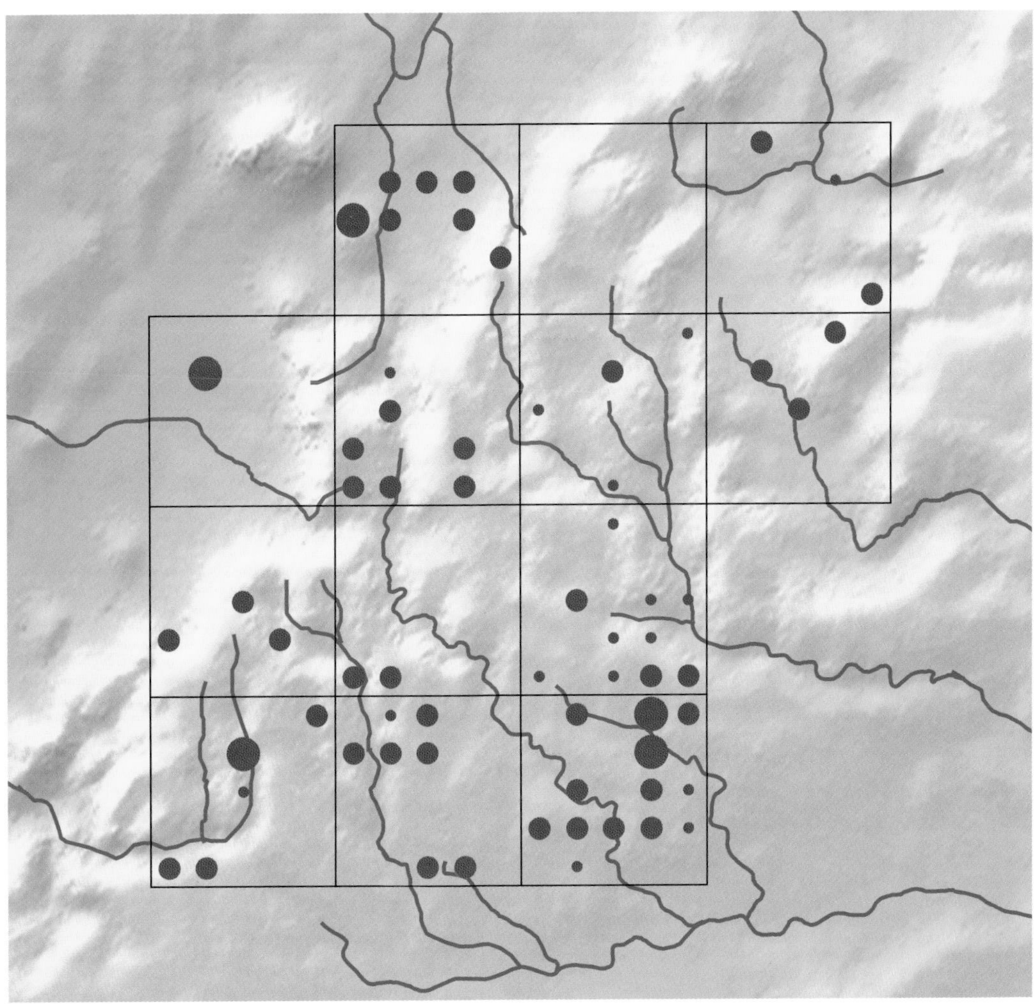

2003–07 survey

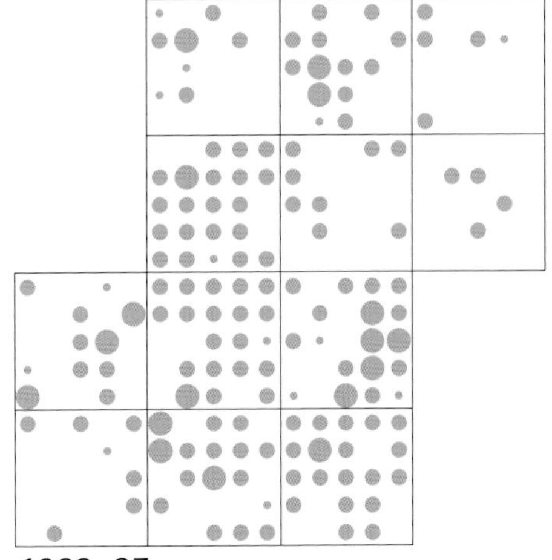

1983–87 survey

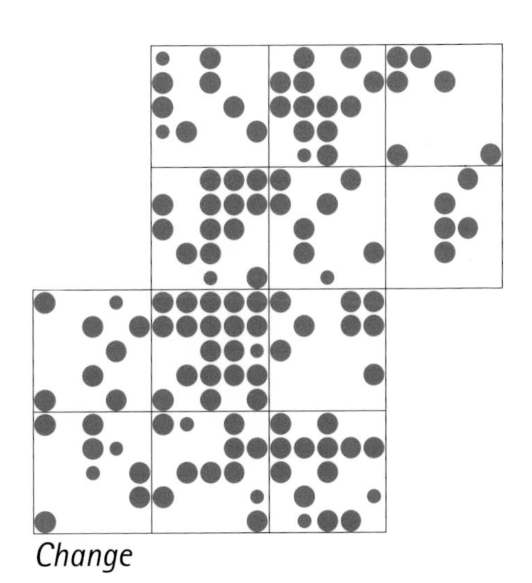

Change

Quail (Common Quail)

Coturnix coturnix

IAIN MAIN

Migrant breeder

UK conservation status: RED

No population trends available

Numbers of tetrads with breeding records:

	2003–07		1983–87
	13 squares	12 squares	12 squares
Possible	**1**	1	0
Probable	**21**	21	19
Confirmed	**1**	1	0
Totals	**23**	**23**	**19**

The difficulty of surveying Quails should not be underestimated. They sometimes call for extended periods, but can be silent for long periods as well and it is impossible to tell whether silence in a field in which calls had previously been heard indicates that the bird has found a mate or has moved on. Proof of breeding can only be achieved by brood sightings, which are rare and only occurred at one location during the survey.

Although not common, this nationally scarce bird is heard regularly in most years in the Cotswolds. There is a great deal of suitable habitat in the area in the form of large cereal fields, both on the High Wold and in the dip slope valleys. They have bred on the high ground and occasionally in the Avon Vale. The number of observations (almost invariably of the tri-syllabic 'whip-ip-ip' call) varies significantly from year to year. Despite the much greater survey effort, the total number of records was very similar to that for the previous Atlas, in which it was stated that: 'In Gloucestershire, the Cotswolds are regarded as a stronghold of this scarce migratory game bird'. It was also noted that 'underestimation of numbers is likely' and that 'The [clusters] of records in the apparently erratic distribution may be the result of more efficient recording in some places ...'.

The distribution maps show the species to be rather thinly scattered from the southwest to the northeast, with no particular pattern evident in the change map. In many cases in the later survey, calling was heard on only one occasion at a particular site, and there were very few areas in which birds were heard in more than one year of the survey. Overall, given the increased observer effort in this survey, it would seem that the number of breeding Quails may be lower in the survey area now than it was in the 1980s, perhaps mirroring what is happening nationally.

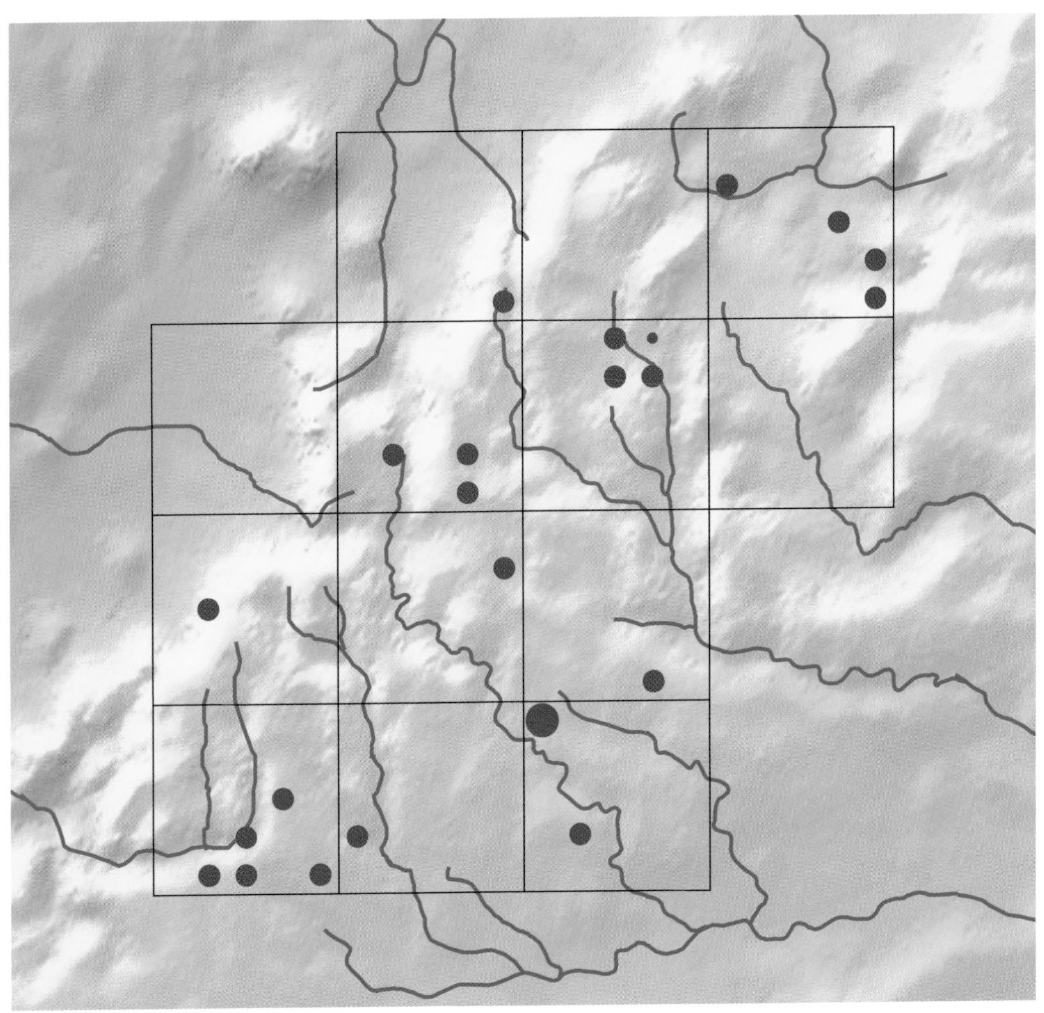
2003–07 survey

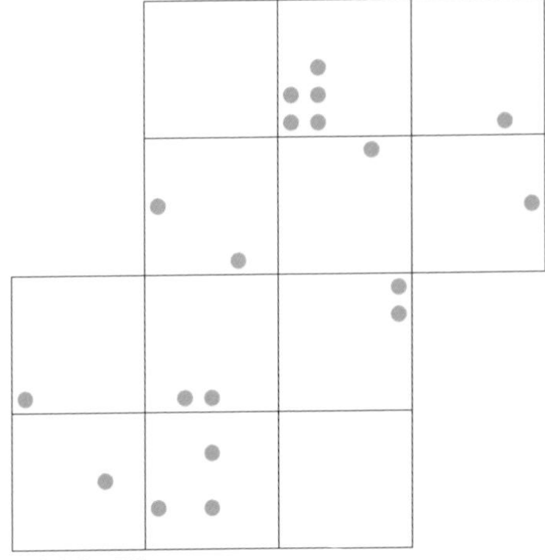
1983–87 survey

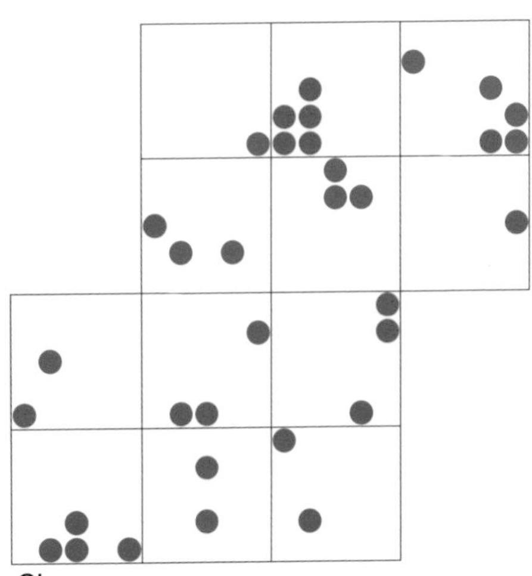
Change

Pheasant (Common Pheasant)

Phasianus colchicus

Pheasants are an important source of income on many Cotswold estates and captive-bred birds are released throughout the survey area in large numbers. Survivors from the shoots have spread throughout the area, and can be found in almost any type of rural habitat, from woodland and steeply sided uncultivated slopes to open arable country. The urban area in and around Cheltenham provides one noticeable gap in the distribution map, although single birds stray into the suburbs. Calling and displaying birds were found virtually everywhere else, often some distance from the nearest shoot or release site.

In the first Atlas report, it was suggested that: 'The relatively low number of confirmed breeding records of this very common bird reflects the lack of interest and rather disdainful attitude which most observers have towards such an artificially maintained species'. In the later survey, the number and percentage of confirmed breeding records declined (although the total number of records increased), despite the increased survey effort, and perhaps a 'less disdainful' attitude. This suggests that the low number of confirmed breeding records may be more to do with the less-than-easy task of seeing young birds before they are able to fly significant distances (which happens quite early in life). It may also be that only a small percentage of the displaying males actually attempt to breed. About a third of the confirmed breeding records were from the finding of broken egg-shells, so the success rate of raising young is not clear. Whatever the truth of this, it certainly seems possible that Pheasant breeding success in the wild in the Cotswolds during this survey was less than it was in the 1980s.

Introduced breeder

UK conservation status: Unlisted

Long-term UK trend (1970–2006): +76%

Recent population trend (1994–2007): +40%

Numbers of tetrads with breeding records:

	2003–07		1983–87
	13 squares	12 squares	12 squares
Possible	**13**	10	34
Probable	**235**	220	154
Confirmed	**69**	67	78
Totals	**317**	**297**	**266**

RICHARD TYLER

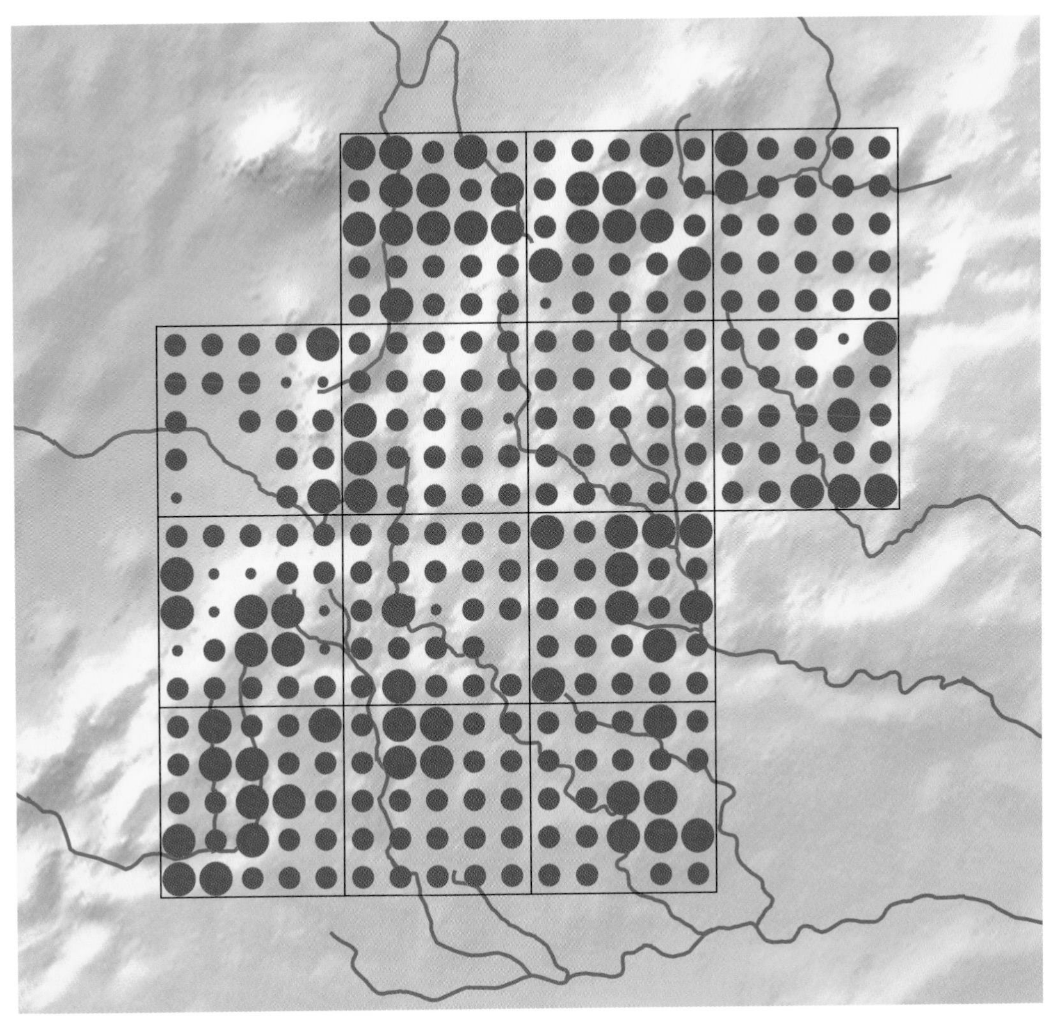

2003–07 survey

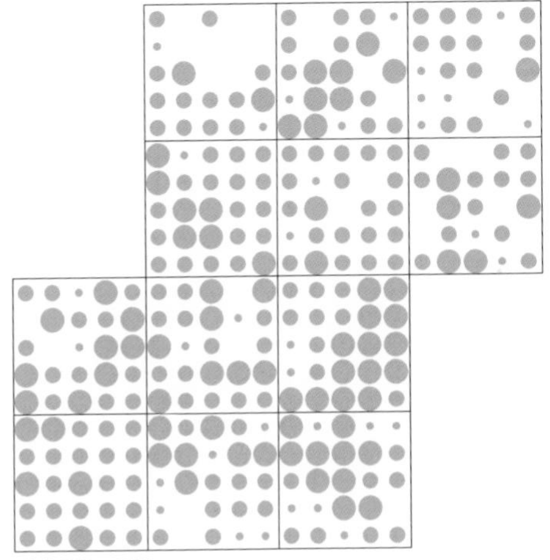

1983–87 survey

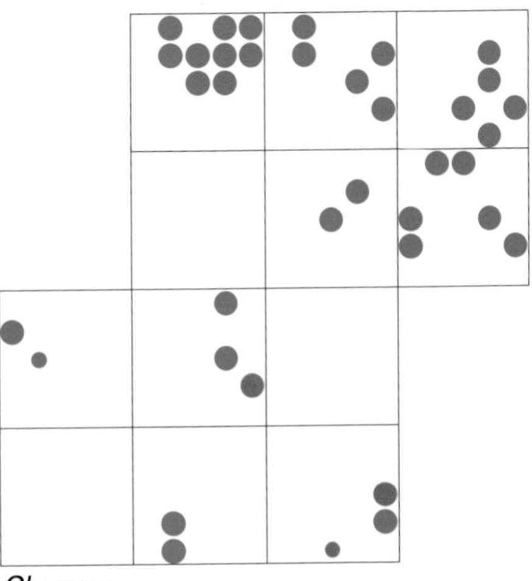

Change

Little Grebe

Tachybaptus ruficollis

Little Grebes can be relatively early nesters. Their whinnying breeding call aided their location and pairs were regularly seen building nests. Confirmation of breeding was obtained by locating occupied nests and observing adults with chicks on the water. Nests were found as early as the end of March during the survey period (although some of these did not survive long enough for young to be hatched) and young birds were seen from the beginning of May.

Little Grebes are less particular than Great Crested Grebes in their breeding habitat requirements; as a consequence they are significantly more widespread in the survey area. They were found on a range of still waterbodies from small ponds to much larger reservoirs and gravel-pits, as well as on the slower-flowing rivers. The greatest concentration was on Dowdeswell reservoir, where the number of adults on occasions was in double figures.

The data indicate a 31% increase in the number of tetrads producing breeding records in any category. This increase could be due, at least in part, to greater survey effort (and better knowledge of Little Grebe habitats). This is supported by the fact that the number of confirmed breeding records rose from about 20% in the first survey to almost 50% in the second. Indeed, the first Atlas commented: 'Bearing in mind the extensive network of unpolluted and undisturbed streams in the area, and the relatively secretive lifestyle of this inconspicuous bird, it is likely to be more widespread than the survey results suggest'. This assessment appears to have been justified.

Apart from an apparent slight decline in the upper reaches of some of the river systems, such as the Coln and the Windrush tributaries, there is no significant pattern to the change in distribution. It may be that a drop in water levels in these areas has had an adverse effect.

Resident breeder

UK conservation status: **GREEN**

Long-term UK trend (1970–2006): +186%

Recent population trend (1994–2007): +21%

Numbers of tetrads with breeding records:

	2003–07		1983–87
	13 squares	12 squares	12 squares
Possible	**11**	11	11
Probable	**23**	23	28
Confirmed	**31**	30	10
Totals	**65**	**64**	**49**

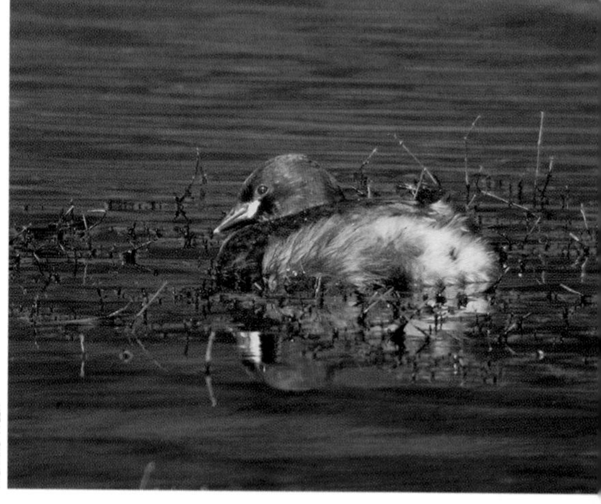

RICHARD TYLER

Species sponsored by Gordon and Jenny Kirk

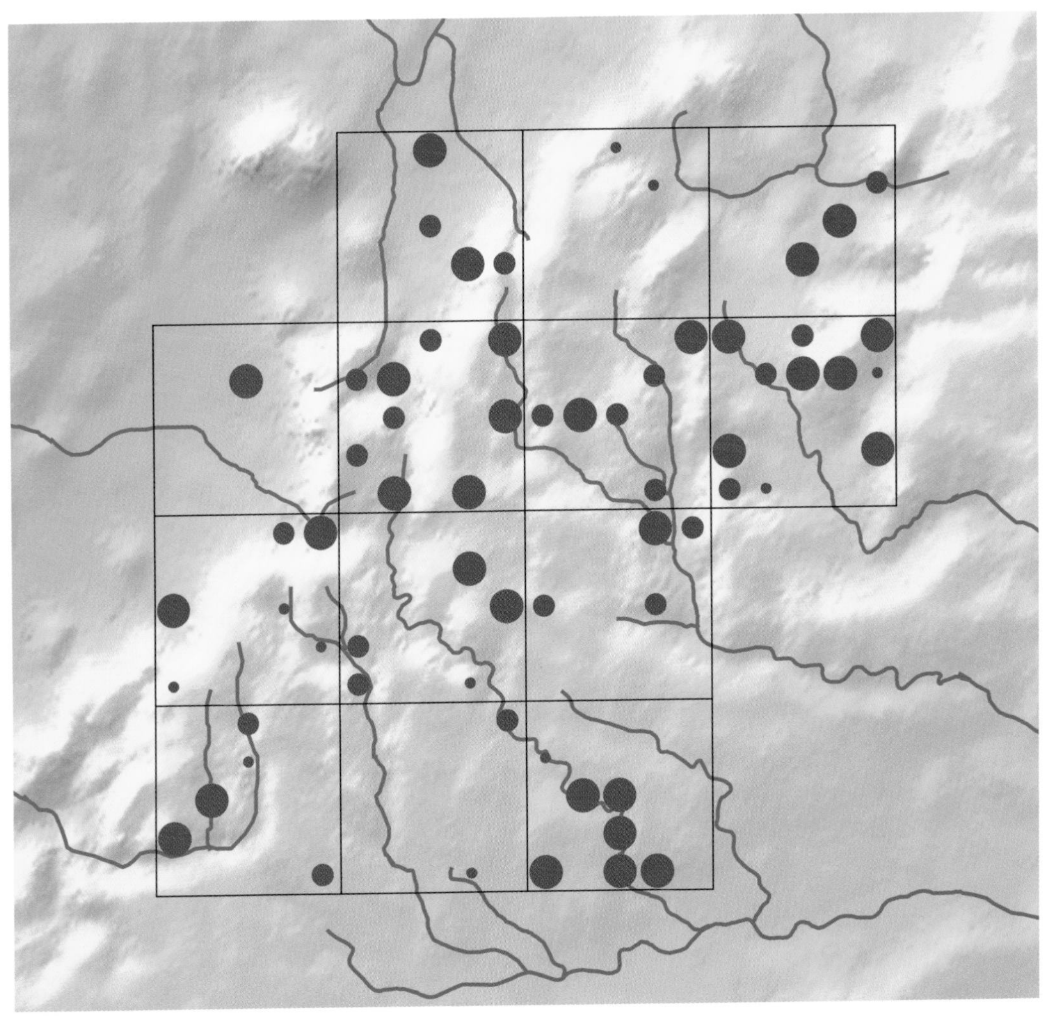
2003–07 survey

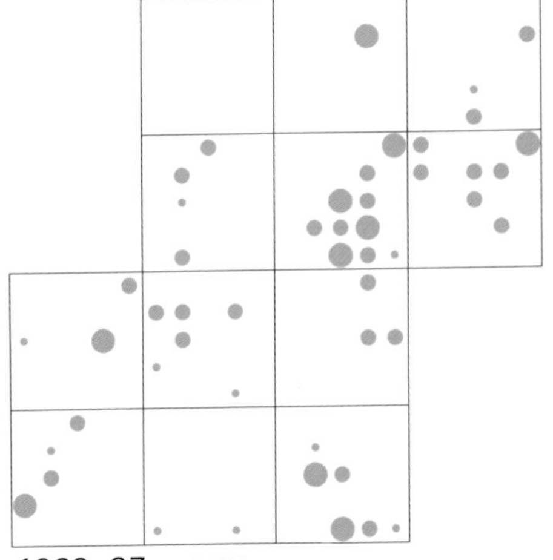
1983–87 survey

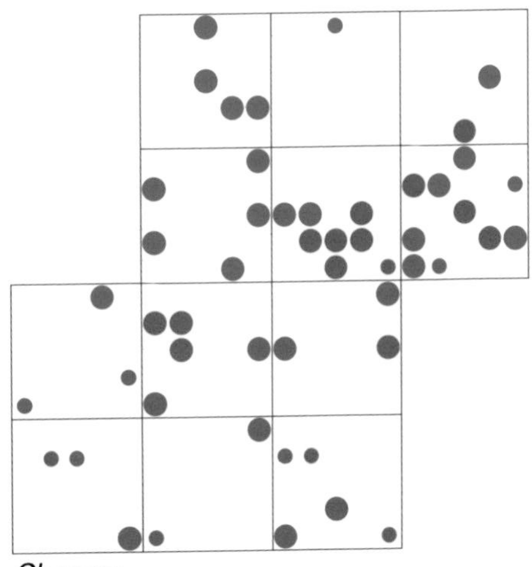
Change

Great Crested Grebe

Podiceps cristatus

ROB BROOKES

Resident breeder

UK conservation status: **GREEN**

Long-term UK trend: Not available

Recent population trend (1994–2007): +18%

Numbers of tetrads with breeding records:

	2003–07		1983–87
	13 squares	12 squares	12 squares
Possible	4	3	1
Probable	1	1	2
Confirmed	7	6	5
Totals	12	10	8

Great Crested Grebes have a more restricted breeding habitat than Little Grebes, since they need relatively large bodies of open water surrounded by plenty of aquatic vegetation. A limited number of sites provide this habitat in the survey area, and most known sites were occupied, principally: Pittville Park in Cheltenham, the gravel-pits at Bourton-on-the-Water and Fairford, the reservoirs at Dowdeswell and Witcombe, the ornamental lake at Cirencester Park, and Rendcomb Park. Misarden Park Lake was not occupied, possibly because it is in the bottom of a deep valley.

The species is generally single brooded, the young having a long period of dependence on the adults, which makes confirmation of breeding very easy. The breeding season is quite extended—eggs can be laid from April to at least August and, in one exceptional case, eggs were laid in December 2006 in Pittville Park. Here, the adults incubated the eggs for nearly two months, but they failed to hatch. Although there is the potential for the birds to breed late in the season, and so be overlooked, it is not believed that any sites were missed.

Nationally, numbers of Great Crested Grebes increased markedly during the twentieth century, owing to extensive gravel extraction programmes. As suitable habitat in the survey area has not changed much since the 1980s survey (the main gravel extraction in the Fairford area taking place just outside the recording area), this survey found that the distribution of the bird is much as it was then, as evidenced by the paucity of dots on the change map. There are probably around 20 breeding pairs in the area.

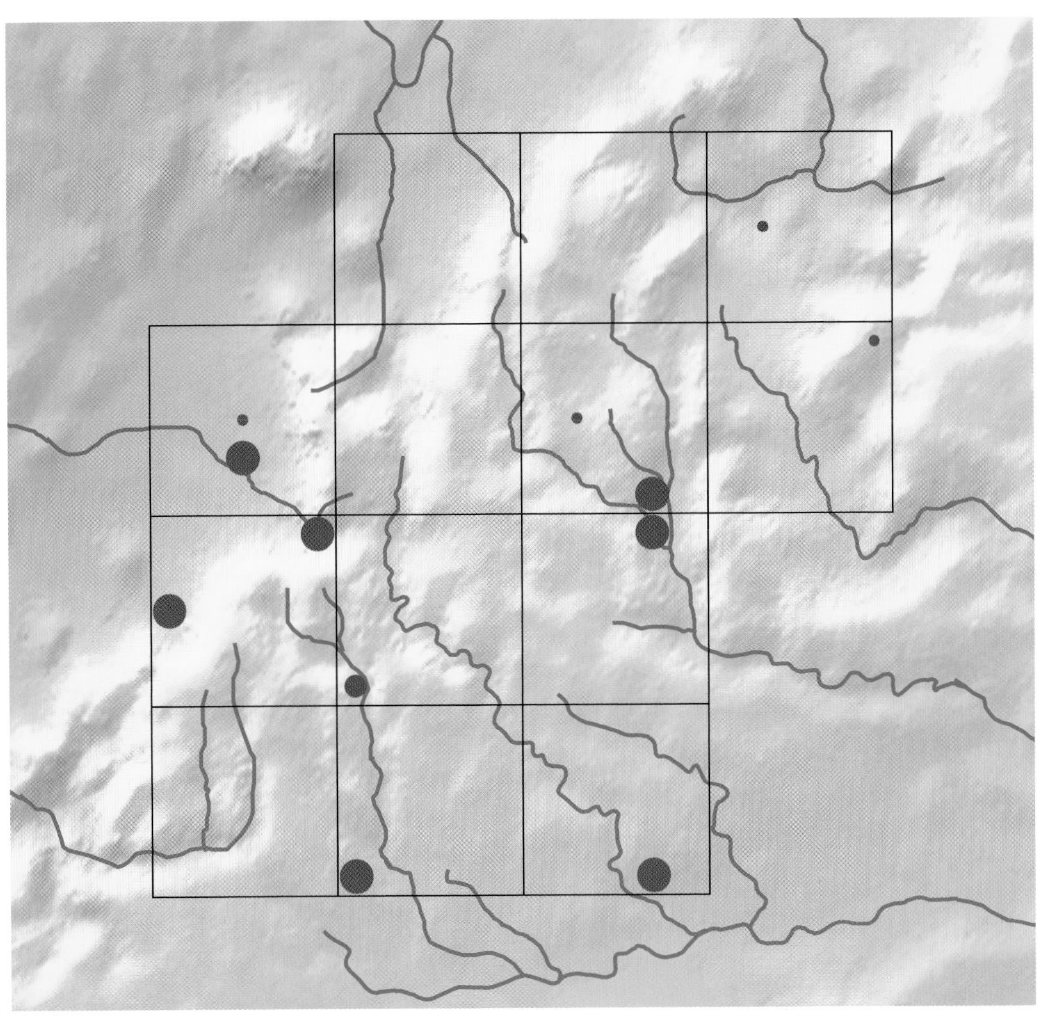

2003–07 survey

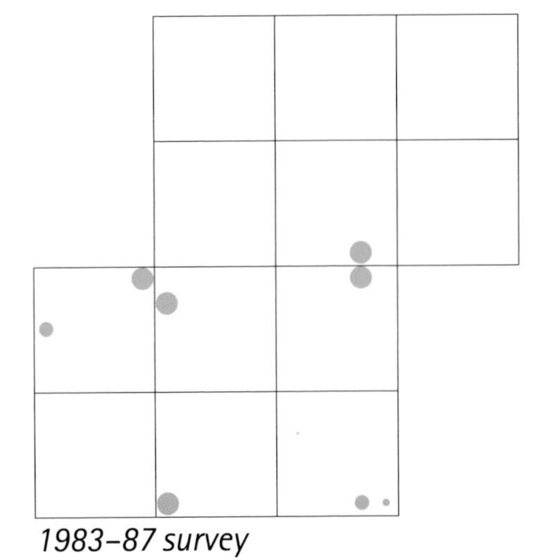

1983–87 survey

Change

Grey Heron

Ardea cinerea

Grey Herons usually nest colonially, most often in tall trees, but also in lower trees and bushes. Despite the large size of both the bird and its nest, actually locating breeding areas (and subsequently finding the nests) is surprisingly difficult. In the Cotswolds, coniferous trees are preferred as nesting sites, which makes the location of nests more difficult, particularly early in the season.

However, locating occupied nests, or observing adults returning to the same site, is still probably the only reliable way of confirming that breeding is occurring in a particular tetrad, as the Grey Heron's feeding range is probably more extensive than that of any other species breeding in the survey area. Therefore, the data for this bird must be treated with extra caution. The wide range of 'possible' records reflects the bird's long-distance feeding trips rather than possible breeding, and some of the breeding activity may have been of birds some distance from the nest site.

In the first survey in the 1980s, only two heronries were confirmed in the survey area, with breeding activity suspected in another two. The greater survey effort in the new survey has led to more sites being located, generally of one to three nests. Whether the increase in confirmed breeding is a result of a real expansion in the bird's distribution (possibly resulting from an increase in the number of breeding pairs), or of the greater survey effort, is unclear. However, it is of interest that about 10 successful nests have been found in hitherto unknown sites, and more investigation of these 'new' areas is warranted. The main breeding population remains at the Lower Harford heronry on the dip slope near the River Windrush and consists of around 10 nests each year.

As noted above, the numbers of possible breeding records in the above table is likely to be too large. Whereas the locations of the main heronries in the survey area are well known, suitable breeding habitat is present in most tetrads. Some recorders interpreted the presence of a feeding bird in suitable habitat as 'possible', whereas others, knowing there to be no heronry nearby and therefore no possible breeding, did not record the bird. For the same reason, the distribution map is probably most useful as an indication of how far Grey Herons move from their breeding sites to feed, rather than an indication of breeding distribution.

Resident breeder

UK conservation status: GREEN

Long-term UK trend (1929–2006): +58%

Numbers of tetrads with breeding records:

	2003–07		1983–87
	13 squares	12 squares	12 squares
Possible	**62**	60	40
Probable	7	7	2
Confirmed	7	7	2
Totals	**76**	**74**	**44**

RICHARD TYLER

Species sponsored by Andy Lewis

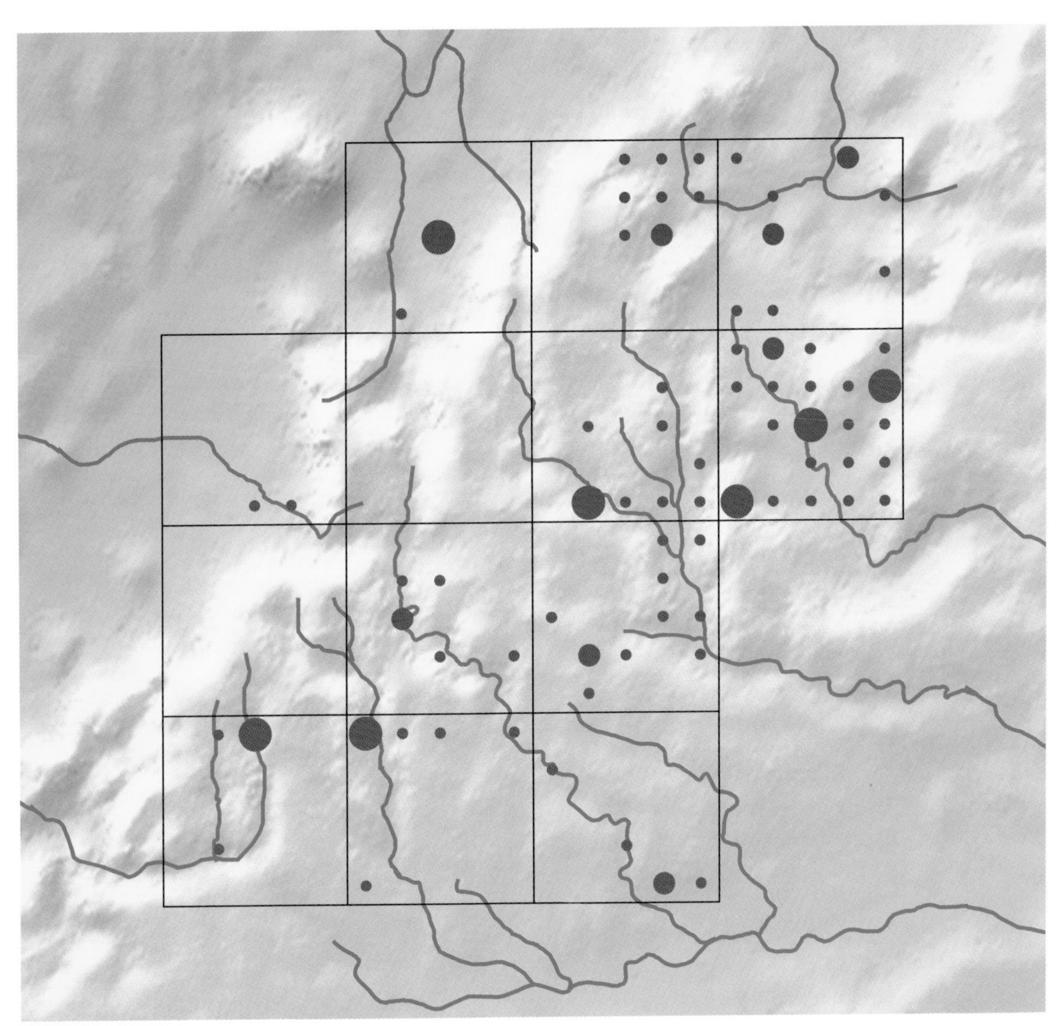

2003–07 survey

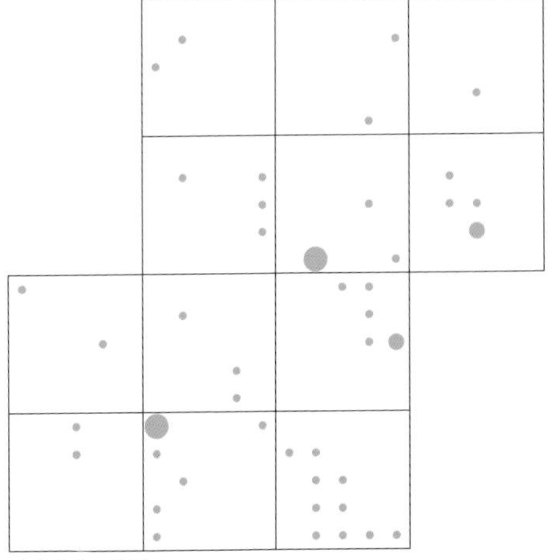

1983–87 survey

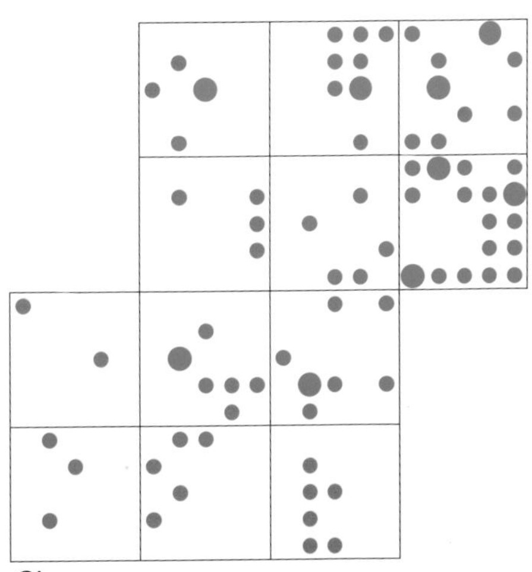

Change

Sparrowhawk (Eurasian Sparrowhawk)

Accipiter nisus

GRAHAM WATSON

Resident breeder

UK conservation status: GREEN

Long-term UK trend (1975–2006): +106%

Recent population trend (1994–2007): -12%

Numbers of tetrads with breeding records:

	2003–07		1983–87
	13 squares	12 squares	12 squares
Possible	**119**	108	83
Probable	**45**	42	23
Confirmed	**36**	33	13
Totals	**200**	**183**	**119**

The Sparrowhawk was one of the most difficult species to survey. As in the first survey in the 1980s, the majority of records were only 'possible', reflecting the fact that encounters with Sparrowhawks tend to be brief. Observation of pairs in display flight over suitable woods was a good indication of breeding activity. Confirmation of breeding was achieved either by finding occupied nests or by hearing young birds calling from a specific area. Since breeding sites were usually in conifers, suitable woods could be identified and nests looked for near the trunk of a tree (on occasion several nests from previous years could be found in close proximity). Observing adults carrying prey was also sometimes a good pointer to the location of nest sites, but was not itself a conclusive indicator of breeding, as Sparrowhawks often feed some distance from where they catch their prey.

For 2003–07, the records are well distributed throughout the survey area, although, as in the 1980s, there was a slight tendency for 'probable' and 'confirmed' records to be from the Cotswold escarpment. The even distribution elsewhere probably reflects the fact that the Cotswolds is well populated with both moderately sized wooded areas and smaller copses and spinneys, all of which provide suitable nesting areas for Sparrowhawks.

Data from the two surveys indicate a 54% increase in the number of occupied tetrads. Given the difficulty in surveying Sparrowhawks, this could be largely due to the more extensive and intensive surveying in 2003–07; increased survey effort may also be responsible for the large increase (150%) in the number of confirmed breeding records. Anecdotal evidence from recorders suggests that there had been no significant sustained change in local Sparrowhawk numbers between the surveys, though there might have been some periodic fluctuations.

Species sponsored by Charles and Jackie Johnson

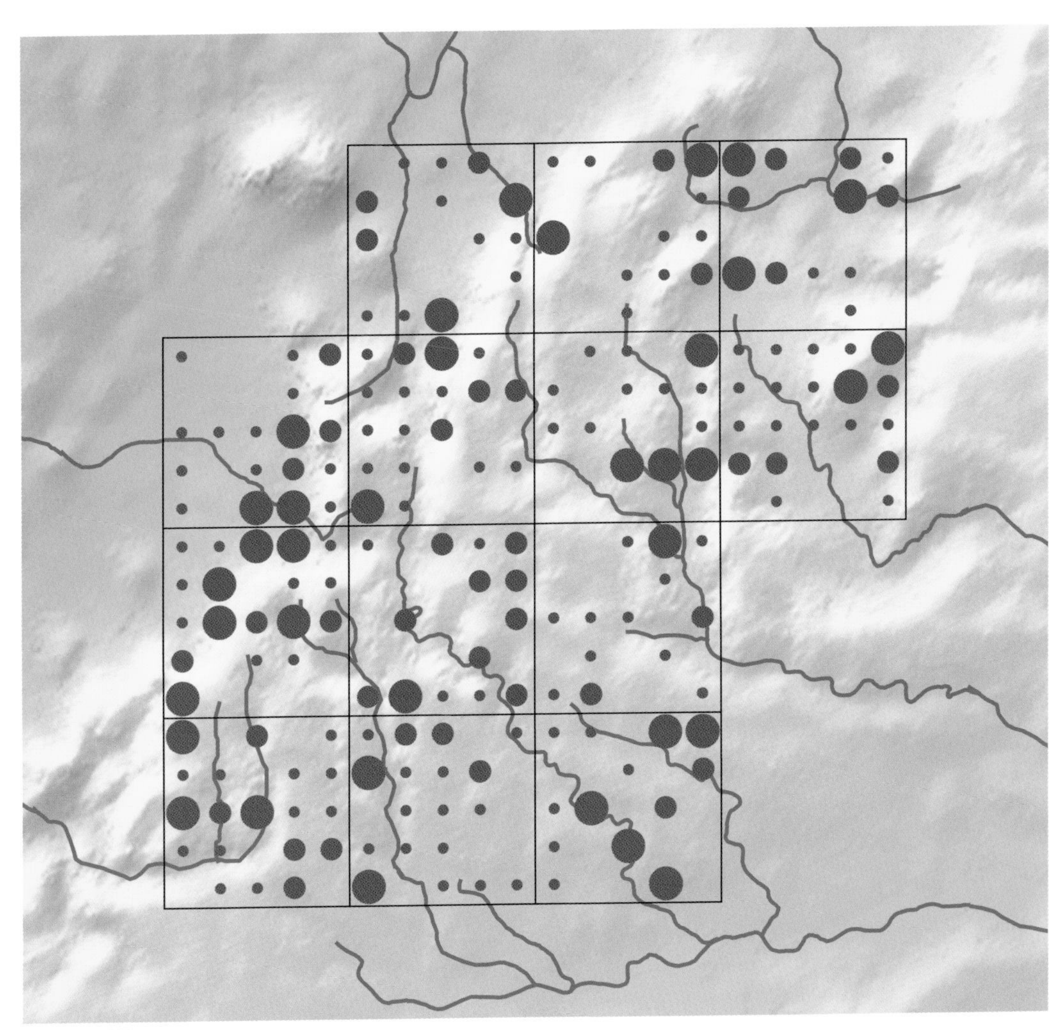

2003–07 survey

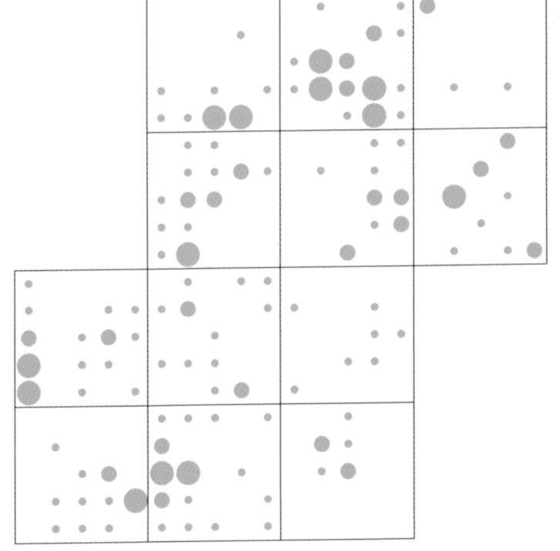

1983–87 survey

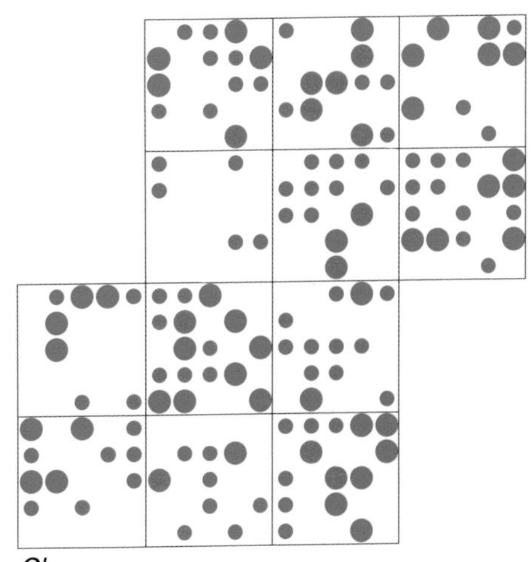

Change

Buzzard (Common Buzzard)

Buteo buteo

Resident breeder

UK conservation status: GREEN

Long-term UK trend (1970–2006): +518%

Recent population trend (1994–2007): +56%

Numbers of tetrads with breeding records:

	2003–07		1983–87
	13 squares	12 squares	12 squares
Possible	**43**	39	47
Probable	**166**	156	20
Confirmed	**101**	94	11
Totals	**310**	**289**	**78**

Buzzards build fairly large nests in trees, and egg-laying in the survey area starts around mid-April, so in many cases it was possible to see the nests before the trees became covered in foliage. Being aerial birds, Buzzards are very easy to spot displaying. However, they do tend to have large territories: deciding whether a bird was a 'native' or from a neighbouring tetrad did cause some indecision in how to record them, and may have led to a slight overestimate in the number of tetrads occupied for breeding purposes. Despite this, it seems that a significant majority (85–90%) of tetrads in the survey area were occupied, compared with one-quarter in the 1983–87 survey. Confirmation of breeding could be most easily achieved by finding an occupied nest early in the season or by hearing the plaintive calls of young in the nest from about mid-July onwards.

Traditionally associated with hilly areas with plentiful woodland, the recent dramatic recolonization of much of central England has seen Buzzards occupy virtually any rural area with a tree to nest in and a good source of food. This has been especially noticeable in the survey area, with many nests being found in trees on arable farmland in the Avon Vale, although they are still most numerous in the wooded areas of the Cotswold escarpment and Stroud Valleys.

The Buzzard has become comfortably the most common and widely distributed raptor of the area. Comments in the first Atlas included: '... some extensive areas of woodland are avoided ...'; 'The paucity of records from the east of the region follows the national pattern ...'; and 'The Atlas area now holds about 15 breeding pairs ...'. In the 1983–87 survey, Buzzards were almost entirely absent from the Stour Valley, the Evenlode Valley and the Thames tributaries, but in this survey they were as common there as elsewhere, and it was often possible to see, from a vantage point in spring, several birds in the air over their breeding territories.

The number of tetrads in which breeding was confirmed has increased nearly 10-fold. The increase in total number of breeding pairs may be as high as 20-fold, with about 300 breeding pairs in the region. The NCOS Winter Random Square Survey (see Preface) suggests a year-on-year increase of 18% from 1995 to 2007.

ROB BROOKES

Species sponsored by Siobhan Barker and Christopher Main

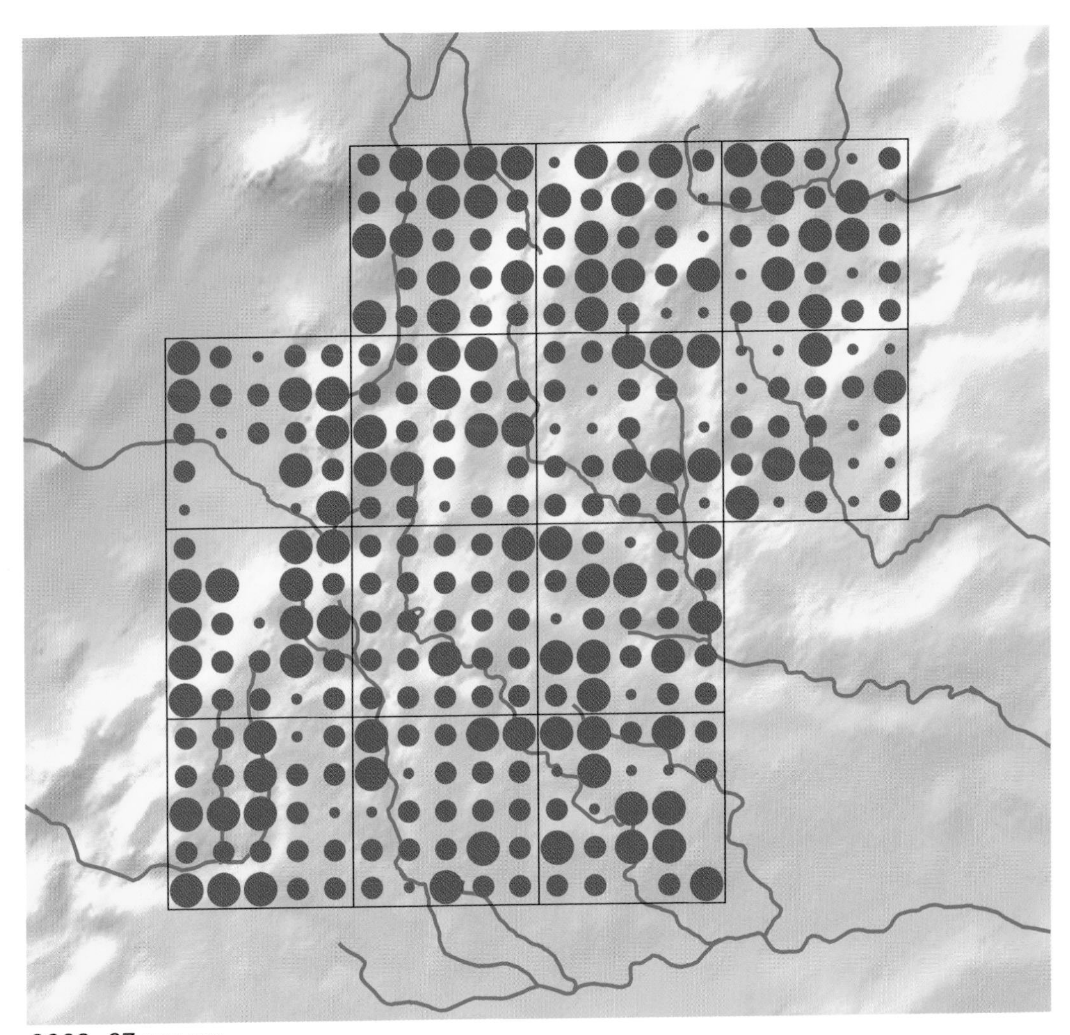

2003–07 survey

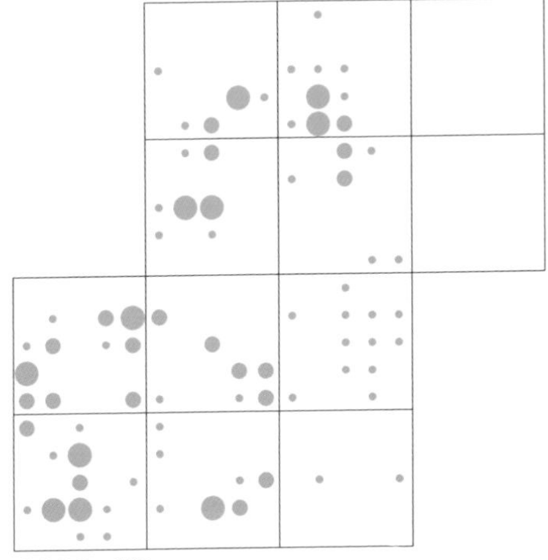

1983–87 survey

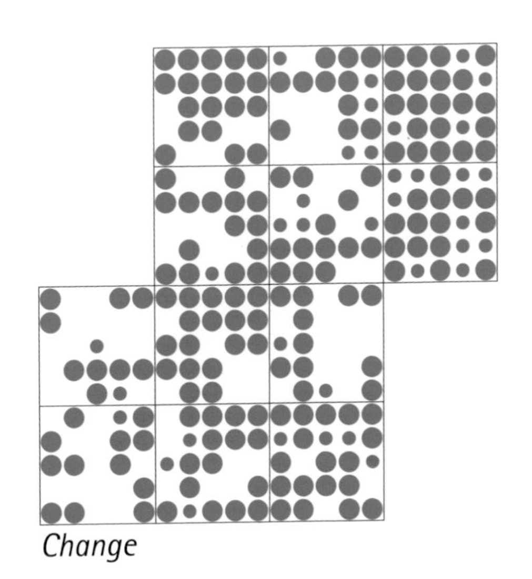

Change

Kestrel (Common Kestrel)

Falco tinnunculus

ROB BROOKES

Resident breeder

UK conservation status: AMBER

Long-term UK trend (1970–2006): -31%

Recent population trend (1994–2007): -29%

Numbers of tetrads with breeding records:

	2003–07		1983–87
	13 squares	12 squares	12 squares
Possible	**149**	135	102
Probable	**87**	85	42
Confirmed	**53**	50	14
Totals	**289**	**270**	**158**

The majority of Kestrel records were of individual birds in suitable breeding territory. Observation of breeding activity was mostly of males and females in close proximity, or of males or pairs involved in their winnowing display flight. Confirmation of breeding was obtained from seeing adults returning repeatedly to a nest site, or family groups hunting together.

In the 1980s survey, Kestrels were observed in just over half of the tetrads surveyed—more than the number of tetrads apparently holding breeding Sparrowhawks. However, it was suggested that the actual number of breeding pairs of Kestrels might be fewer, because ease of observation facilitated recording. The same may have been true in 2003–07, though the disparity between the numbers of records for the two species is significantly greater, with the Kestrel again being the more commonly recorded, being found in over 85% of tetrads. However, confirmation of breeding of both species requires patient observation, and the percentages of confirmed records for the two are similar.

The conclusion reached over the status of Sparrowhawks is that there has not been any significant change in distribution since the first Atlas survey period, and the same may be true of Kestrels; although the increase in number of occupied tetrads appears to be significant, it could be due to the greater survey effort. In the first Atlas, it was stated: 'With the changes of land use from pasture to arable, prime hunting areas have diminished and with them potential nesting sites in over-mature trees.' Since that time, this change has not been reversed, and so it is encouraging that the number and distribution of Kestrels appears to be fairly stable. The opinion of observers is that Kestrels occur in higher numbers in arable areas than in pasture, and it may be that margins around arable fields and adjoining minor roads provide an adequate food supply. In general, the distribution of breeding activity records (categories 2 and 3) is fairly uniform across the survey area.

Species sponsored by Peter Ormerod

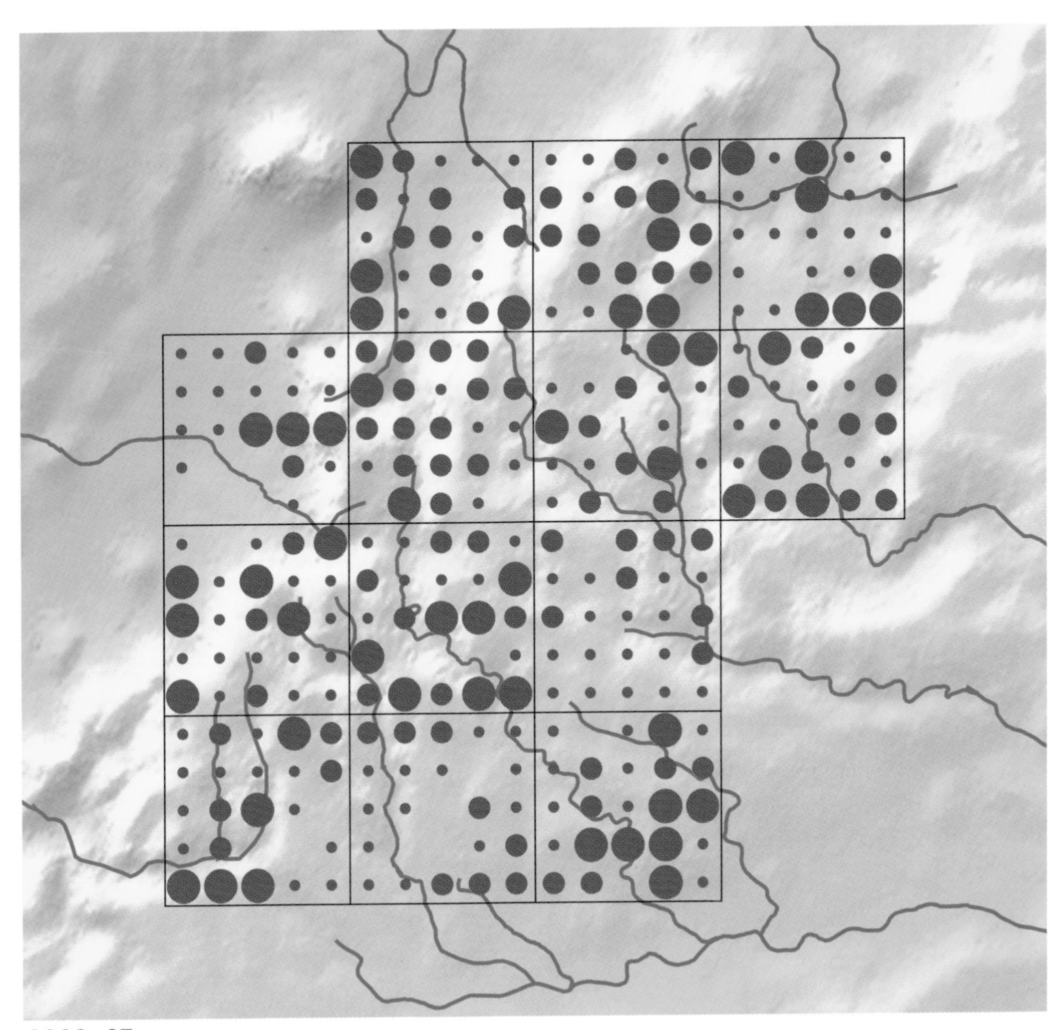

2003–07 survey

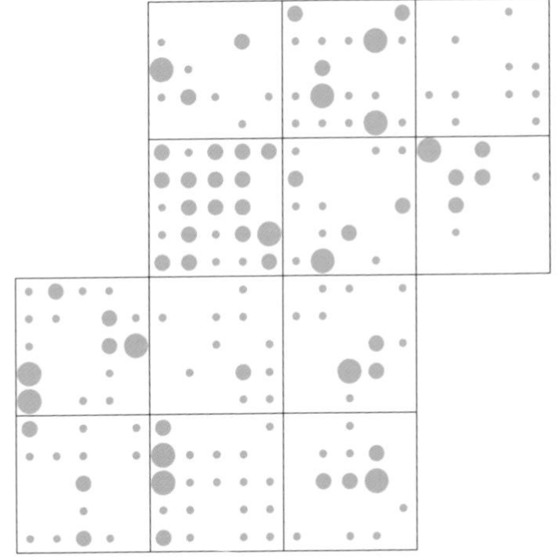

1983–87 survey

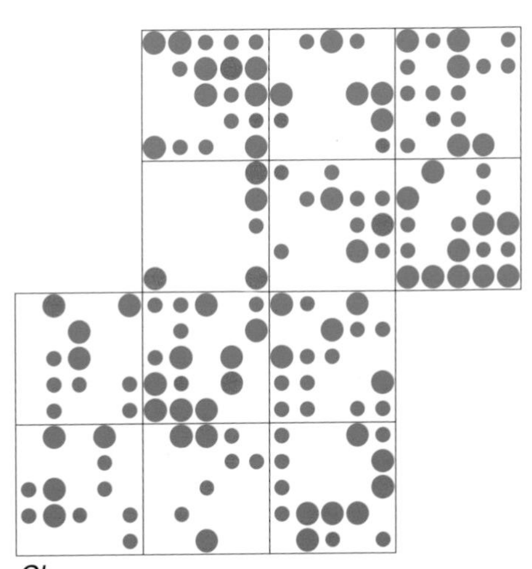

Change

Hobby (Eurasian Hobby)

Falco subbuteo

The Hobby is undoubtedly one of the most difficult species to survey effectively. It hunts over a wide area, and frequently carries prey more than a kilometre to the nest site. Most sightings were of single birds, with no specific indication of breeding activity. Hobbies' nests (typically disused crows' nests) tend to be in isolated trees in open country and, as the birds are also subject to protection legislation, care had to be taken not to disturb them at or near a nest site.

They were sparsely distributed throughout the area, in farmland up to about 200 m, but avoiding deep valleys and extensive woodland, and were less frequent in the Avon and Severn Vales. The nest sites chosen were usually in the vicinity of villages with high hirundine populations, and this is where many of the sightings of hunting birds occurred.

Sightings tended to be concentrated around the first two weeks of May when the birds arrive to set up territory, and in the second half of summer when hirundine populations attract adults seeking food for their young. It is believed that many of the single sightings could indicate breeding pairs in the vicinity, but only the 'probable' and 'confirmed' dots on the distribution map are likely to be a reasonable representation of the species' status.

Migrant breeder

UK conservation status: **GREEN**

Long-term UK trend: Not available

Recent population trend (1994–2007): +14%

Numbers of tetrads with breeding records:

	2003–07		1983–87
	13 squares	12 squares	12 squares*
Possible	**25**	25	25
Probable	**4**	4	3
Confirmed	**9**	9	1
Totals	**38**	**38**	**29**

*For confidentiality reasons the 1983–87 Atlas mapped all records as category 1.

Comparison of the two surveys suggests that there has been little change in distribution, or in the total number of occupied tetrads. However, nine confirmed breeding locations were found in this survey, compared with only one in the previous survey. Although this probably reflects observers' better knowledge of the species and its habits, as well as the greater survey effort, it is possible that it also indicates a slightly increased breeding population compared with that found in the 1980s.

GRAHAM WATSON

Species sponsored by Debbie Colbourne and Julian Miles

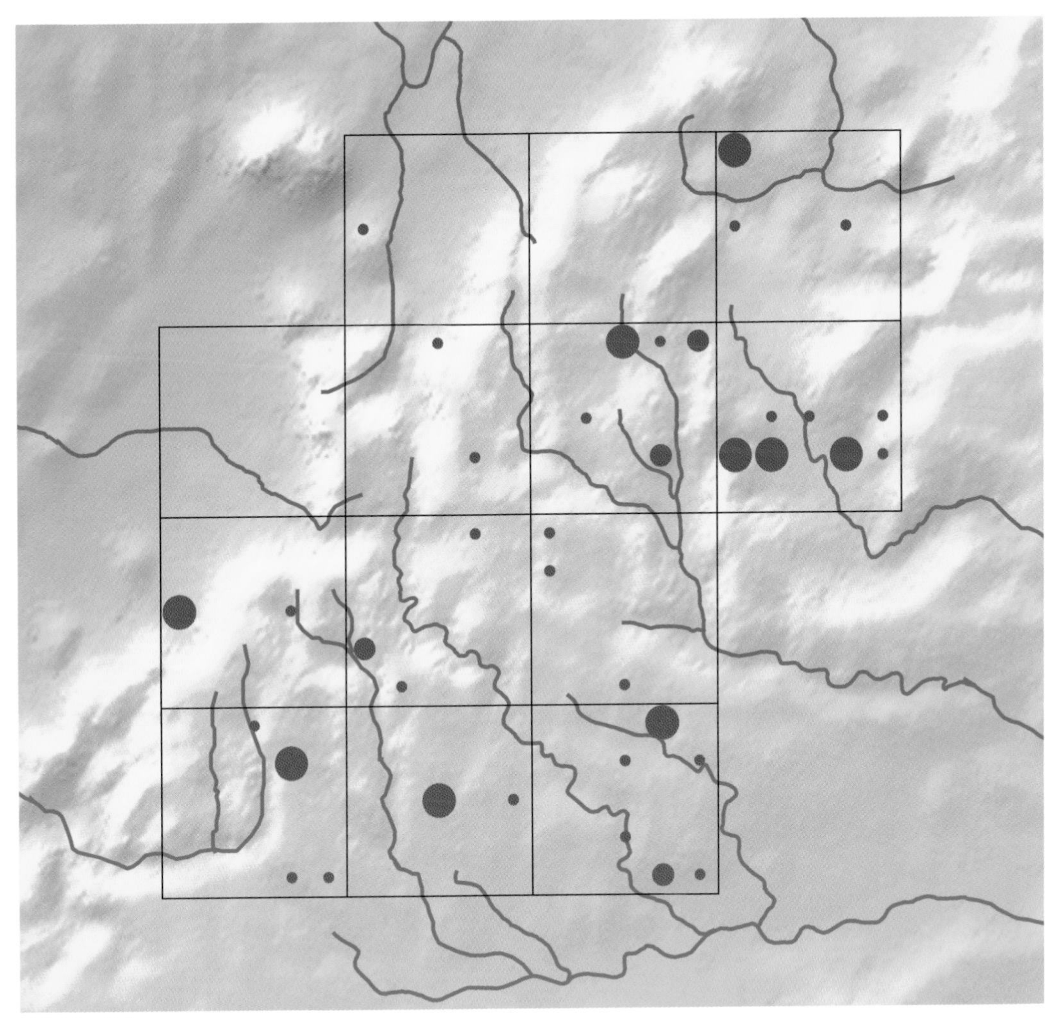
2003–07 survey

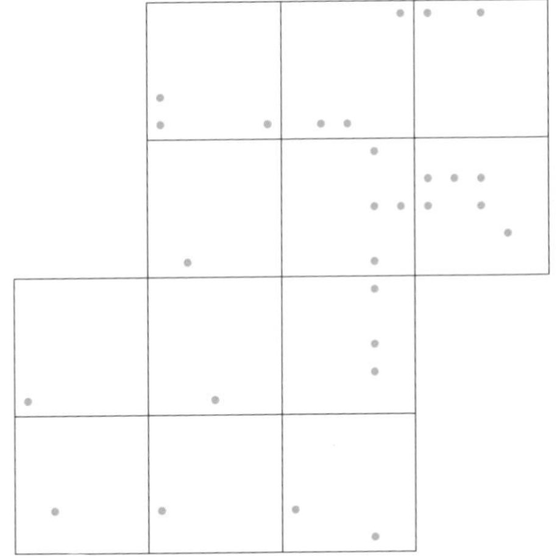
1983–87 survey

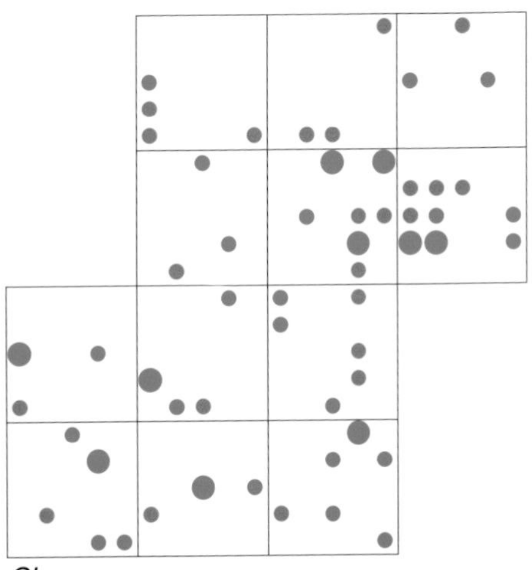
Change

Moorhen (Common Moorhen)

Gallinula chloropus

ROB BROOKES

Resident breeder

UK conservation status: **GREEN**

Long-term UK trend (1970–2006): -4%

Recent population trend (1994–2007): +16%

Numbers of tetrads with breeding records:

	2003–07		1983–87
	13 squares	12 squares	12 squares
Possible	**37**	36	39
Probable	**45**	44	32
Confirmed	**124**	113	77
Totals	**206**	**193**	**148**

The Moorhen is a common and widespread breeding bird in the survey area. It is found along the main river valleys as well as on large lakes and small ponds (including rural 'garden' ponds), and even regularly in narrow field margin drainage ditches. Its wide distribution confirms that the superficially dry Cotswolds environment contains many tetrads not completely waterless.

Observation of adults in pairs and/or building nests made locating territories very easy—in fact, it was unusual to find a suitable piece of water without a pair of Moorhens in residence. Confirmation of breeding was also often quite straightforward, through observation of birds sitting on nests or of adults with broods of newly hatched young. Nests are usually very well hidden in reeds standing in water, but occasionally exposed sites will be chosen, although these will often suffer predation. The high percentage of confirmation is probably an indication of their relative abundance as well as the wide distribution of the species.

The 1983–87 survey also found the Moorhen to be widespread, and the change map shows that its breeding distribution was similar in both surveys. The moderate increase in the total number of records, and the small increase in number of confirmed breeding records, can be attributed to the increased survey effort in 2003–07.

Species sponsored by Peter Ormerod

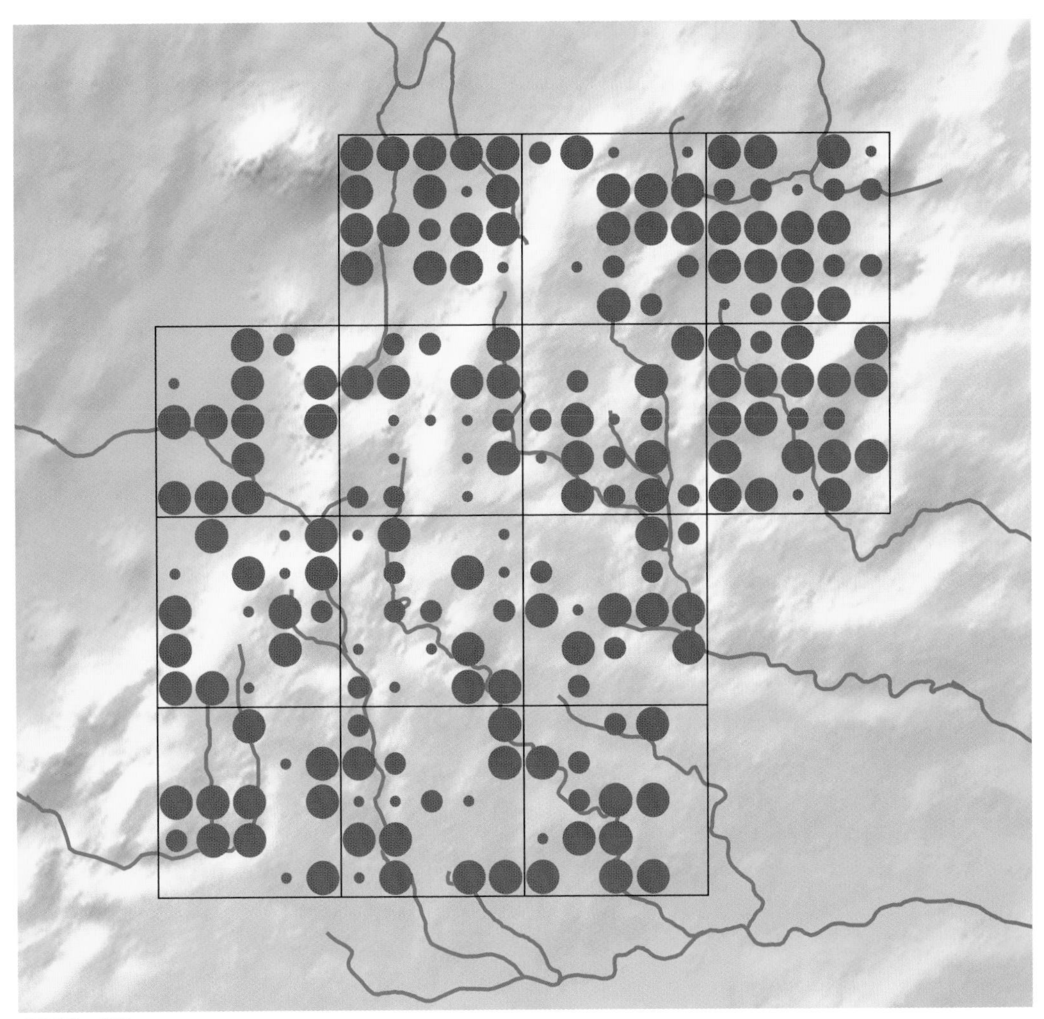

2003–07 survey

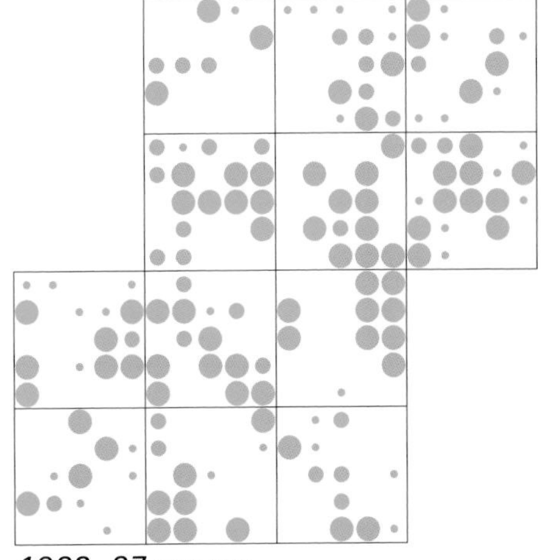

1983–87 survey

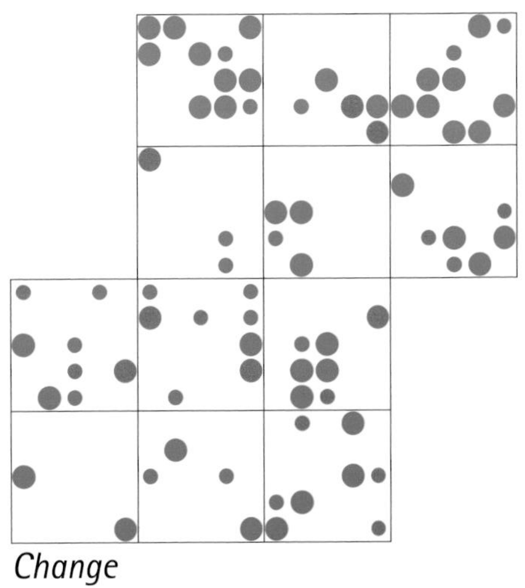

Change

Coot (Common Coot)

Fulica atra

Coots are generally more restricted than Moorhens in their breeding requirements. The first Atlas suggested that, in the survey area, they 'are unlikely to breed on bodies of water less than 0.5 hectare in area ...', although their colonization of the slow-moving rivers to the south and east of Bourton-on-the-Water (the Windrush and the Churn at Fairford) was noted. Interestingly, in the new survey, Coots were also found on relatively small ponds, and the number of suitable sites in the generally dry habitat of the Cotswolds means that it is still a common and quite widely distributed species (although it appears to have been lost from some tetrads in the Windrush river system around Bourton-on-the-Water).

Location of Coot breeding areas was relatively straightforward, building and occupation of nests being frequently observed, as was the sighting of broods of young birds on open ponds. This meant that a very high proportion (about 75%) of the records were of confirmed breeding.

As with the Moorhen, the greater survey effort compared with that in the 1980s meant that there were more breeding records in total,

Resident breeder

UK conservation status: **GREEN**

Long-term UK trend (1970–2006): +77%

Recent population trend (1994–2007): +32%

Numbers of tetrads with breeding records:

	2003–07		1983–87
	13 squares	12 squares	12 squares
Possible	**8**	8	6
Probable	**19**	18	21
Confirmed	**77**	74	52
Totals	104	100	79

and more confirmed breeding records. Although the data suggest that there has been little change in the distribution of either species, the change map for the Coot does hint at a modest increase in the number of occupied sites in the Stour and Evenlode Valleys; clearly some sites that were occupied in the earlier survey were unoccupied in this survey, and vice versa.

RICHARD TYLER

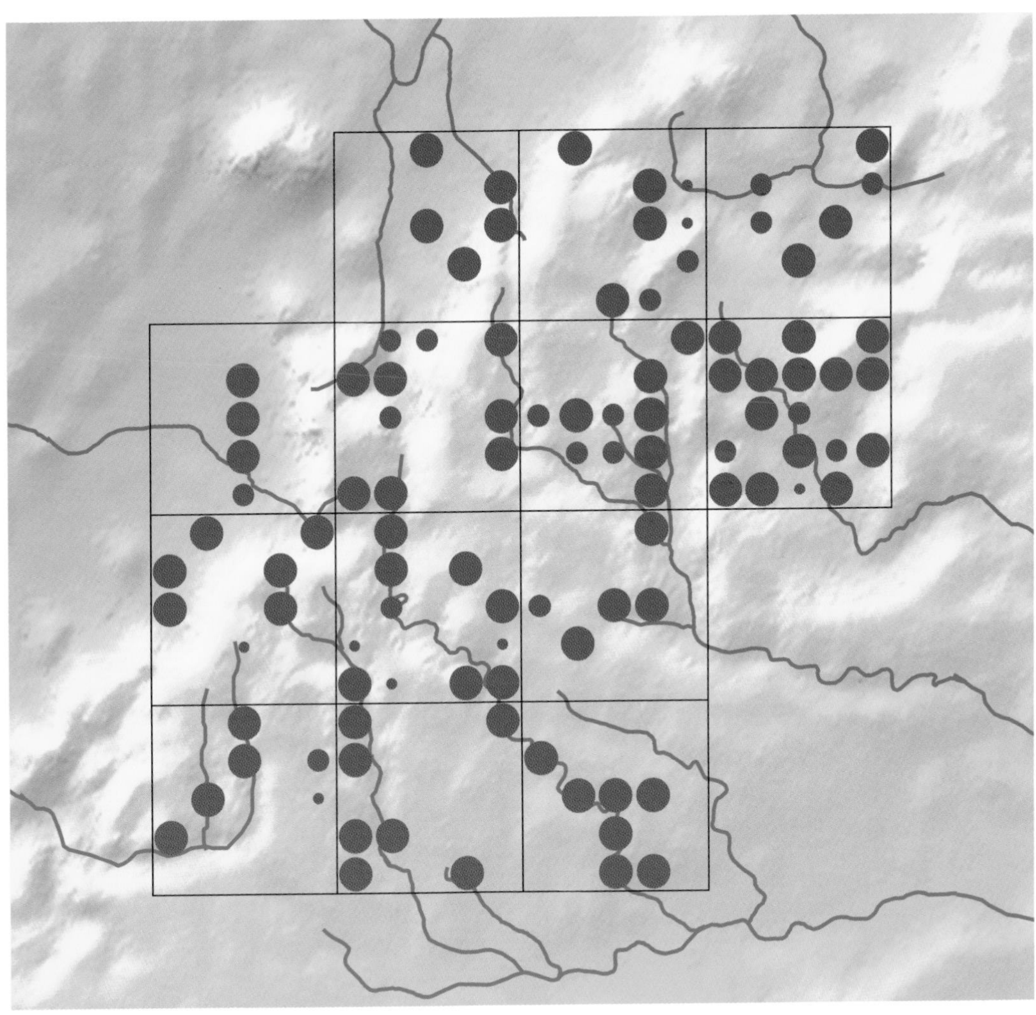
2003–07 survey

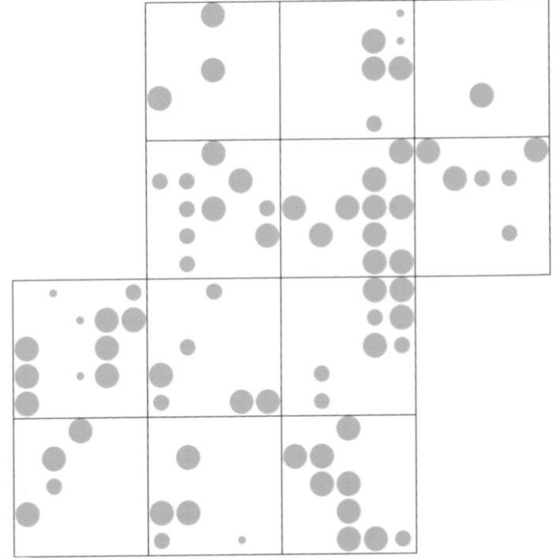
1983–87 survey

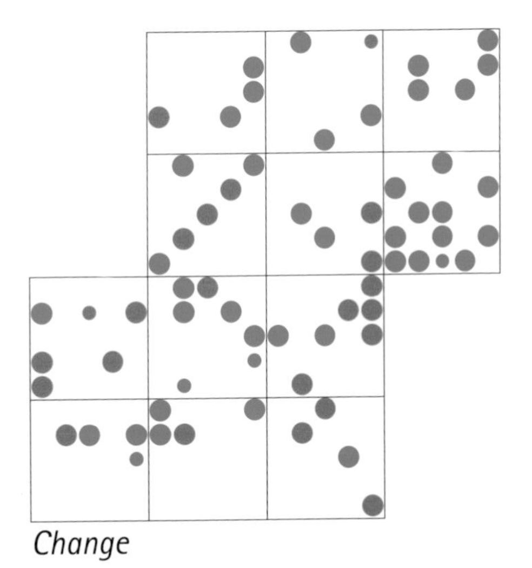
Change

Lapwing (Northern Lapwing)

Vanellus vanellus

The preferred breeding habitat for Lapwings is open, flat or gently undulating land with relatively short vegetation; in the Cotswolds, this has been provided by spring-sown cereals and set-aside, the latter being particularly valuable. It has been suggested that south-facing locations are favoured. The vocal and spectacular display flights of adults make breeding activity easy to spot, although confirmation of breeding requires the young birds to be seen (which is not easy), unless incubating adults can be found. Displaying birds were found in pasture and arable settings both in the Severn and Avon Vales and on higher ground, often in areas of set-aside. In many cases, birds were seen to display for a week or two (and therefore were recorded as probable breeders) and then desert the area. There was only one confirmed breeding record in the Vale areas, and very few records from the extreme southwest of the recording area, around the Stroud Valleys.

The Lapwing is now a much less common breeding bird in the Cotswolds than it was in the 1980s. Following the 1983–87 survey, around 200 pairs were estimated to breed; in the present study, improved coverage notwithstanding, it is believed the number attempting to breed may have been almost halved. Although the favoured areas (revealed by the pattern of distribution) are much the same as they were, distribution density has clearly declined, with Lapwings being present in significantly fewer tetrads. However, the number of breeding territories found, the number of areas in which displaying birds were seen, and the total number of confirmed breeding territories have probably been greater than was expected at the outset of the survey. Recent financial encouragement to farm in an environmentally friendly fashion may have resulted in more areas becoming available for Lapwings to breed, although the suspension of set-aside payments to farmers may act to counter this.

Migrant/resident breeder

Gloucestershire BAP species

UK conservation status: AMBER

Long-term UK trend (1970–2006): -47%

Recent population trend (1994–2007): -18%

Numbers of tetrads with breeding records:

	2003–07		1983–87
	13 squares	12 squares	12 squares
Possible	**19**	19	30
Probable	**59**	56	69
Confirmed	**34**	33	45
Totals	112	108	144

ROB BROOKES

Species sponsored by Ian Boyd

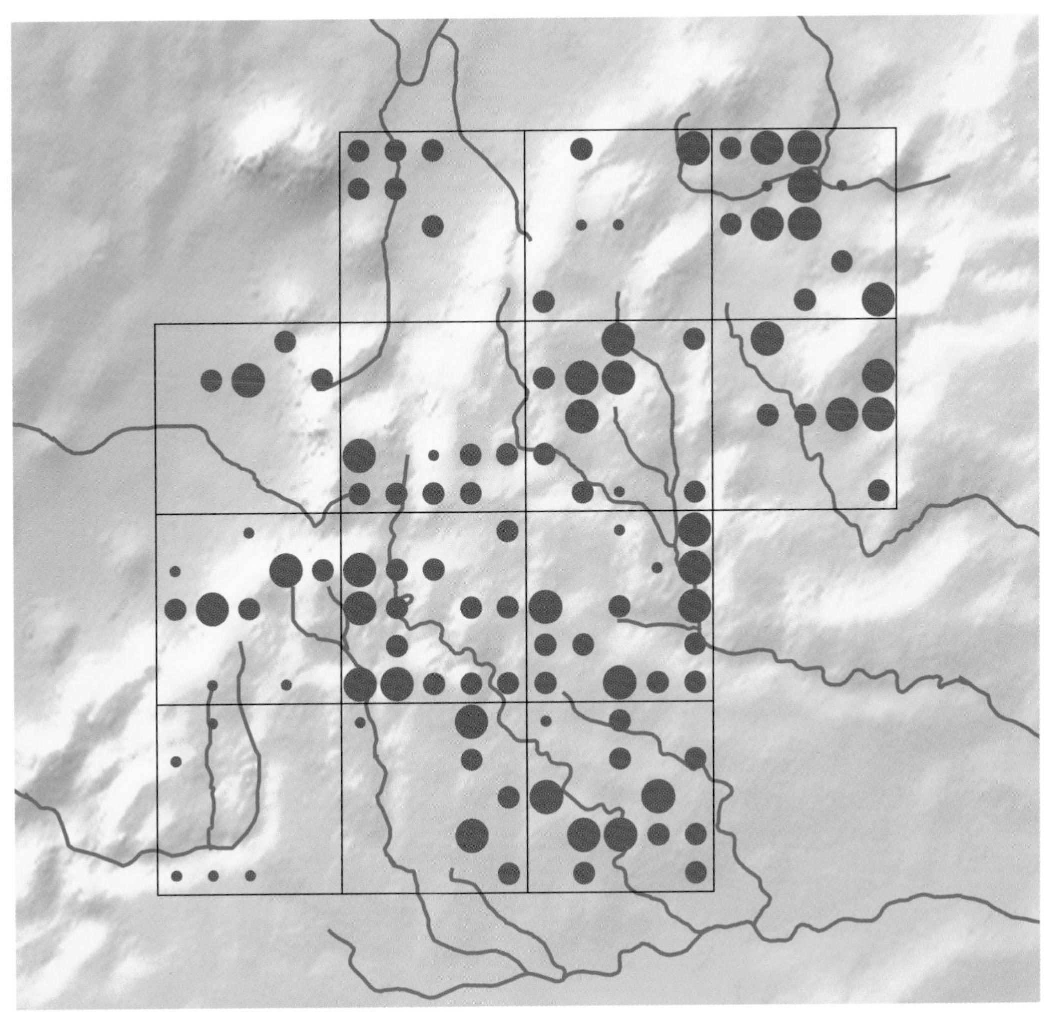

2003–07 survey

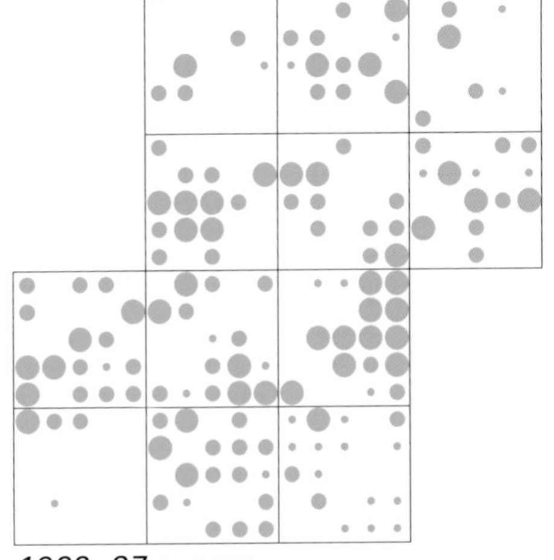

1983–87 survey

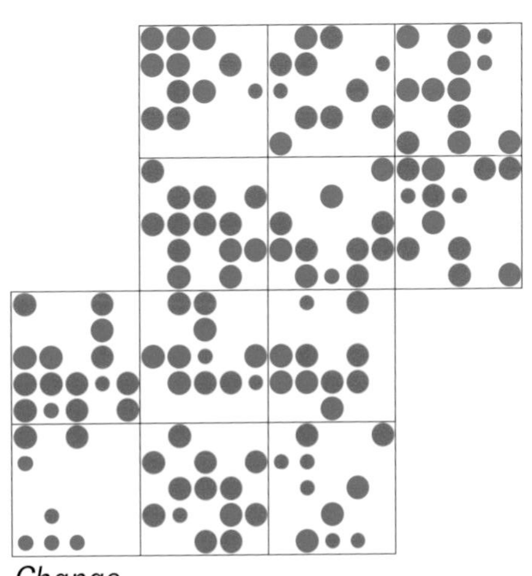

Change

Curlew (Eurasian Curlew)

Numenius arquata

DAVE PEARCE

Migrant/resident breeder

UK conservation status: AMBER

Long-term UK trend (1970–2006): -54%

Recent population trend (1994–2007): -36%

Numbers of tetrads with breeding records:

	2003–07		1983–87
	13 squares	12 squares	12 squares
Possible	3	3	1
Probable	7	7	4
Confirmed	2	2	1
Totals	12	12	6

The Curlew has always been a very localized breeder in the survey area. Its distinctive call and associated display flight make the location of established territories relatively easy, and it is unlikely that many breeding sites were missed. However, confirmation of breeding is difficult: since nesting birds are wary and quiet, observing family parties with young birds is the most likely way to achieve this.

Although the total number of breeding records is small, as it was in the 1980s, the change in breeding sites is of interest. In the 1980s there was a very well established breeding population, on damp, lowland grassland between Tewkesbury and Toddington, the eastern edge of which lies within the recording area. By the mid- to late 1990s, this area had been deserted, although breeding was confirmed for a few years just to the northeast between Childswickham and Aston Somerville, again in low-lying grassland. During the early part of the 2003–07 survey, there were records of isolated pairs of birds in this area, but no indication of breeding.

However, breeding did take place in the area between Hawling and Salperton in some years in the 1990s, and was confirmed during the 2003–07 survey. This site differs from the others in being upland grassland. Breeding activity was also noted in another upland grassland area near Cleeve Common, as well as in several lowland grassland locations in the east of the region near Bledington in the Evenlode Valley. Breeding was confirmed at one of these sites.

Species sponsored by Peter Ormerod

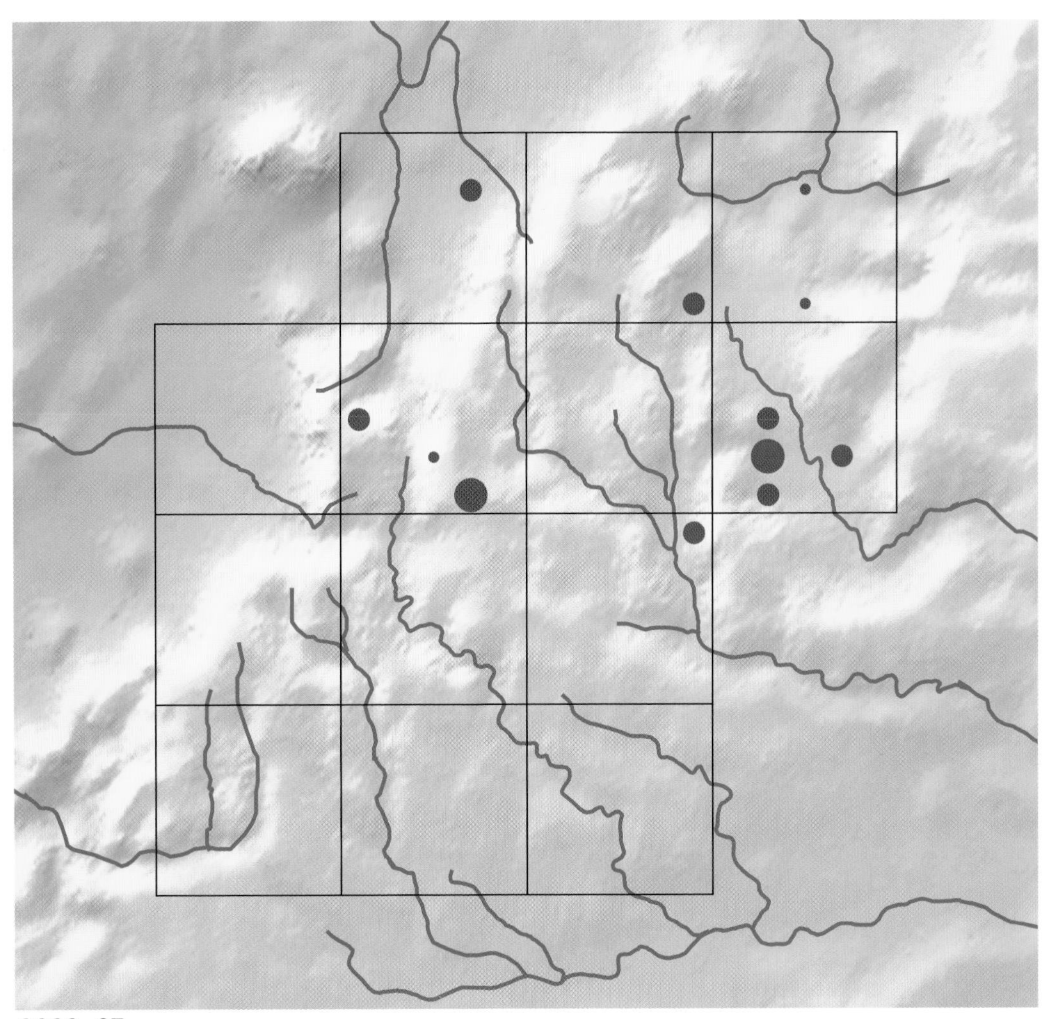

2003–07 survey

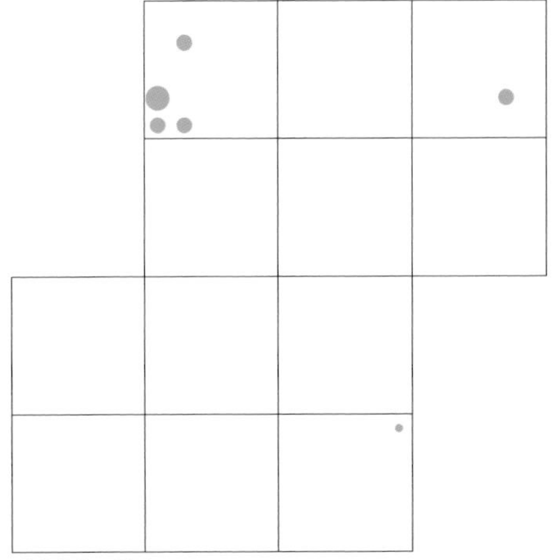

1983–87 survey

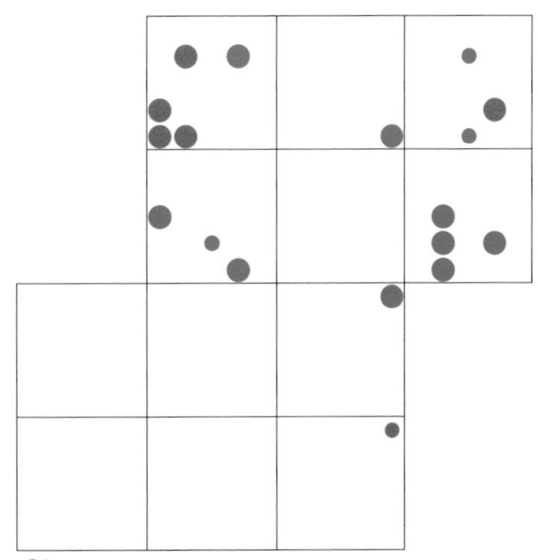

Change

Feral Pigeon

Columba livia

Feral Pigeons are familiar birds and their well-known call and colonial habits made the location of breeding populations easy. However, confirmation of breeding was not so easy, as the young are fed on regurgitated pigeon 'milk' and do not have any immediately obvious plumage features. Breeding confirmation was best achieved by seeing birds sitting on nests (although this was frequently inside buildings).

Throughout the survey area, Feral Pigeons were found breeding in most towns and many farms, although their occurrence in villages is best described as patchy. There is no doubt that they were also breeding widely at the time of the first survey in the 1980s, although the species was not included in the published Atlas, because of the difficulty in distinguishing between truly feral birds and free flying birds from local lofts. This difficulty still remains, and the results of this survey are more a reflection of how different field workers treated the species than of their actual distribution, as they were recorded in virtually every tetrad in some 10 km squares, but were almost unrecorded in others.

Introduced breeder

UK conservation status: **GREEN**

Long-term UK trend: Not available

Recent population trend (1994–2007): -16%

Numbers of tetrads with breeding records:

	2003–07		1983–87
	13 squares	12 squares	12 squares
Possible	**34**	32	
Probable	**44**	38	
Confirmed	**21**	18	
Totals	**99**	**88**	

ROB BROOKES

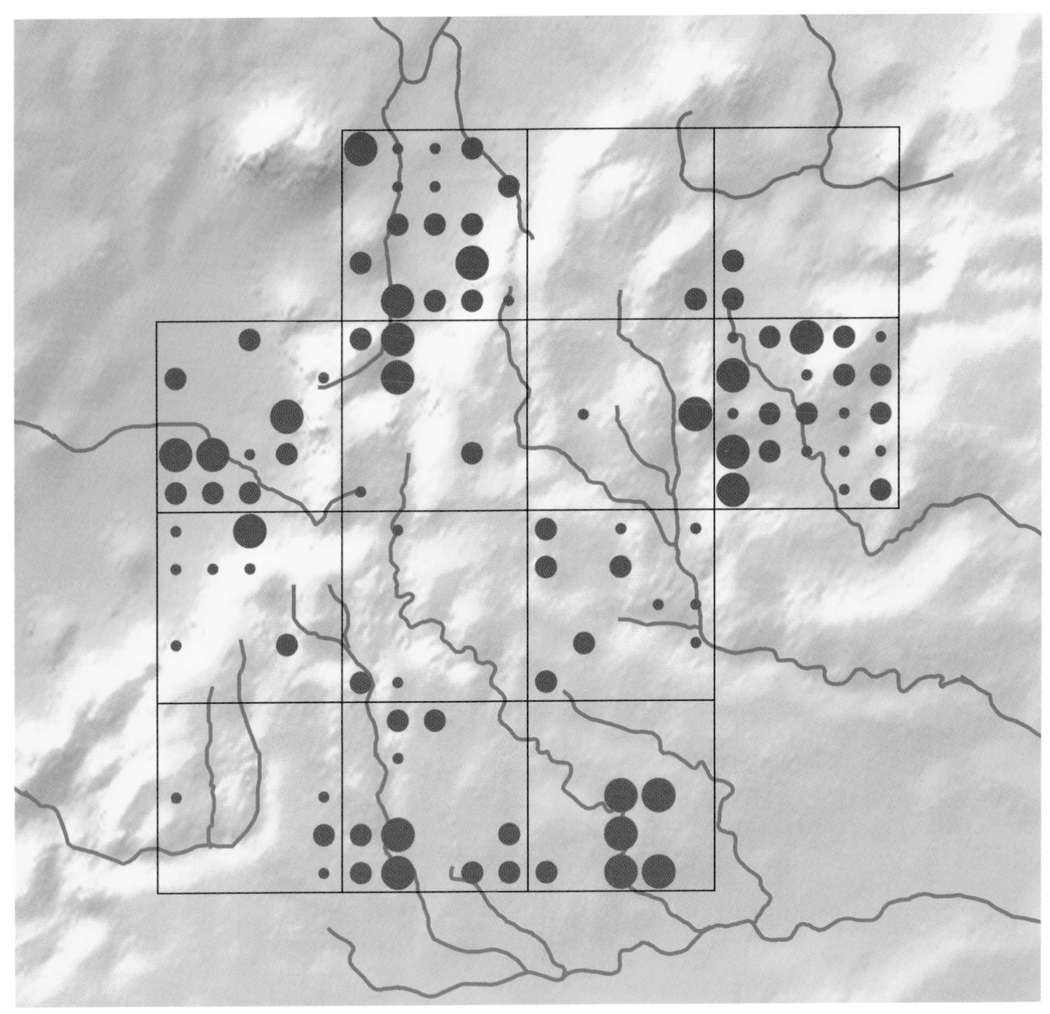

2003–07 survey

Stock Dove (Stock Pigeon)

Columba oenas

ROB BROOKES

Resident breeder

UK conservation status: AMBER

Long-term UK trend (1970–2006): +75%

Recent population trend (1994–2007): -1%

Numbers of tetrads with breeding records:

	2003–07		1983–87
	13 squares	12 squares	12 squares
Possible	21	20	32
Probable	232	213	154
Confirmed	43	42	64
Totals	**296**	**275**	**250**

Stock Doves have an extended breeding season, and their distinctive, far-reaching call can be heard from early spring well into midsummer. This provided an easy way of detecting breeding intent, as did the carrying of nest material. However, adults do not openly carry food to young and the young themselves are not particularly distinctive; therefore, confirmation of breeding proved much more difficult, and was mainly achieved by observing adults entering nest holes.

Typical habitat for Stock Doves is parkland, open pasture and farmland, which is found throughout much of the Cotswolds, particularly the lower parts of the Cotswold scarp. They nest in holes in trees, quarries and Cotswold stone barns and similar buildings, and also in rock faces on Crickley Hill and Cleeve Common. They are widely distributed across the region, but in most parts they are not particularly numerous.

The only significant gap in the breeding distribution appears to be in the north of the area, in part of the mainly arable High Wold, roughly in the area between Broadway, Paxford, Longborough and Stanway. The 10% increase in total breeding records could be a consequence of the greater survey effort, but the reduction in the number of confirmed breeding records (by approximately one-third) probably reflects a real decrease in breeding density.

Species sponsored by Tony and Pam Perry

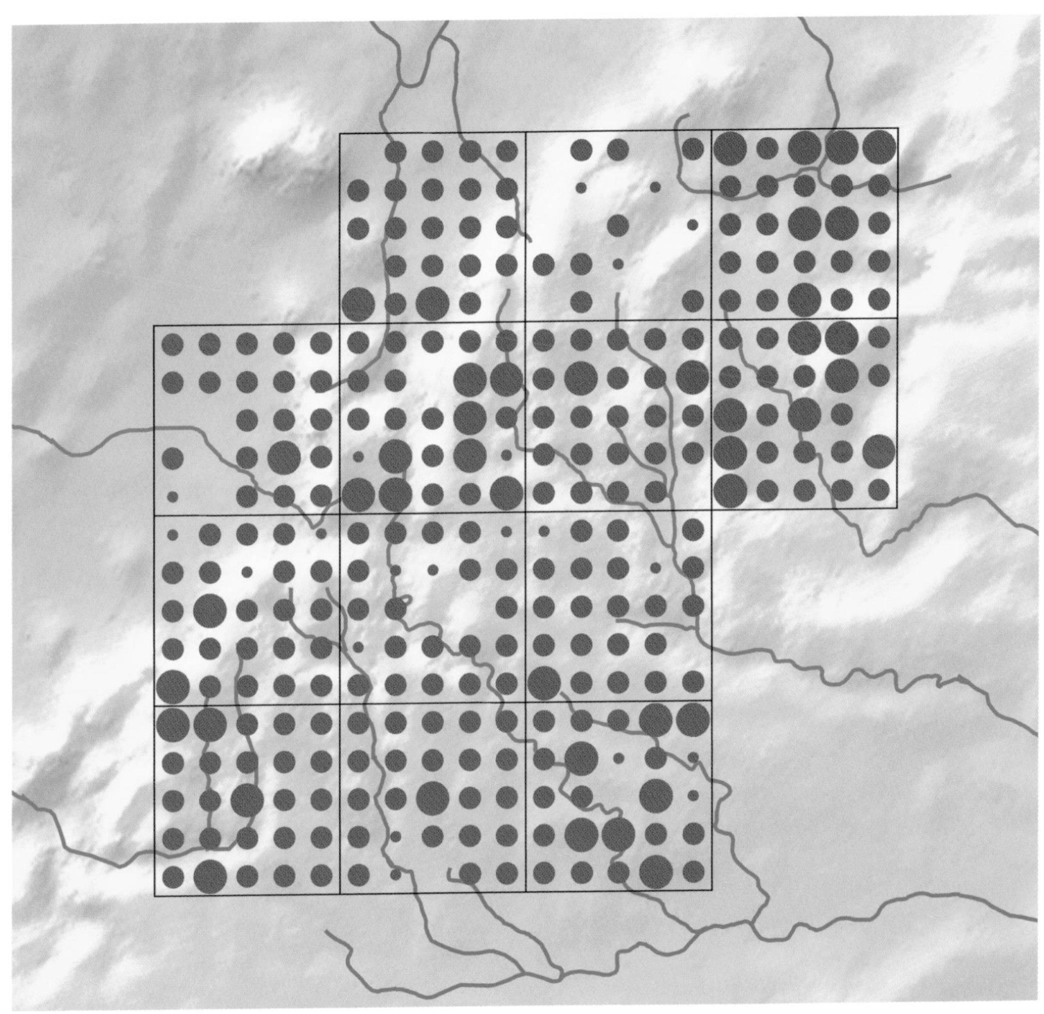

2003–07 survey

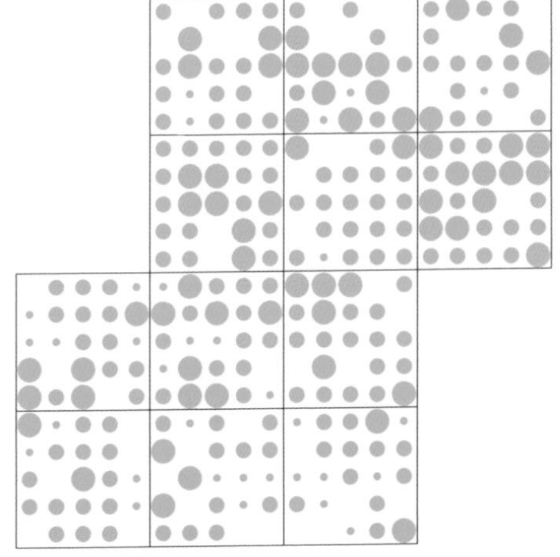

1983–87 survey

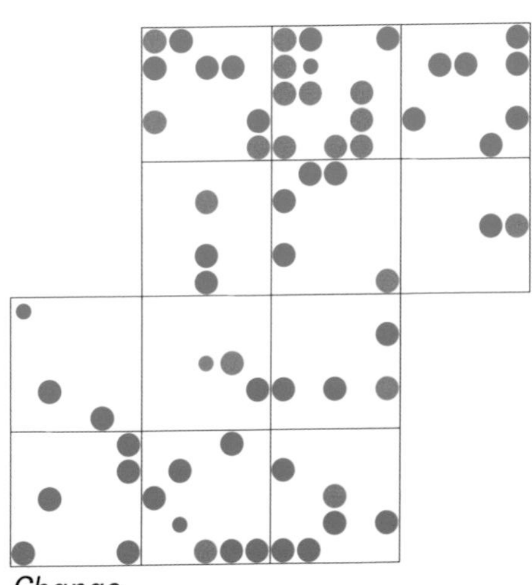

Change

Woodpigeon (Common Wood Pigeon)

Columba palumbus

Although quite often a single-brooded species, the Woodpigeon has an extended breeding season, breeding having been recorded in every month of the year. In the Cotswolds, however, breeding follows the typical rural pattern of peaking in late summer and running into early autumn (August to late September), although nests were also found in April and May, and nest building was noted as early as February in urban areas. Confirmation of breeding was generally achieved by finding an occupied nest (a platform of twigs built in the canopy of small trees or within larger shrubs) which in high summer is well hidden, or young birds close to their nest site. The young are fed on pigeon 'milk', so the adults cannot be seen carrying food, and active feeding of young is not easy to detect. The timing of the peak of breeding activity (at the end of the survey season) may explain the slightly low percentage of confirmed breeding records.

The survey results confirm that the Woodpigeon is truly a ubiquitous bird of the recording area, breeding being recorded in all tetrads. Although confirmation of breeding was only achieved in about 45% of tetrads (somewhat higher than found in the first survey), the sheer numbers of birds, including those found singing and displaying, throughout the study area suggests that it breeds widely in all tetrads. Although a distribution survey cannot detect changes in the overall breeding density, the general impression of observers was that Woodpigeons had become more numerous and widespread in the years between the two surveys. Observers also noted a move into urban areas, and gardens in particular.

Resident breeder

UK conservation status: GREEN

Long-term UK trend (1970–2006): +110%

Recent population trend (1994–2007): +22%

Numbers of tetrads with breeding records:

	2003–07		1983–87
	13 squares	12 squares	12 squares
Possible	2	2	14
Probable	171	161	170
Confirmed	152	137	115
Totals	**325**	**300**	**299**

RICHARD TYLER

Species sponsored by Matthew Main

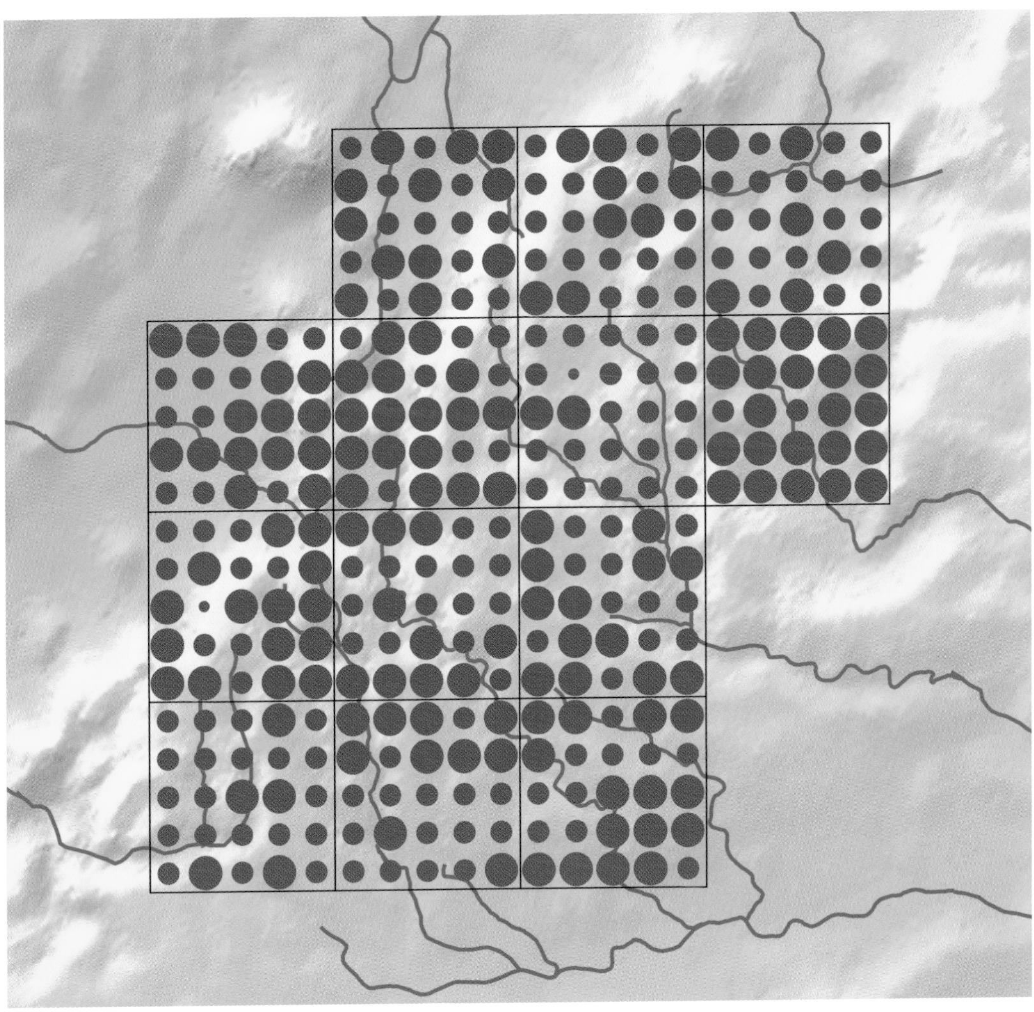

2003–07 survey

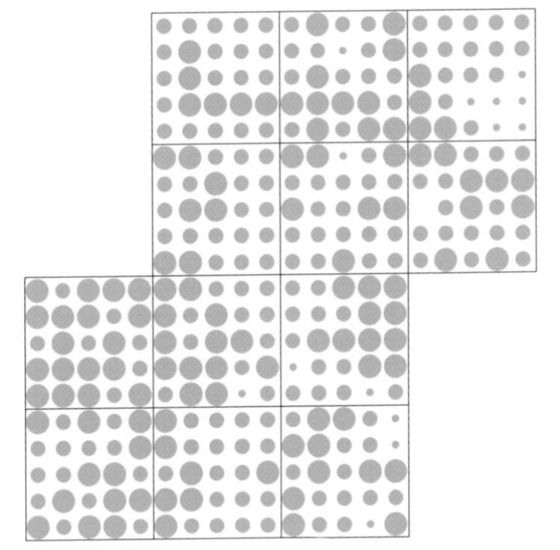

1983–87 survey

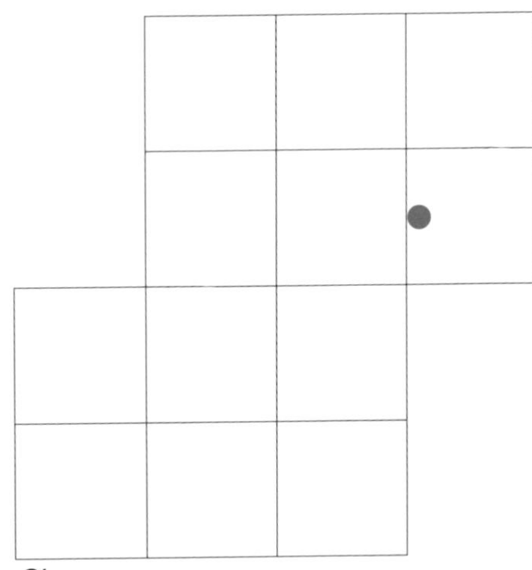

Change

Collared Dove (Eurasian Collared Dove)

Streptopelia decaocto

RICHARD TYLER

Resident breeder

UK conservation status: **GREEN**

Long-term UK trend (1975–2006): +387%

Recent population trend (1994–2007): +27%

Numbers of tetrads with breeding records:

	2003–07		1983–87
	13 squares	12 squares	12 squares
Possible	**19**	18	35
Probable	**191**	175	155
Confirmed	**58**	53	32
Totals	**268**	**246**	**222**

Although common and widely distributed, Collared Doves are much more tied to human habitation (towns, villages and farmyards) than Woodpigeons. Collared Doves have a very extended breeding period, singing and displaying from very early in the year. Occupied nests were found well into October, and it is known as a species that will breed all year round, where its use of the abundant urban food supply is advantageous. The three-note 'song' is extremely familiar, making the presence of breeding birds easy to detect, and the map is probably a true picture of its distribution. However, confirmation of breeding usually requires finding a nest in a dense conifer (often in a private garden) or recognizing newly fledged young from their lack of a collar and slightly duller, browner plumage; this is not easy, and only about 20% of breeding records were confirmed.

Collared Doves first bred in the Cotswolds in 1963 at Sherborne and Coln St Aldwyns, and by the time of the first survey in the 1980s, they were established in about three quarters of the tetrads of the survey area. Allowing for the increased intensity of coverage in the 2003–07 survey, there does not appear to have been much change in total tetrad occupancy over the intervening 20 years. However, anecdotal opinion suggests that there may have been a peak in population density in the Cotswolds in the years between the two surveys, and that numbers have declined slightly since. The change map suggests a shift in distribution, with an increase in the south and northeast of the region, and some losses in the central areas. These changes do not appear to be related to any particular habitat type.

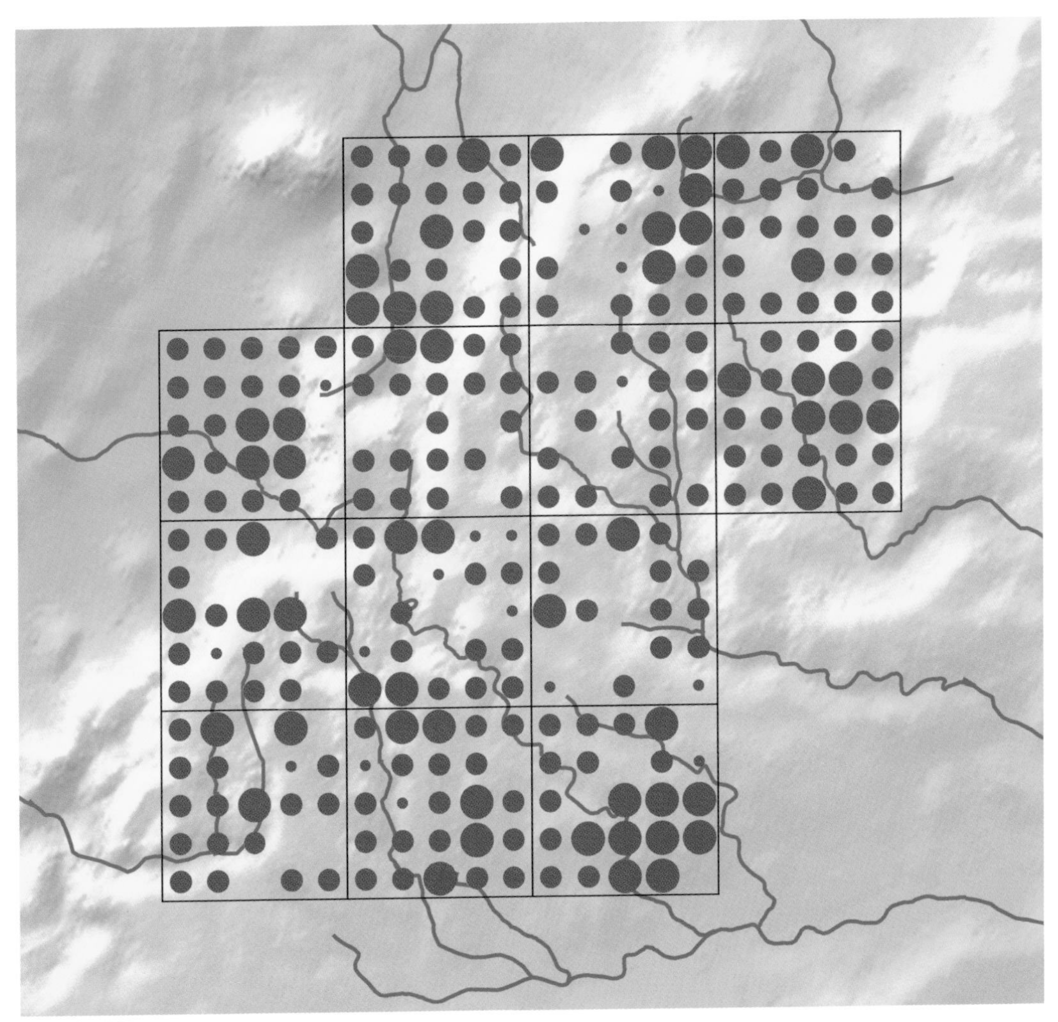
2003–07 survey

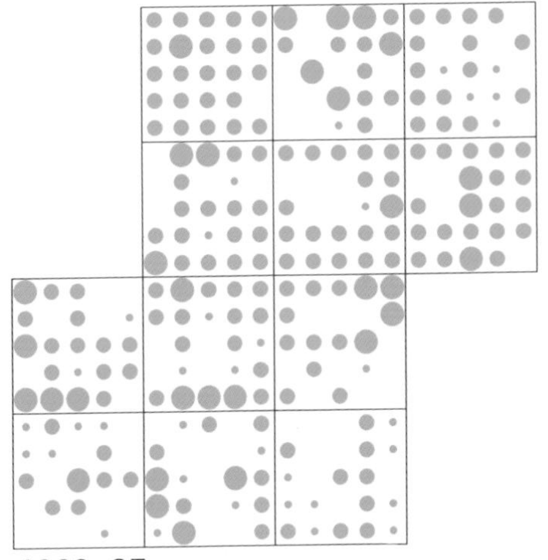
1983–87 survey

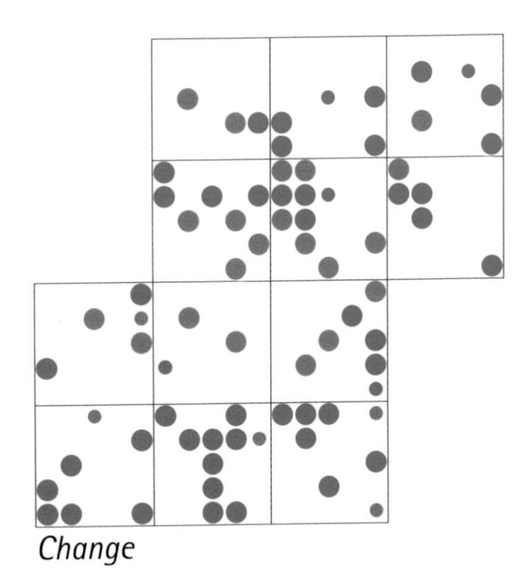
Change

Turtle Dove (European Turtle Dove)

Streptopelia turtur

RICHARD TYLER

Migrant breeder

Gloucestershire BAP species

UK conservation status: RED

Long-term UK trend (1970–2006): -86%

Recent population trend (1994–2007): -66%

Numbers of tetrads with breeding records:

	2003–07		1983–87
	13 squares	12 squares	12 squares
Possible	**2**	2	14
Probable	**23**	22	77
Confirmed	**0**	0	1
Totals	**25**	**24**	**92**

The purring call of Turtle Doves is far-reaching, although often difficult to pinpoint. Detecting the presence of territorial birds was therefore easy with a little patience. However, confirmation of breeding was difficult and, despite significant effort in some traditional territories, was not achieved during this survey.

Of all the species breeding in the Cotswolds survey area, few have declined as dramatically as the Turtle Dove. Despite the greater survey effort this time, they were recorded in fewer than 8% of tetrads over the five-year period, and in some of these in one year only. This compares with over 30% of tetrads in the first Atlas, when the bird was described as 'nowhere a familiar bird' and that it was 'sparsely distributed and numbers varied considerably from year to year'.

It is believed to have been still declining during the survey; interestingly, a similar suggestion was made in the first Atlas. This decline does not appear to be habitat-related in the Cotswolds. Even the cluster of tetrads in the northwest, in the Avon Vale on and around a section of the disused railway track from Cheltenham to Stratford-upon-Avon, may now be reduced to a couple of pairs of birds. This is despite the habitat appearing to be ideal: relatively sheltered, with low trees, bushes and open areas, and unchanged over many years. Apart from this population, and a few pairs in the Stour Valley, the species appears to be virtually absent from the scarp and Severn and Avon Vales, as well as the Stroud Valleys and the western parts of the dip slope. The decline or extinction in the wooded areas south of Cheltenham is particularly noticeable. This all seems to accord with the national decline in numbers and the contraction of its range.

Species sponsored by Mick and Jo Jones

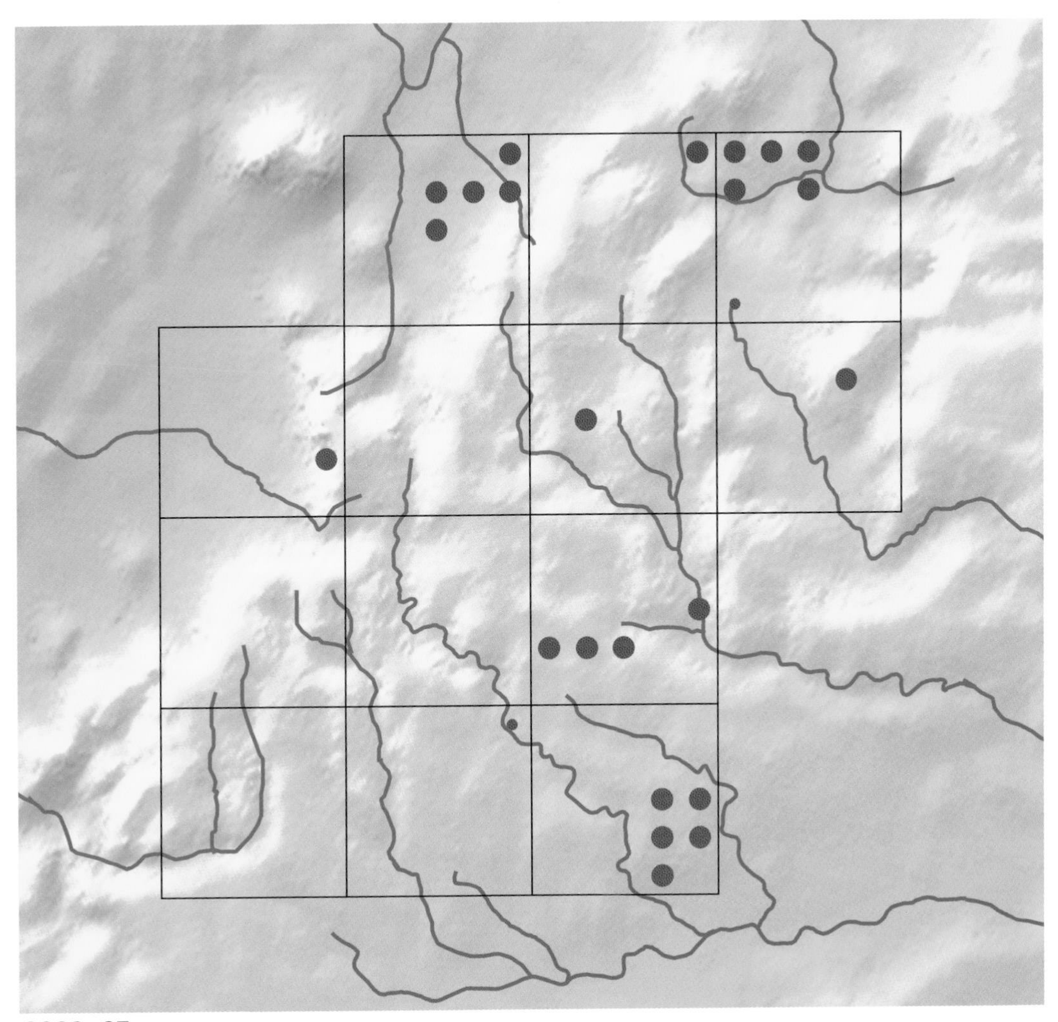
2003–07 survey

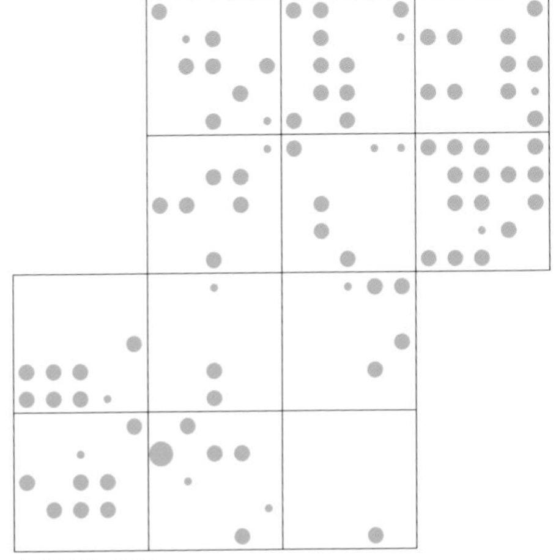
1983–87 survey

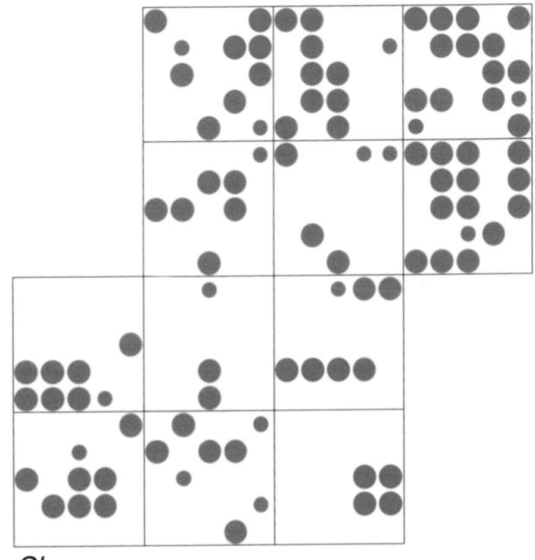
Change

Cuckoo (Common Cuckoo)

Cuculus canorus

Cuckoos have shown one of the biggest declines in distribution and numbers of all the species in the Cotswolds since the early 1990s. Their easily recognizable and far-carrying call means that there is a significant chance that they have been overrecorded–most of the records are 'probable', through hearing the song. There is also the possibility that the data may include some passage migrants. However, the same comment was made in the first Atlas, so a direct comparison of the data may still yield useful information. This suggests that the number of occupied tetrads has been nearly halved, despite the increased survey effort.

In the first Atlas, it was suggested that most occupied tetrads held no more than one or two pairs, giving a total population of 200–300 pairs. The distribution density was probably even thinner this time, and the total number of pairs is likely to be well below 100, and possibly below 50. The most reliable places for seeing and hearing Cuckoos remain Cleeve Common and the hillside above Prestbury. Their hosts there are most likely to be Meadow Pipits whereas in most of the rest of the survey area, where Cuckoos are encountered increasingly rarely, Dunnocks are thought to be the host species.

In the first survey, there were six records of confirmed breeding (although only one of these involved finding an occupied nest–that of a Dunnock–while the rest were of young birds in tetrads where territory holding had already been observed). In this survey, there were no confirmed breeding sites, despite the increased effort, and no reports of young birds. This further supports the feeling that breeding density, as well as distribution, has decreased significantly.

Migrant breeder

UK conservation status: AMBER

Long-term UK trend (1970–2006): -47%

Recent population trend (1994–2007): -37%

Numbers of tetrads with breeding records:

	2003–07		1983–87
	13 squares	12 squares	12 squares
Possible	5	5	9
Probable	114	106	183
Confirmed	0	0	6
Totals	119	111	198

RICHARD TYLER

Species sponsored by Peter Ormerod

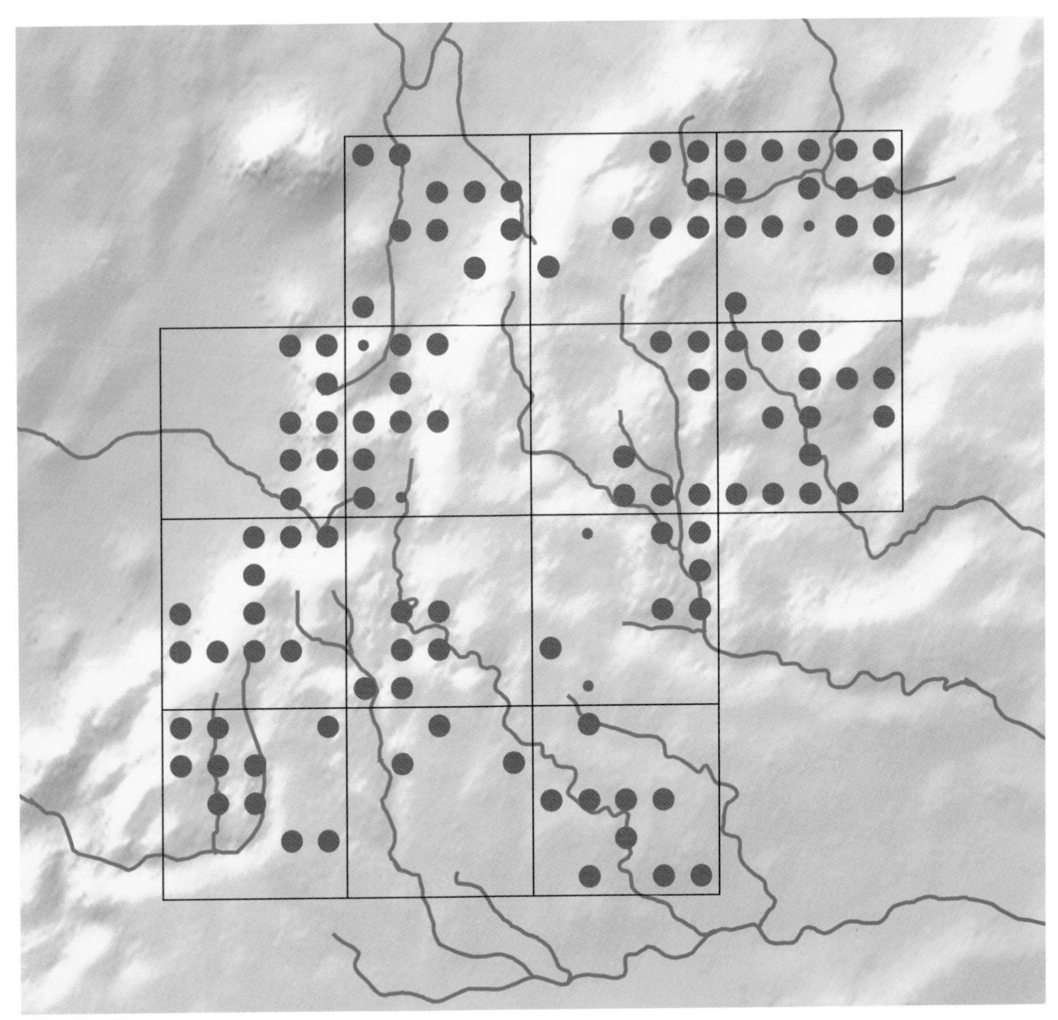
2003–07 survey

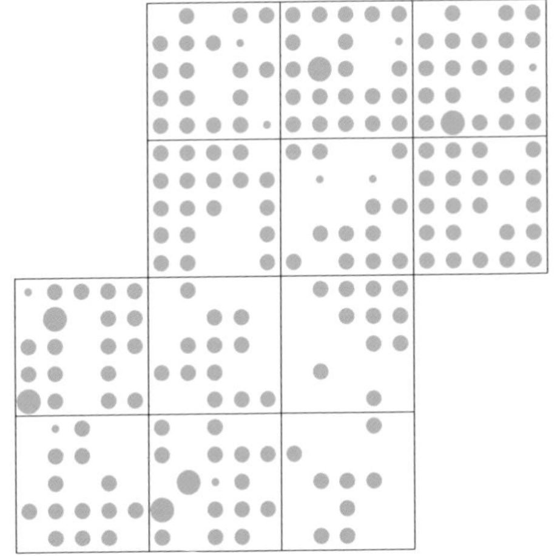
1983–87 survey

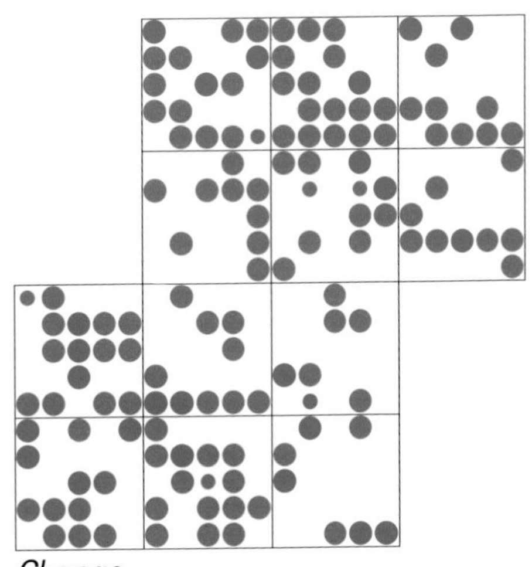
Change

Barn Owl

Tyto alba

This survey has highlighted what seems to be a major expansion in the Barn Owl population in the Cotswolds. In the 1980s, definite breeding activity was only noted in four tetrads (though it was also thought possible in another 10), all in the vicinity of the River Churn to the north of Cirencester. All but one of the other sightings were associated with the Thames tributaries in the extreme south of the recording area. In this survey, there was a dramatic increase in both total number of occupied tetrads and confirmed breeding sites, and Barn Owls were found to be widely distributed over a large part of the North Cotswolds, although interestingly they were not recorded in their former stronghold along the River Churn. They also appeared to be almost completely absent from the Cheltenham area and the surrounding vale.

Barn Owls require rough ground or woodland edges for feeding. Although large areas suitable for feeding have probably declined since the 1980s, there has been a move towards leaving wider field margins. It may be that the study area has always contained territory suitable for Barn Owls, and that population increases outside the Cotswolds have allowed the species to move in and increase within the area. The erection of nestboxes in various locations across the country has also almost certainly had a beneficial effect, and captive release schemes may also have contributed.

Resident breeder

UK conservation status: AMBER

No population trends available

Numbers of tetrads with breeding records:

	2003–07		1983–87
	13 squares	12 squares	12 squares
Possible	**27**	27	10
Probable	**16**	15	1
Confirmed	**22**	22	3
Totals	**65**	**64**	**14**

ROB BROOKES

The first Atlas commented: 'It is likely that some breeding pairs have been overlooked and that the species is under-recorded in the study area. However, the breeding population in any one year is almost certainly fewer than 10 pairs'. The greater survey effort in the 2003–07 survey may have exaggerated the apparent increase in distribution seen, but the move of the species into large areas where it was previously unrecorded would indicate that this increase is real. Although the annual variation in numbers makes it difficult to estimate the size of the population in the survey area (the species is very prone to major breeding population increases and decreases, determined by the availability of prey), an average approaching 50 pairs each year would not be an unreasonable guess.

Species sponsored by Ian and Val Tucker

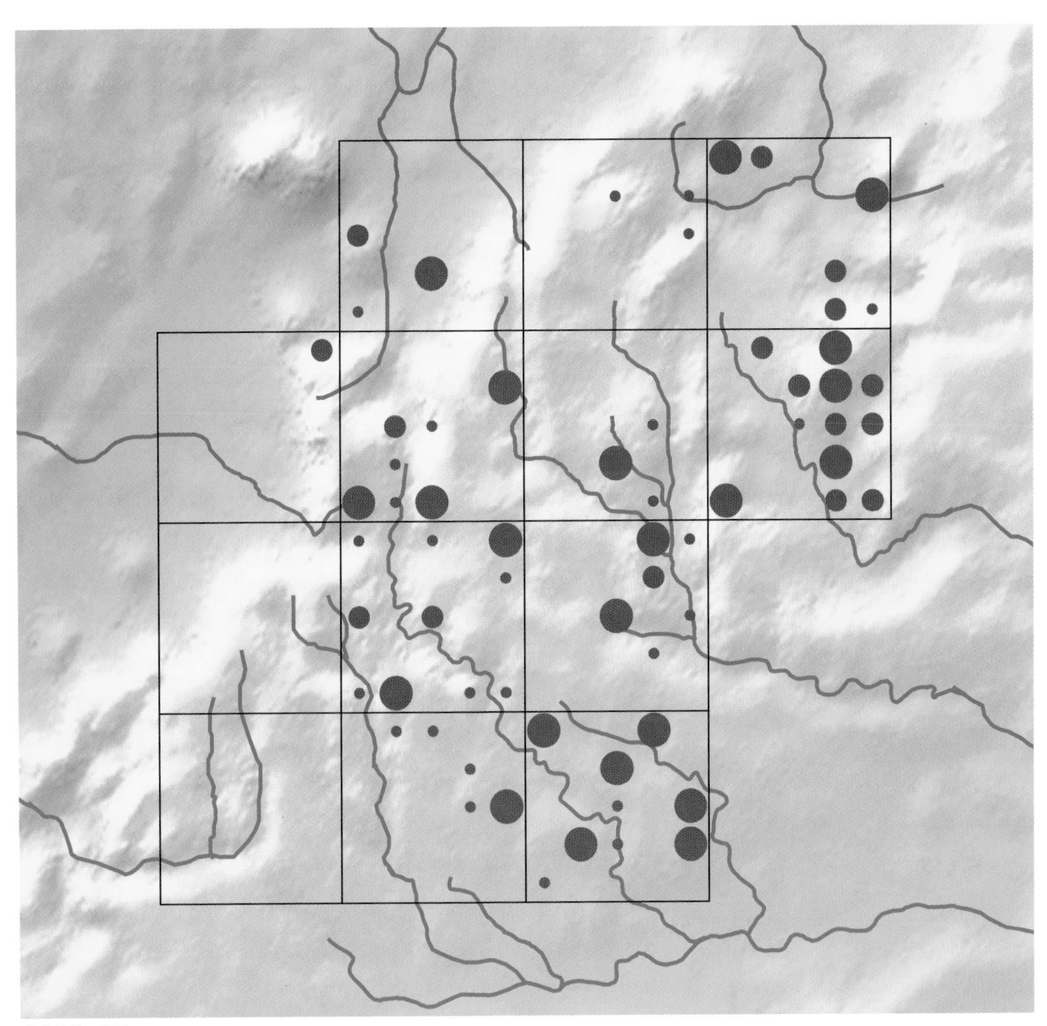

2003–07 survey

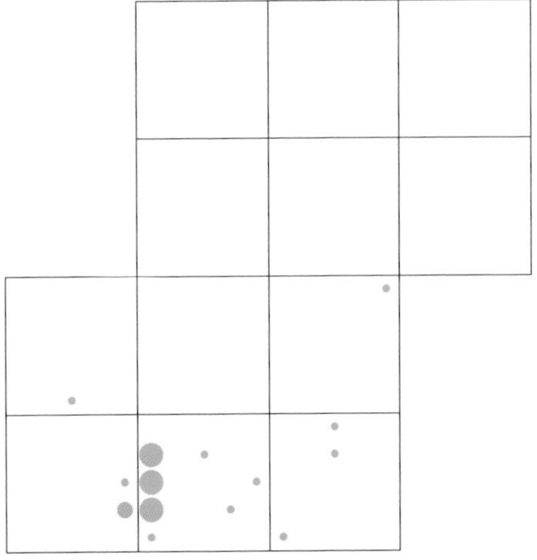

1983–87 survey

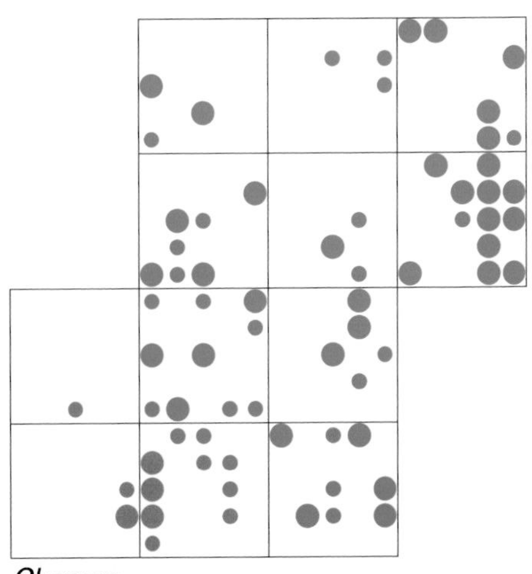

Change

Little Owl

Athene noctua

ROB BROOKES

Introduced breeder

UK conservation status: Unlisted

Long-term UK trend (1970–2006): -24%

Recent population trend (1994–2007): -26%

Numbers of tetrads with breeding records:

	2003–07		1983–87
	13 squares	12 squares	12 squares
Possible	**40**	39	68
Probable	**58**	57	30
Confirmed	**28**	25	27
Totals	**126**	**121**	**125**

Little Owls could readily be seen hunting during the long spring and summer evenings, and confirmation of breeding could often be obtained by observing them leaving or entering nest holes. Although they remain a widespread species in the study area, being found in any type of agricultural land containing trees with holes for nesting, they are somewhat thinly distributed. As in the survey in the 1980s, the boles of old pollarded willows and ash trees provided a good number of nest sites, but they were also noted nesting in farm outbuildings and quarry faces. They were described in the first Atlas as being 'more common in the lower grasslands than in the arable uplands.' This would seem to be still broadly the case, although breeding density is believed to have declined markedly.

The numbers of occupied tetrads in the two surveys are similar. However, there may have been a more marked decline that has been masked by the increase in survey effort. The change map provides a clue to this, with over half of the occupied tetrads in the first survey now having been lost, and replaced by nearly as many new ones. It could reasonably be postulated that many of the new sites in this survey were occupied, but overlooked, in the first survey, and that there has, in fact, been something like a 30% reduction in the number of occupied tetrads.

Comparison of the data with that for the Tawny Owl might support this hypothesis. In the survey in the 1980s Tawny Owls were found in fewer tetrads than were Little Owls. This was attributed to the fact that Little Owls are easier to find and survey. In this survey, the two distribution maps are now quite similar. Since there is no evidence of an increase in Tawny Owl numbers, it appears that Little Owl numbers have probably declined since the first survey.

Species sponsored by Rozza Birch

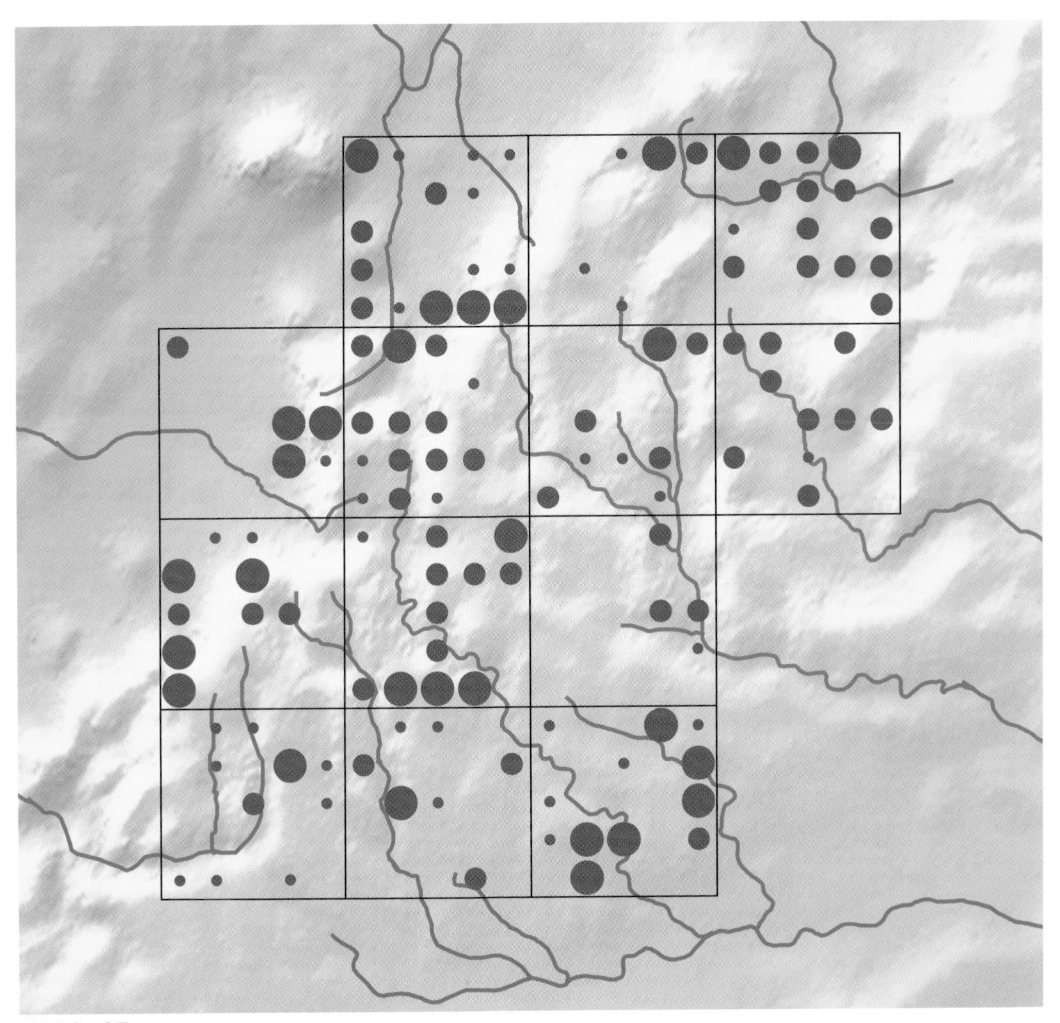

2003–07 survey

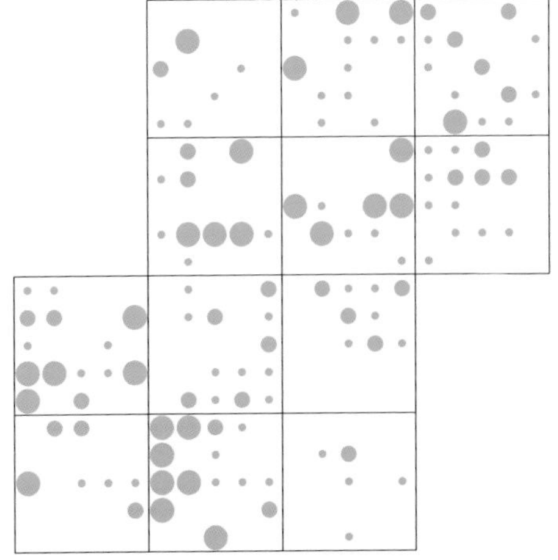

1983–87 survey

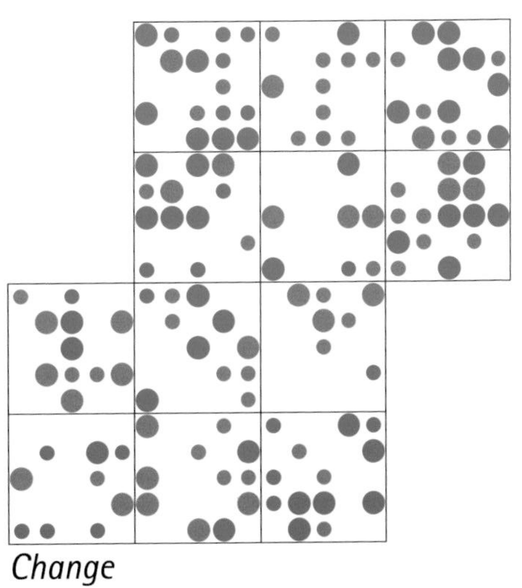

Change

Tawny Owl

Strix aluco

The Tawny Owl is a more difficult species to survey than would at first seem likely. They are very vocal at the start of their breeding season, which is late winter, but they are much less so during spring and summer, although the begging calls of young birds can be heard during this period. This problem was found in the survey for the first Atlas (in which it was commented that they were 'undoubtedly under-recorded'), and so specific effort this time was made to survey some out-of-the-way sites in February and early March each year before the main field work began.

Tawny Owls generally require concentrations of mature trees in which suitable holes for nesting can be found, although there were instances of large open-fronted nestboxes being used. The distribution map of this species is quite similar to that of the Little Owl (although there is no great similarity in their habitat requirements): it is absent from the rather tree-less arable areas in the Severn Vale north of Cheltenham, and somewhat thinly distributed down the central upland arable areas of the dip slope.

The extra survey effort may account for the 34% increase in number of occupied tetrads that was found this time compared with the first survey. The general feeling of observers in the area is that Tawny Owl numbers may have declined since the early 1990s, suggesting that the distribution recorded in the first survey was a significant underestimate. However, the extent to which the distribution maps accurately represent the actual status of the species is probably more uncertain than for any other species in this survey.

Resident breeder

UK conservation status: GREEN

Long-term UK trend (1970–2006): -23%

Recent population trend (1994–2007): -9%

Numbers of tetrads with breeding records:

	2003–07		1983–87
	13 squares	12 squares	12 squares
Possible	**17**	15	30
Probable	**81**	79	48
Confirmed	**33**	31	15
Totals	**131**	**125**	**93**

ROB BROOKES

Species sponsored by Sarah and Dennis Gornall

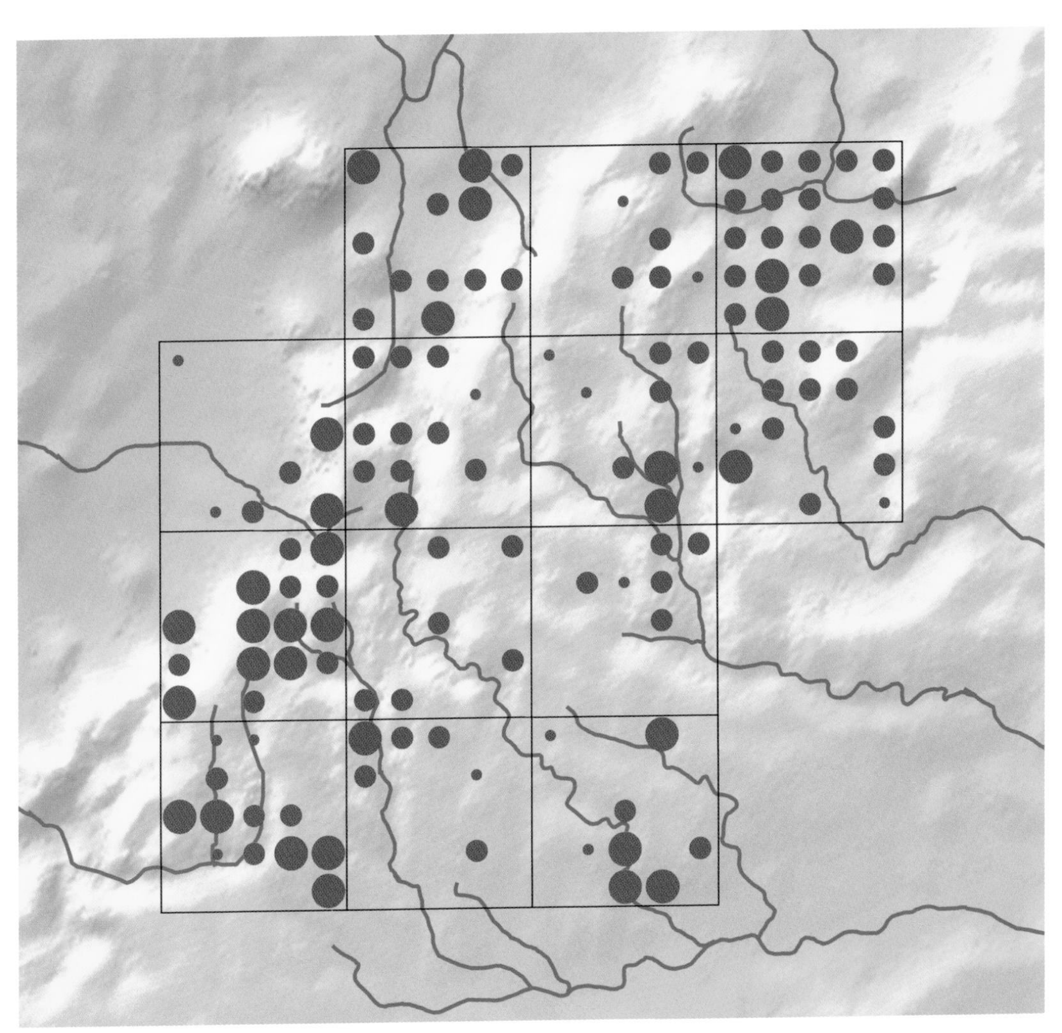

2003–07 survey

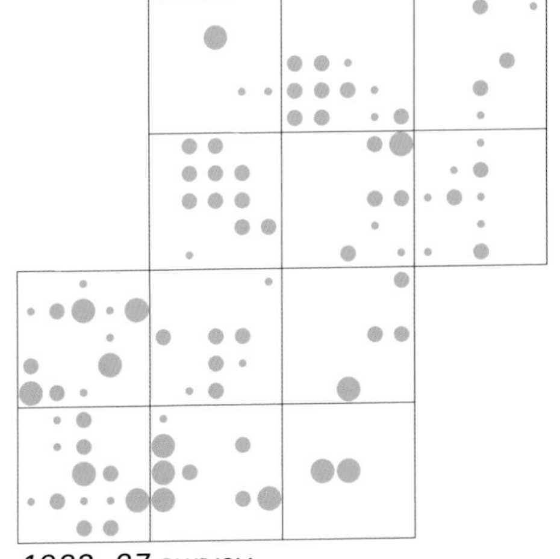

1983–87 survey

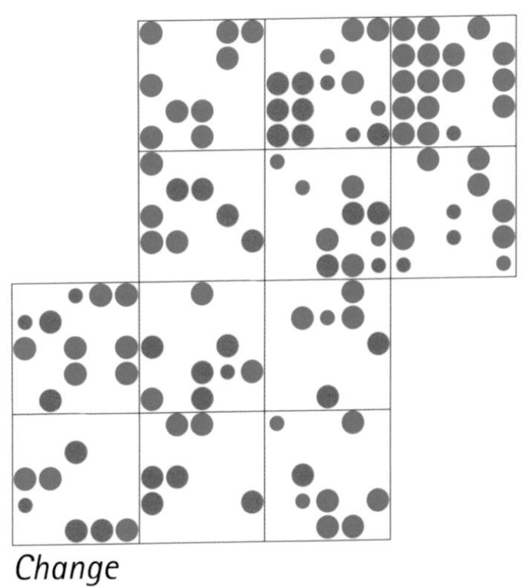

Change

Swift (Common Swift)

Apus apus

ROB BROOKES

Migrant breeder

UK conservation status: GREEN

Long-term UK trend: Not available

Recent population trend (1994-2007): -41%

Numbers of tetrads with breeding records:

	2003–07		1983–87
	13 squares	12 squares	12 squares
Possible	**44**	40	93
Probable	**66**	65	64
Confirmed	**77**	71	32
Totals	**187**	**176**	**189**

The Swift proved to be one of the most difficult species to survey accurately. Although numerous and very easy to locate, they can feed many kilometres from their nesting sites or spend all summer on the wing as non-breeders, thus making them hard to classify. Agreeing a consistent method amongst observers of determining whether Swifts were showing breeding activity ('probable' breeding) was problematic and may be one reason why some 10 km squares had many records whilst adjoining ones had few. Seeing a Swift entering a hole in a roof in late spring was a fairly cast-iron indication of breeding, but this required luck or patience. More usual was to see Swifts approaching a potential nest site, and then veering away–this often indicated that other Swifts were in residence, and could be taken as evidence of probable breeding.

In the survey area, as elsewhere in the UK, Swifts are entirely dependent on human constructions for breeding, and the traditional limestone tiled roofs of Cotswolds villages provide good breeding sites, although re-roofing with modern alternatives may not be as satisfactory. Swifts seen in open countryside should be regarded as feeding rather than breeding. This may not have been consistently adhered to in the first Atlas survey in the 1980s, when 'probable breeding' was thought to have been overestimated.

Although Swifts were seen throughout the summer over most villages in the recording area, there was often no definite evidence of breeding activity. Although the total number of breeding records was slightly lower during this survey than in the first survey, the number of confirmed breeding areas was more than doubled. It is probably safe to assume that those areas in which probable and confirmed breeding were noted represent the major Swift colonies of the region, but the tetrads recorded as possible breeding should be treated with some caution. These may be breeding areas in which the observer was overcautious, but equally they may be of non-breeding birds in areas that were good for feeding.

Species sponsored by The Brookes family

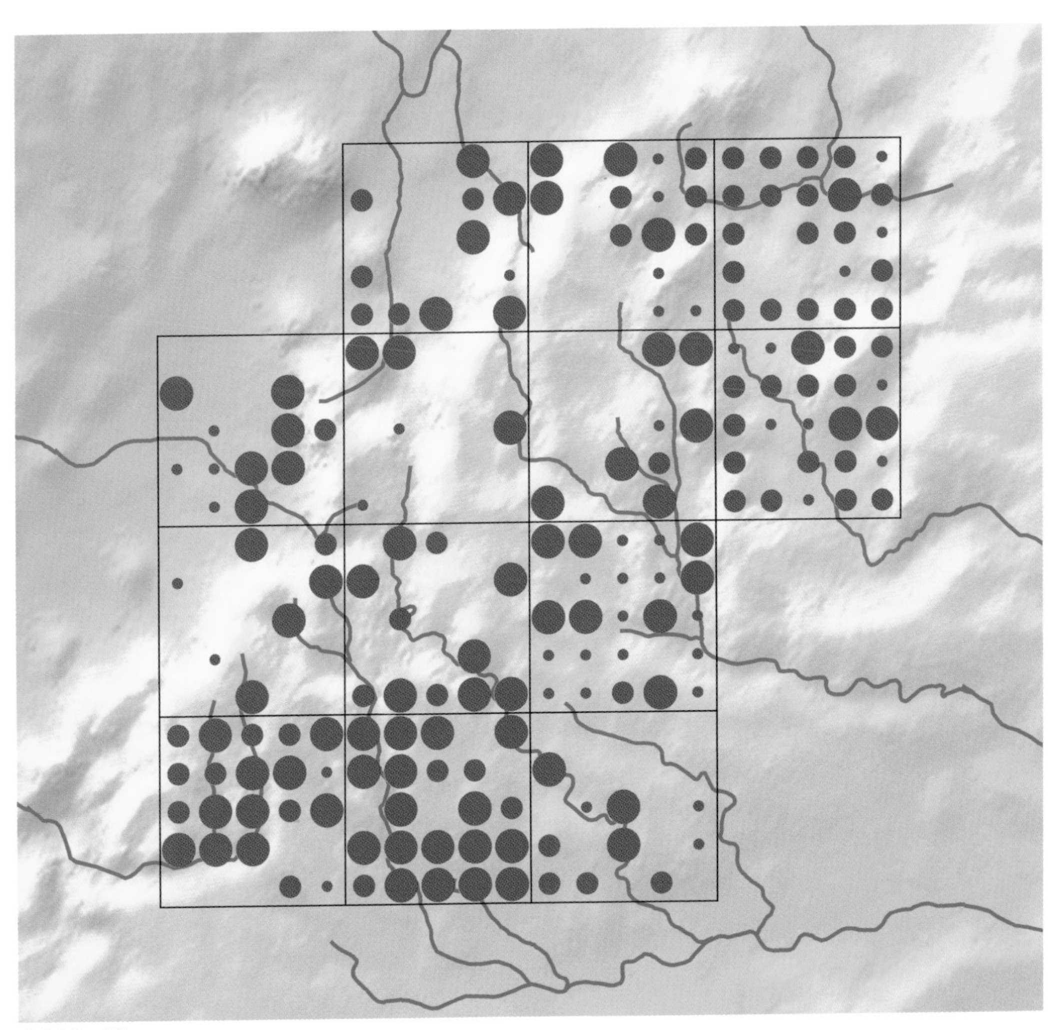

2003–07 survey

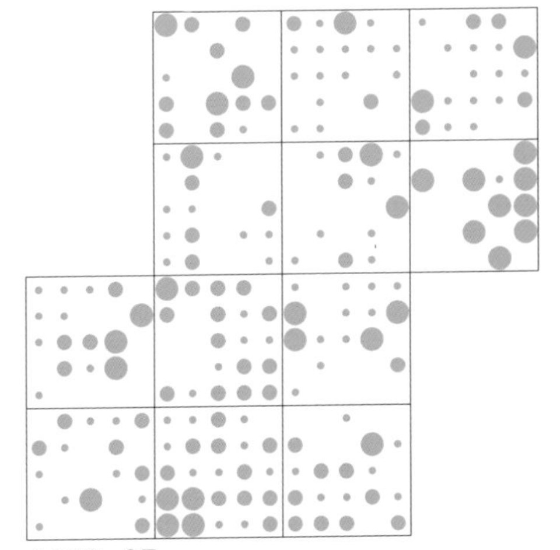

1983–87 survey

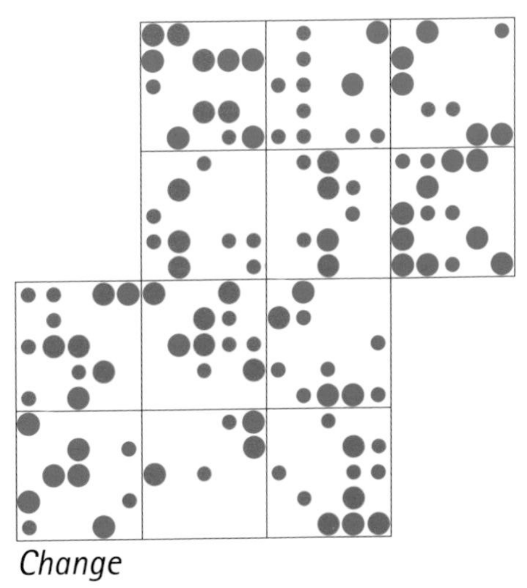

Change

Kingfisher (Common Kingfisher)
Alcedo atthis

The loud call of Kingfishers, their dazzling plumage and their habit of flying up and down small rivers carrying food to their young meant that they were difficult to miss if they were breeding in an accessible area. They require banks in which they can excavate tunnels and nest chambers, and the river systems of the Coln, Churn, Windrush, Evenlode and Isbourne amongst others provide a reasonable supply of such places.

In this survey the total number of Kingfisher records was markedly higher (by about 60%) than in the 1980s survey. Although the number of possible breeding records declined, the number of tetrads in which both probable and confirmed breeding activity were noted was significantly greater. However, it is doubtful whether this can be taken in isolation as showing a real change in the distribution of the species. In the first survey, Kingfishers were found on all of the main river systems in the area, and this was again the case. They were also seen to visit quite small garden ponds in urban areas. It was also suggested in the first Atlas that the species had almost certainly been underrecorded, owing to the inaccessibility to observers of suitable stretches of water in private grounds. Whilst lack of access to certain areas may still have been a problem in the later survey, observers were probably much more aware of the Kingfisher's potential breeding locations because of the considerable amount of local knowledge that had been accumulated in the years between the two surveys. This increased familiarity certainly helped in locating actual nest sites and confirming breeding. It is likely that the distribution map accurately represents the spread of the species, although numbers do fluctuate, and Kingfishers may not have bred in each area every year.

Migrant/resident breeder

UK conservation status: AMBER

Long-term UK trend (1975–2006): +7%

Recent population trend (1994–2007): +2%

Numbers of tetrads with breeding records:

	2003–07		1983–87
	13 squares	12 squares	12 squares
Possible	19	18	21
Probable	15	14	3
Confirmed	15	15	5
Totals	**49**	**47**	**29**

RICHARD TYLER

Species sponsored by Constance Perry

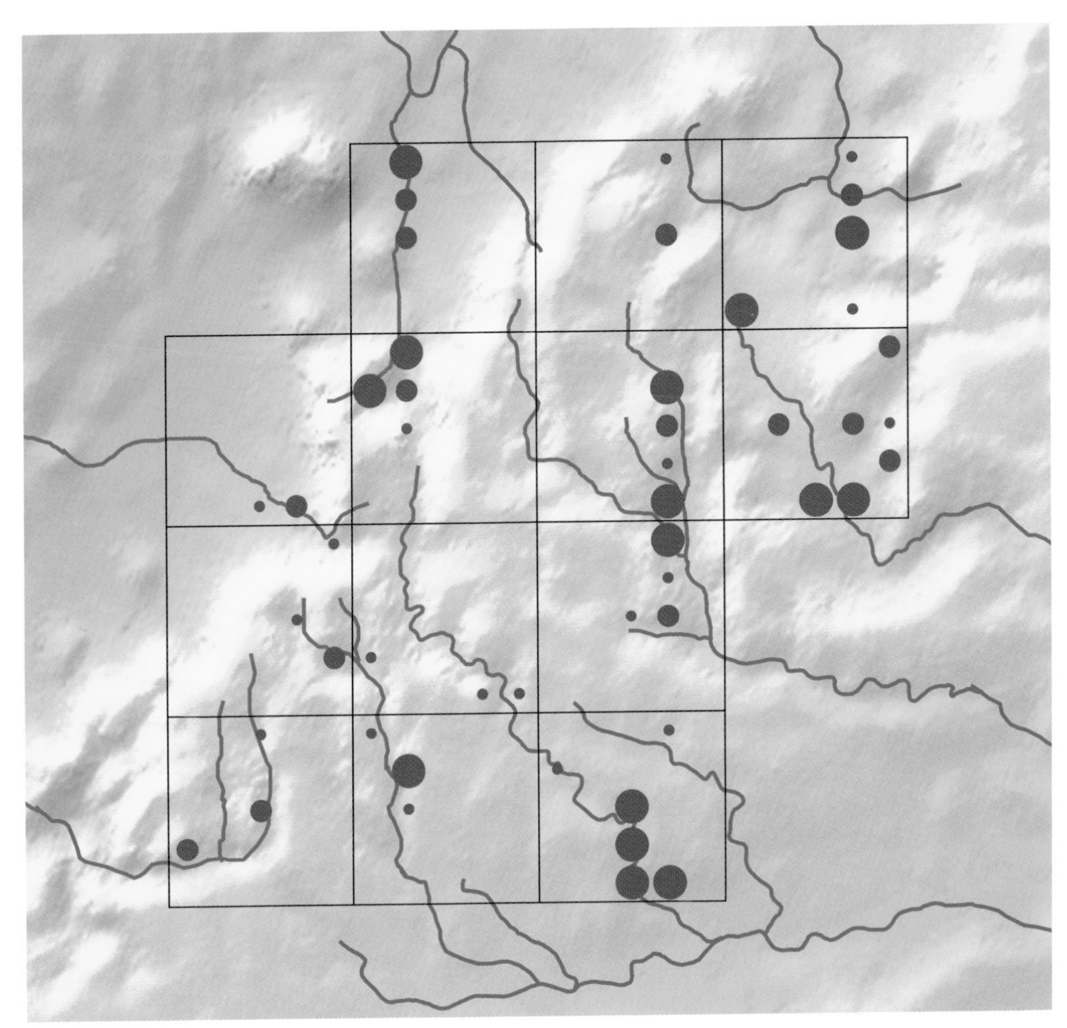

2003–07 survey

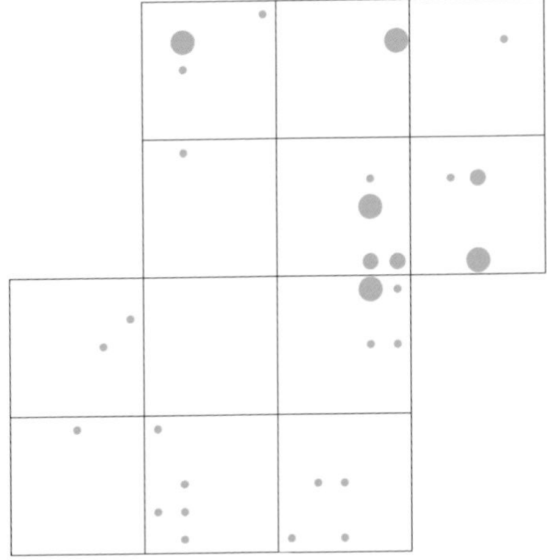

1983–87 survey

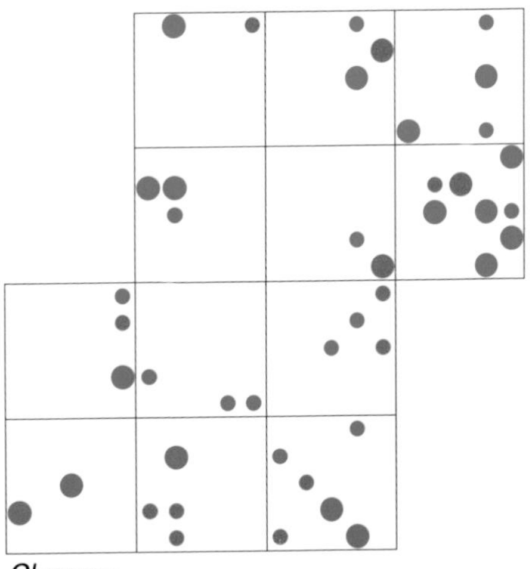

Change

Green Woodpecker

Picus viridis

DAVE PEARCE

Resident breeder

UK conservation status: AMBER

Long-term UK trend (1970–2006): +119%

Recent population trend (1994–2007): +31%

Numbers of tetrads with breeding records:

	2003–07		1983–87
	13 squares	12 squares	12 squares
Possible	**28**	27	20
Probable	**171**	158	117
Confirmed	**73**	65	15
Totals	**272**	**250**	**152**

The long drawn-out 'yaffling' territorial call of the Green Woodpecker is very recognizable and carries far, making locating birds easy. Although birds call throughout the year, yaffling only occurs during the breeding season and so could be taken to indicate probable breeding. However, their nests are much more difficult to find than those of the Great Spotted Woodpecker, and so confirmation of breeding relies more on seeing newly fledged young.

Green Woodpeckers are often found in areas of old pasture and grassland, which provide them with their staple food (ants) and mature trees in which to excavate breeding holes. These dual requirements can be satisfied throughout most of the survey area, but they are particularly numerous on the less cultivated scarp slopes. They are also suburban birds, as can be seen, for example, by their presence in all tetrads in the Cheltenham area, where they feed in gardens, parks and playing fields. There are few areas in which the species appears to be completely absent but one such is located to the east and west of Aldsworth in the Thames Vale, where the landscape is primarily cereal fields surrounded by dry stone walls.

While not quite as widely distributed as Great Spotted, Green Woodpeckers are common birds in the Cotswolds. The increase in the total number of records by almost two-thirds indicates an expansion into new tetrads (highlighted by the change map) than can be accounted for by the greater survey effort. There was also a fourfold increase in the number of confirmed breeding tetrads, which is noteworthy in view of the relative difficulty of proving breeding. The relative abundances of the two species are discussed in more detail in the species account for the Great Spotted Woodpecker.

Species sponsored by John and Viv Phillips

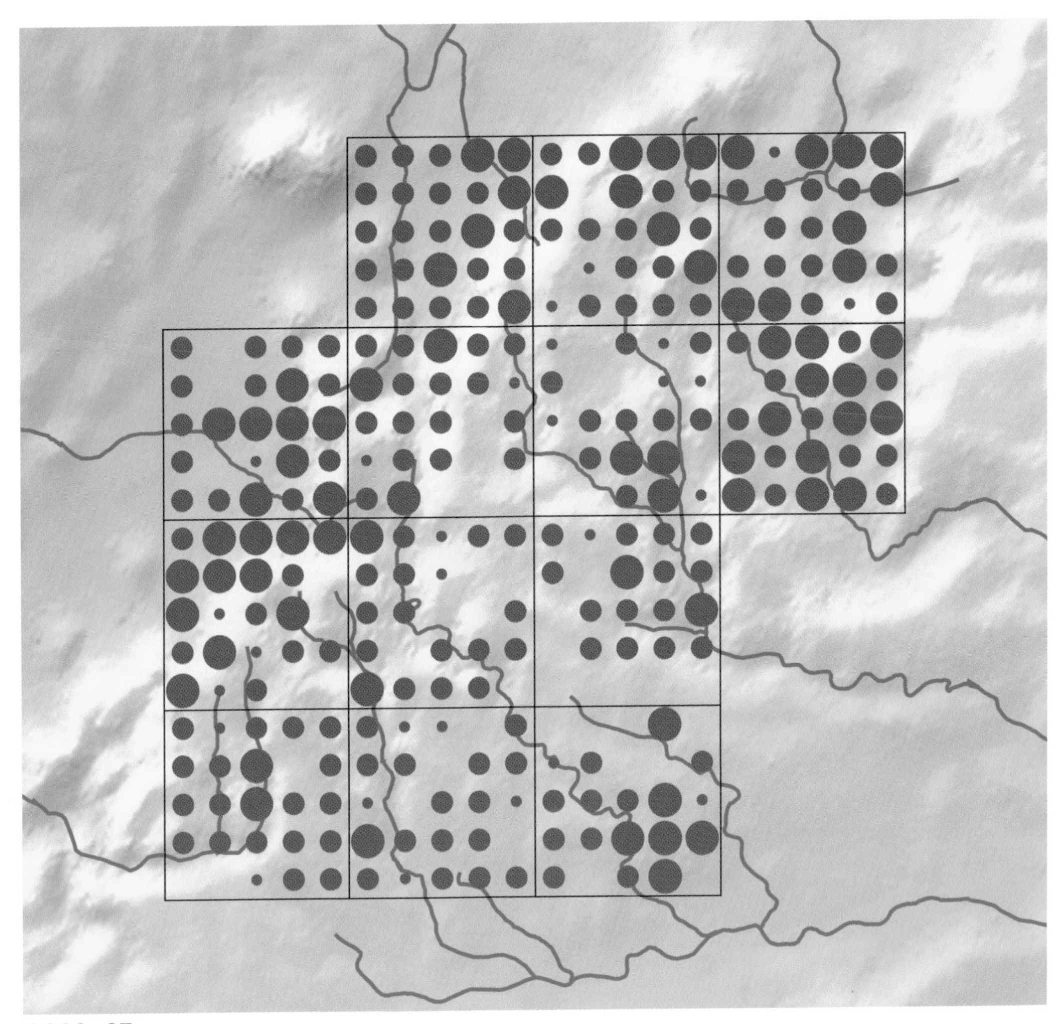

2003–07 survey

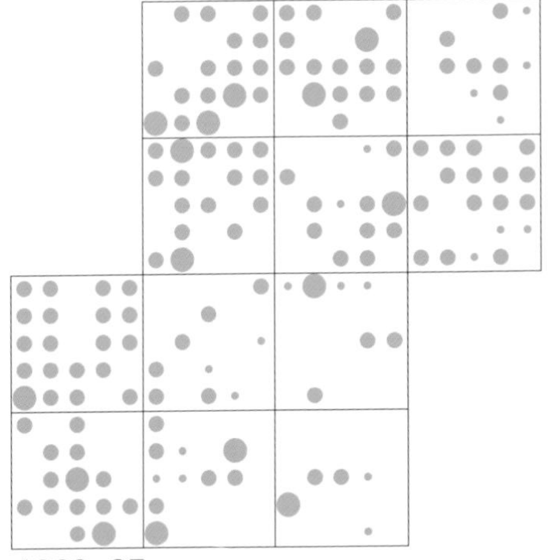

1983–87 survey

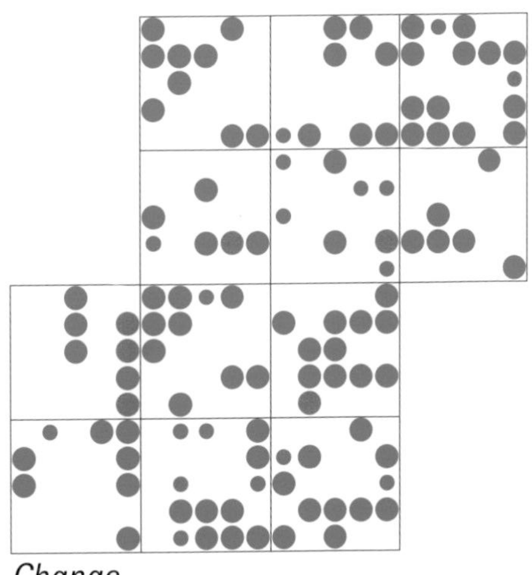

Change

Great Spotted Woodpecker

Dendrocopos major

GRAHAM WATSON

Resident breeder

UK conservation status: GREEN

Long-term UK trend (1970–2006): +314%

Recent population trend (1994–2007): +150%

Numbers of tetrads with breeding records:

	2003–07		1983–87
	13 squares	12 squares	12 squares
Possible	**49**	44	49
Probable	**133**	127	85
Confirmed	**116**	106	37
Totals	**298**	**277**	**171**

As the territorial drumming of the male Great Spotted Woodpecker is regularly heard from New Year's Day through to early May, the location of birds with breeding intentions is made very easy. Young in the nest hole are very vocal, and provided a large number of confirmed breeding records.

There are no particular areas in which Great Spotted Woodpeckers are noticeably absent. Although they have a need for trees in which to nest, even the wide expanses of arable farmland found in the Avon Vale and parts of the High Wold contain some small groups of mature trees that provide these.

Results from this survey suggests that there has been a significant expansion in the distribution of the Great Spotted Woodpecker in the period since the first Atlas. Although this has probably been exaggerated by the greater survey effort this time, the two-thirds rise in total number of occupied tetrads can be taken as evidence of a real increase. The general impression of observers in the area is of a growth in numbers, and it is certainly the most numerous and widely distributed of the woodpeckers in the survey area. Being a largely sedentary species, like the Green Woodpecker, data from the Winter Random Square Survey, which the NCOS has been conducting since 1994, give an indication of the relative abundance of the two species. It suggests that the Great Spotted Woodpecker is about one-third more numerous than the Green Woodpecker, though this should only be taken as a very rough indicator of relative breeding numbers.

It has been suggested that the success of the Great Spotted Woodpecker may have been helped in some way by the decline of the Starling from some rural areas, as the latter tended to oust woodpeckers from their newly constructed nest holes.

Species sponsored by Charlotte and Jessica Nash

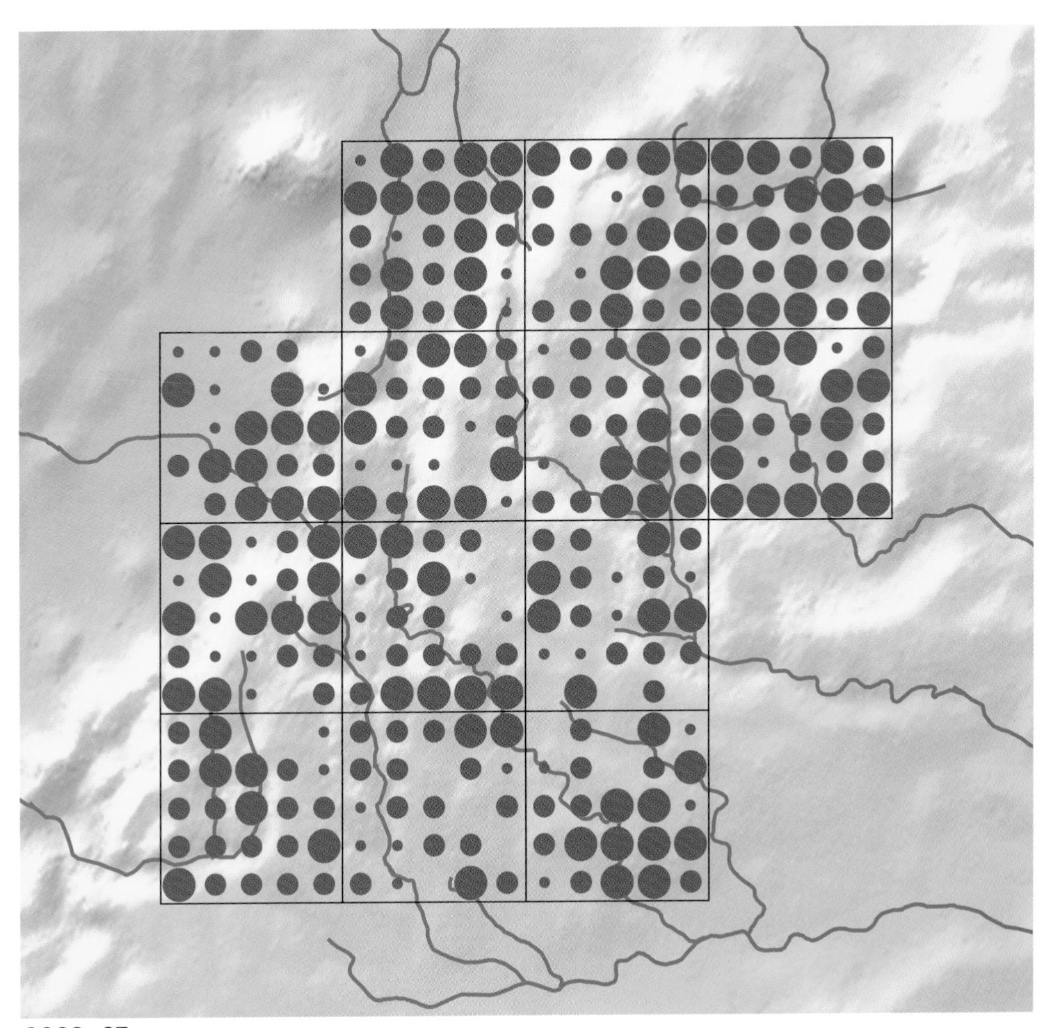

2003–07 survey

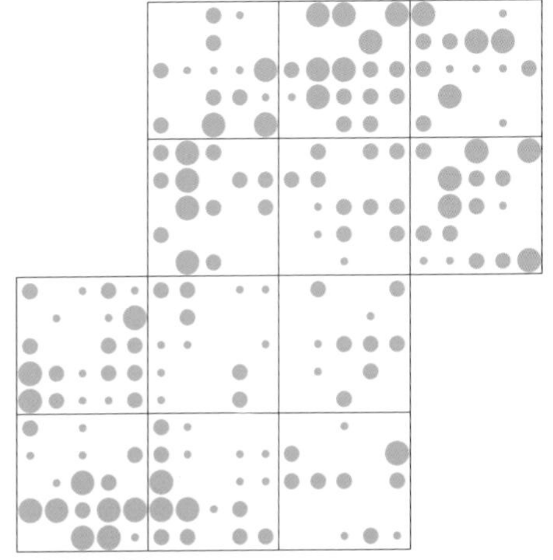

1983–87 survey

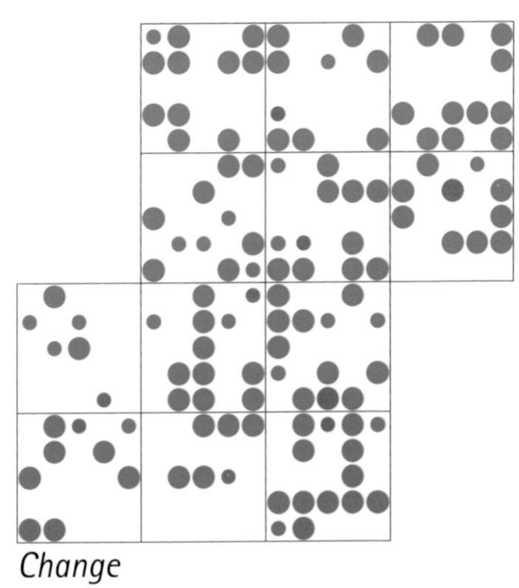

Change

Lesser Spotted Woodpecker

Dendrocopos minor

Given its small size and somewhat unobtrusive habits, the Lesser Spotted Woodpecker is a difficult species to detect, and requires either specific local knowledge, or much effort and a bit of luck. Both its distinctive call and its drumming can be heard in March and April, but the latter is quieter and less frequent than that of the Great Spotted Woodpecker. Whilst adults are subsequently more muted, young birds will call loudly from nest holes, which are often excavated high in a tree.

Unlike both of the larger woodpeckers, the distribution and numbers of Lesser Spotted Woodpeckers declined significantly in the recording area in the years between the two surveys. Although they have never been particularly common birds, Lesser Spotted Woodpeckers are now distinctly scarce. In the 1980s survey, the majority of the records came from the northern half of the recording area but most of these appear to have gone, and the current distribution map shows a thin and randomly scattered population. The loss of mature elms and a decline in the number of orchards might both be causes.

Given the greater survey effort, it may be that the actual thinning out in distribution is more than the maps imply. Some records were of birds heard calling on a single occasion–although they are a relatively sedentary species, they are likely to be more mobile early in the breeding season when looking for a suitable territory or mate. It may have been underrecorded in the first survey–the notes of the time commented that 'the difficulties of recording must be borne in mind when interpreting the findings'. It is likely, however, that this survey gives a fair picture of its current status and only 12 records of a woodland resident would suggest a species that is in danger of disappearing from the region.

Resident breeder

UK conservation status: RED

Long-term UK trend (1975-2006): -73%

Recent population trend: Not available

Numbers of tetrads with breeding records:

	2003–07		1983–87
	13 squares	12 squares	12 squares
Possible	**3**	3	10
Probable	**6**	5	16
Confirmed	**3**	3	1
Totals	**12**	**11**	**27**

RICHARD TYLER

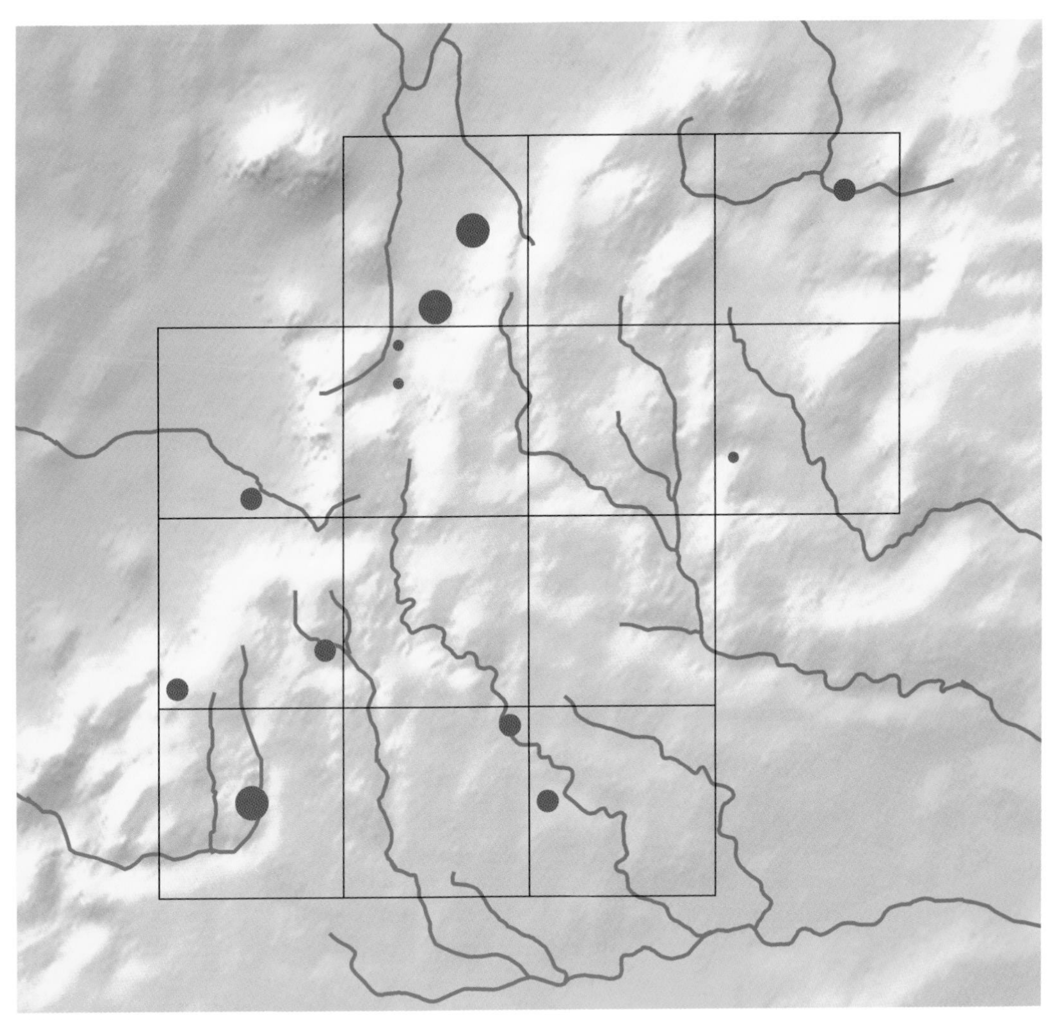
2003–07 survey

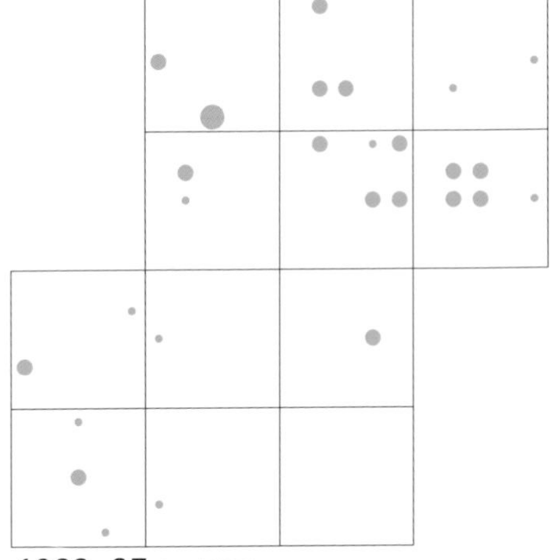
1983–87 survey

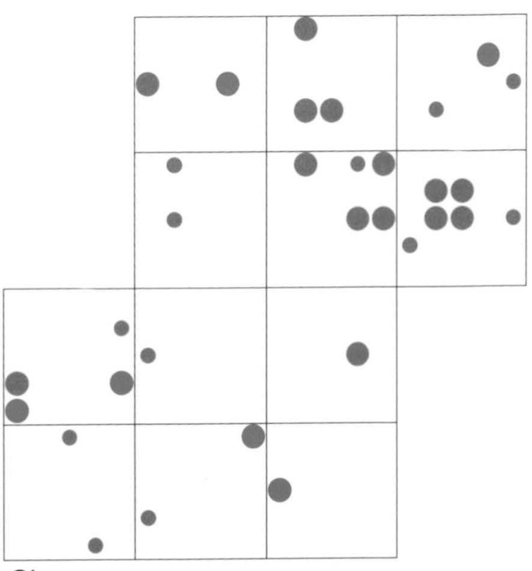
Change

Skylark (Sky Lark)

Alauda arvensis

ROB BROOKES

Resident breeder

Gloucestershire BAP species

UK conservation status: **RED**

Long-term UK trend (1970–2006): -53%

Recent population trend (1994–2007): -13%

Numbers of tetrads with breeding records:

	2003–07		1983–87
	13 squares	12 squares	12 squares
Possible	**0**	0	5
Probable	**233**	218	241
Confirmed	**76**	73	26
Totals	**309**	**291**	**272**

Skylarks are very easy birds to survey on a presence/absence basis—their song is very familiar and easy to pick up, and they are unlikely to be found singing in areas in which they are not attempting to set up or hold territory. Confirmation of breeding is more difficult: although they openly carry food to young in the nest, this is often in the middle of a cereal field and some distance from the closest observation point. This means that patience and a little detective work are required to be absolutely sure. That confirmation was obtained in 76 tetrads (73 of them in the 12 common 10 km squares, 180% more than in the previous survey) suggests that the species is still breeding widely in the Cotswolds.

Nationally, the Skylark has been highlighted by the RSPB as a previously common bird that is in serious decline in some habitats, primarily farmland. The area covered by this survey contains a large percentage of farmland—both arable (where the species breeds) and pasture (where it generally does not), as well as upland rough ground such as Cleeve Common, which it also favours. Despite its decline nationally, the results of this survey indicate that it is a still a widely distributed species in the Cotswolds, being found in all but 16 of the 325 tetrads surveyed. Although the Cotswolds contains a mosaic of habitats, this suggests that the ones favourable to the Skylark are distributed fairly uniformly.

The greater survey effort probably accounts for the slightly higher number of occupied tetrads. However, the change map shows very few areas from which the species has been lost.

Species sponsored by Duncan and Rebecca Dine

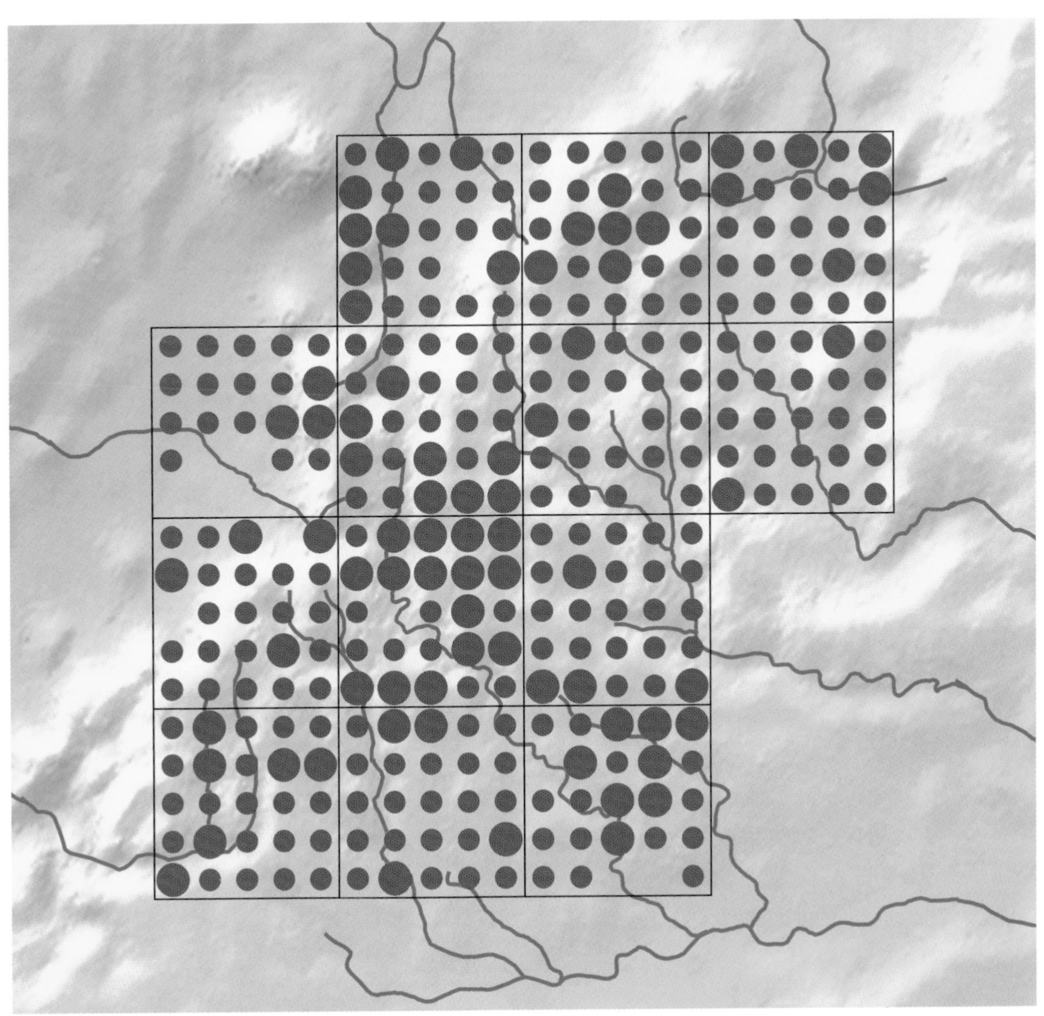

2003–07 survey

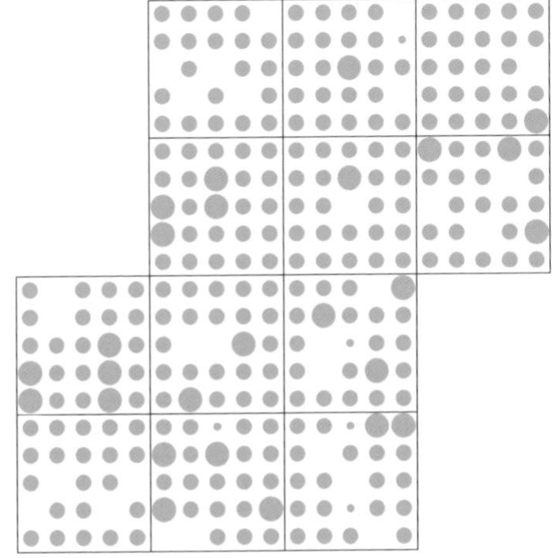

1983–87 survey

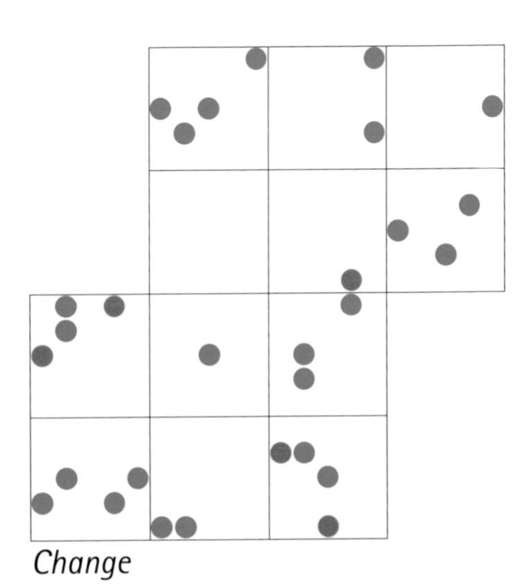

Change

Swallow (Barn Swallow)

Hirundo rustica

ROB BROOKES

Migrant breeder

UK conservation status: AMBER

Long-term UK trend (1970–2006): +25%

Recent population trend (1994–2007): +25%

Numbers of tetrads with breeding records:

	2003–07		1983–87
	13 squares	12 squares	12 squares
Possible	**13**	12	40
Probable	**43**	40	103
Confirmed	**250**	235	129
Totals	**306**	**287**	**272**

The Swallow is a familiar bird throughout the Cotswolds. It is also relatively easy to survey and locate its breeding sites, as it is usually clear from its flight paths where it is attempting to nest. As a result, the percentage of tetrads in which breeding was confirmed was high at over 80% (compared with just under 50% in the 1980s survey) and it is unlikely that many occupied tetrads were missed.

Swallows are predominantly birds of rural situations. In the survey area, they nest in farm buildings, horse stables and similar structures—even sculleries of rural houses—and the wide distribution of the species is an indicator of the diversity of habitat in the region at the tetrad level. The only area where there appears to be a significant absence of breeding Swallows is in and around urban Cheltenham.

The previous Atlas stated: 'Nearly every farm has one or two pairs, and some may support up to a dozen but rarely more'. Sizeable swallow colonies are still very unusual, with up to five pairs being more the norm found in this survey. Although some traditional nesting sites have been lost through the conversion of farm buildings to houses, this may have been partially offset by the building of new stables in the Cotswolds. It is not clear whether there has been any change in population density as a result, but the change map shows that there has been very little alteration in distribution.

Species sponsored by Tony and Pam Perry

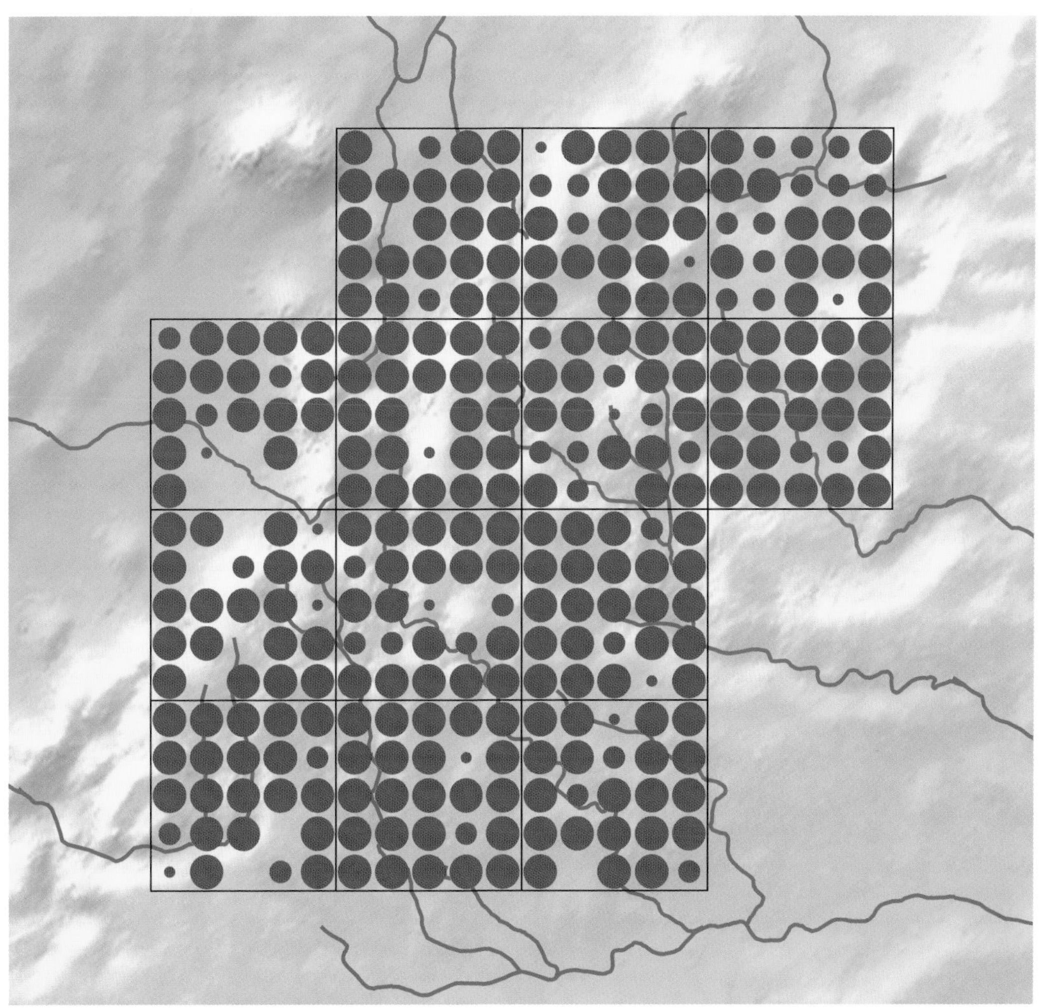
2003–07 survey

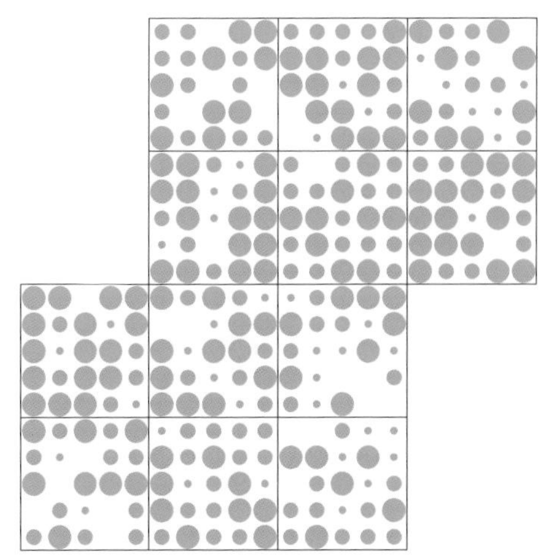
1983–87 survey

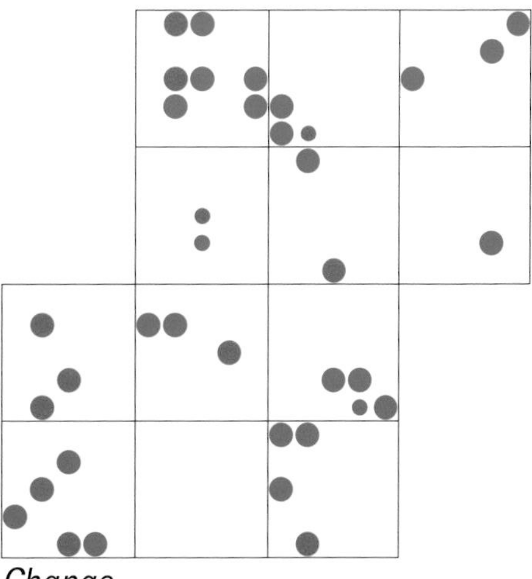
Change

House Martin

Delichon urbicum

ROB BROOKES

Migrant breeder

UK conservation status: AMBER

Long-term UK trend (1970–2006): -35%

Recent population trend (1994–2007): +9%

Numbers of tetrads with breeding records:

	2003–07		1983–87
	13 squares	12 squares	12 squares
Possible	**19**	16	29
Probable	**20**	17	36
Confirmed	**215**	203	148
Totals	**254**	**236**	**213**

Arriving in mid-April and often remaining well into October, the House Martin is a very familiar sight around Cotswolds villages. Most nests were found under the eaves of both old and new buildings, although House Martins also nest in window reveals. Their conspicuous breeding sites together with the protracted breeding season resulted in a very high proportion of confirmed breeding records and it is unlikely that any tetrads with significant colonies were missed. Movement of colonies from one location to nearby ones (often new housing estates) was noted, evidence of the species' ability to rapidly exploit new sites.

In the survey area, the distribution of the House Martin is slightly more restricted than that of the Swallow, as it is entirely dependent on housing and similar buildings. However, the average colony size is probably greater–up to 20 pairs at a single location being not unusual, whereas it was uncommon for there to be more than about five pairs of Swallows in a single location. Breeding sites are mainly in rural villages, and some isolated houses; House Martins have a more localized distribution in large urban areas.

The change map does not suggest any major change in distribution, although there does appear to have been a modest rise in the number of occupied tetrads in the east of the region in the Evenlode Valley. Whether this is a genuine increase or the result of a greater survey effort is not clear.

Species sponsored anonymously

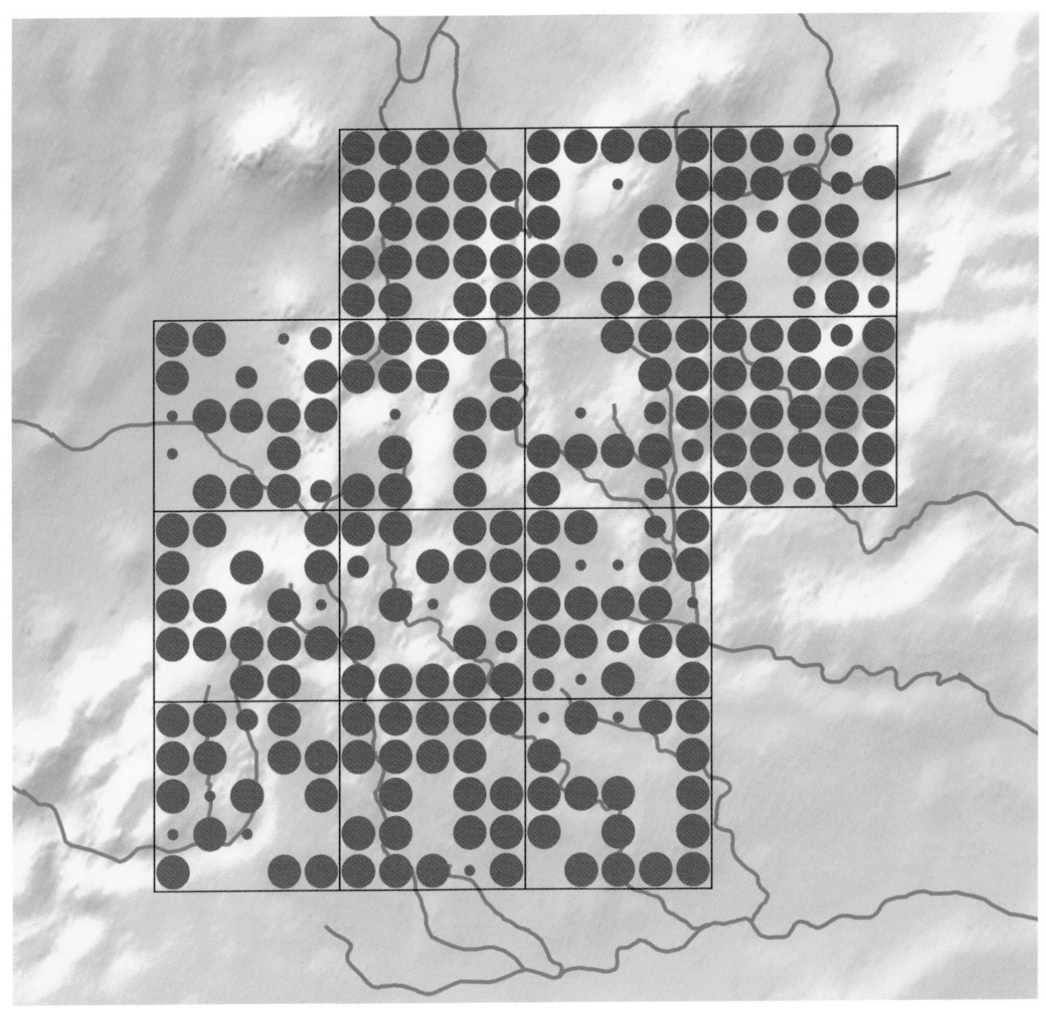

2003–07 survey

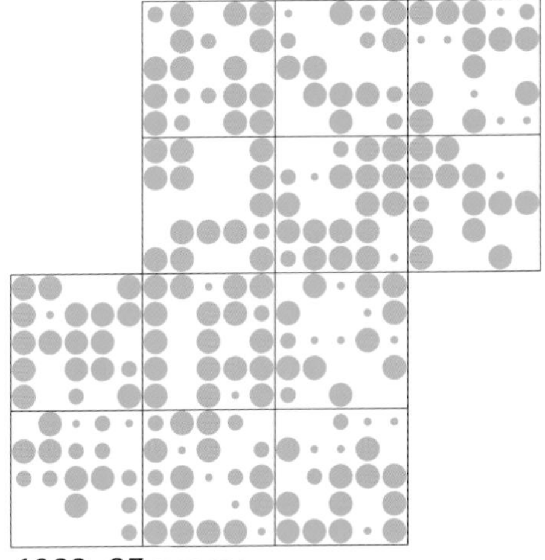

1983–87 survey

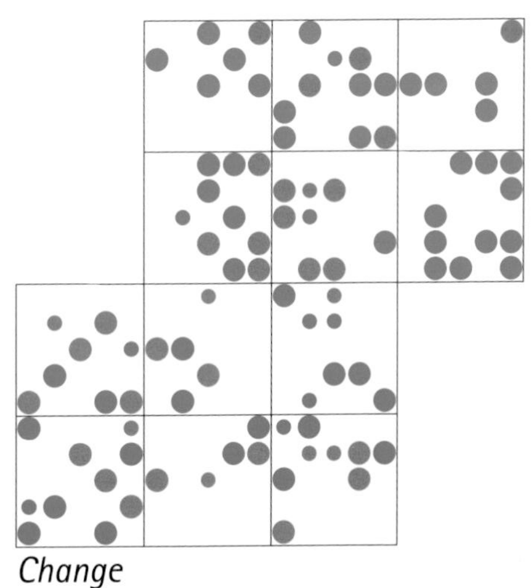

Change

Tree Pipit

Anthus trivialis

RICHARD TYLER

Migrant breeder

UK conservation status: AMBER

Long-term UK trend (1970–2006): -75%

Recent population trend (1994–2007): -11%

Numbers of tetrads with breeding records:

	2003–07		1983–87
	13 squares	12 squares	12 squares
Possible	1	1	4
Probable	31	28	70
Confirmed	7	6	13
Totals	**39**	**35**	**87**

The song of the Tree Pipit is distinctive, comprising a series of phrases and being much more varied than the trill of the Meadow Pipit. Combined with its display flight, this made locating the species straightforward. Tree Pipits also openly carry food to young and so, in theory, confirmation of breeding should not have been too difficult. However, breeding was only confirmed in seven tetrads.

Since the first survey in the 1980s, there has been a major contraction in the species' range, and overall the number of breeding records has decreased by more than 60%. Tree Pipits prefer to breed at the tops of slopes and although large areas of the Cotswold scarp would seem to offer suitable habitat, the distribution map does show a very localized distribution there.

The change map shows very clearly that the Tree Pipit has retreated from the majority of its sites in the north and east of the region (on the High Wold and in the dip slope valleys). The smaller population in the far southwest (in the Stroud Valleys) has also been lost. These were often in new plantations or clear-felled woodland, a habitat which appears to have been largely abandoned. The population now seems to be concentrated in the uncultivated scarp slopes to the southeast of Cheltenham, including several tetrads in the 10 km square SO92, which was not surveyed in the 1980s.

Anecdotal evidence suggests that most of the contraction in range occurred in the early 1990s, by which time Tree Pipits had already deserted their sites in the north. Although the habitat here seems much as it was in the 1980s, there may have been subtle changes that have made it unsuitable for Tree Pipits. However, it may also be a reflection of the national decline. In many of the tetrads in which Tree Pipits were found, several singing birds were located. However, the low number of confirmed breeding records is a cause for concern in a species that quite visibly carries food for young; this may reflect poor breeding success, which may be responsible for a contraction in distribution.

Species sponsored by Geoff Bailey

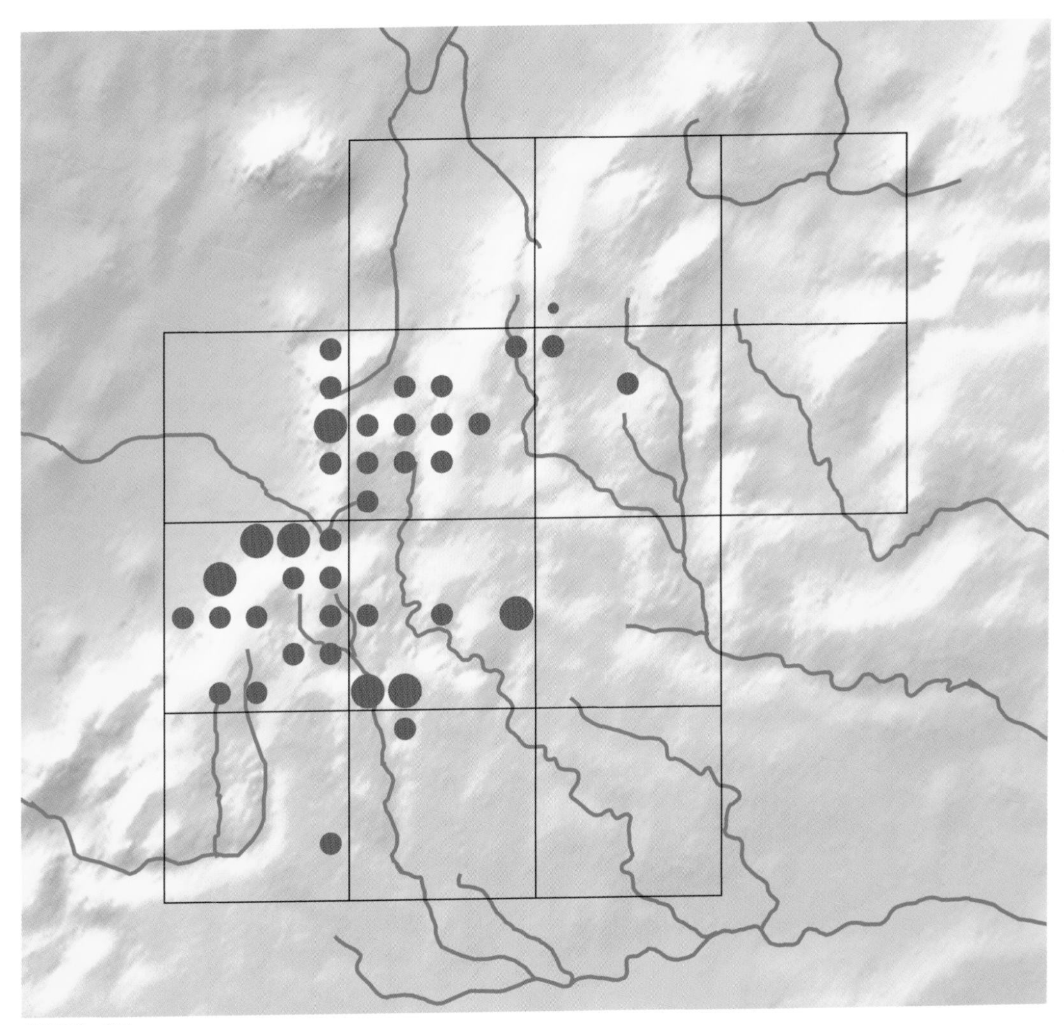

2003–07 survey

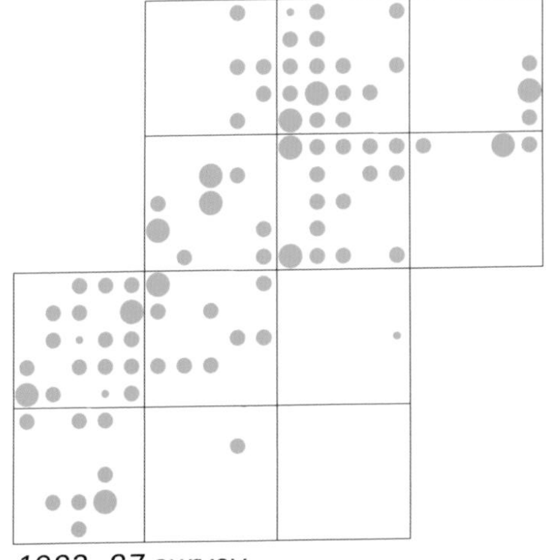

1983–87 survey

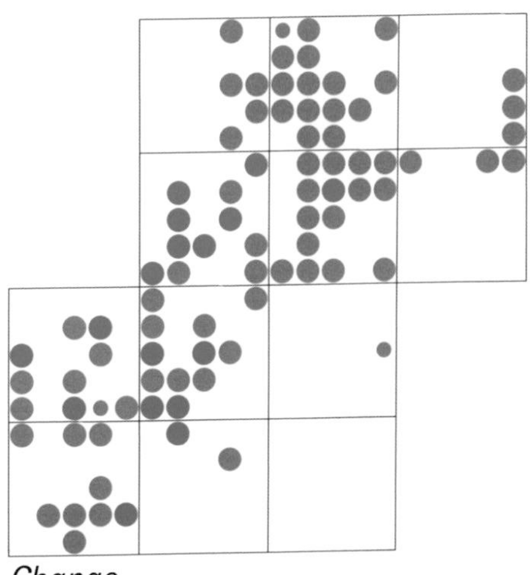

Change

Meadow Pipit

Anthus pratensis

ROB BROOKES

Migrant/resident breeder

UK conservation status: AMBER

Long-term UK trend (1970–2006): -41%

Recent population trend (1994–2007): -16%

Numbers of tetrads with breeding records:

	2003–07		1983–87
	13 squares	12 squares	12 squares
Possible	7	5	6
Probable	8	6	10
Confirmed	3	2	7
Totals	18	13	23

The song and display flight of the Meadow Pipit are quite distinctive. It gives a clear and extended trilling call as it ascends and then descends in a parachute glide. The only likely confusion species is the Tree Pipit, which is visually similar though its song is much more varied. It is possible that observers could confuse the two species in breeding territory, exacerbated by the fact that they may be found in close proximity in some parts of the Cotswolds.

Meadow Pipits are regular spring passage migrants in the region, and care was taken not to include passage records in the data. In theory, breeding is not difficult to confirm, as the species can readily be seen carrying food to nestlings on the ground.

The distribution map clearly indicates the restricted distribution of Meadow Pipits—they require exposed rough pasture, and this type of habitat is mainly confined to the hills at the top of the Cotswold scarp—most records of breeding activity were on Cleeve Common (where they are locally common) and the adjoining areas of Leckhampton and Birdlip. The change map suggests a contraction in range, with the species no longer being found in the open farmland on the High Wold to the east of Cleeve Common. As with the Tree Pipit, there were several occupied tetrads in SO92, which was not surveyed in the 1980s.

Given the ease of surveying breeding activity of this species, it is likely that the distribution map gives a very accurate picture of its breeding range.

Species sponsored by Andrew Cleaver

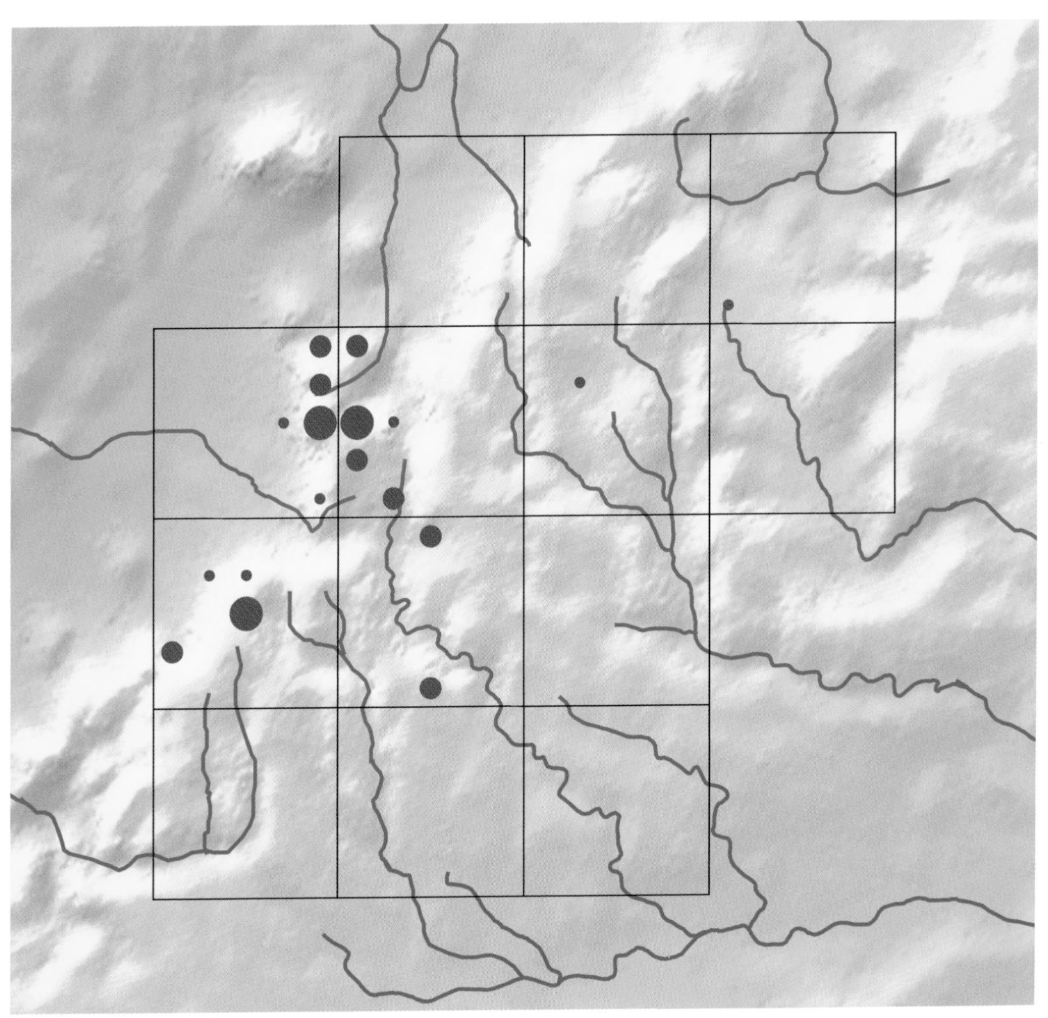

2003–07 survey

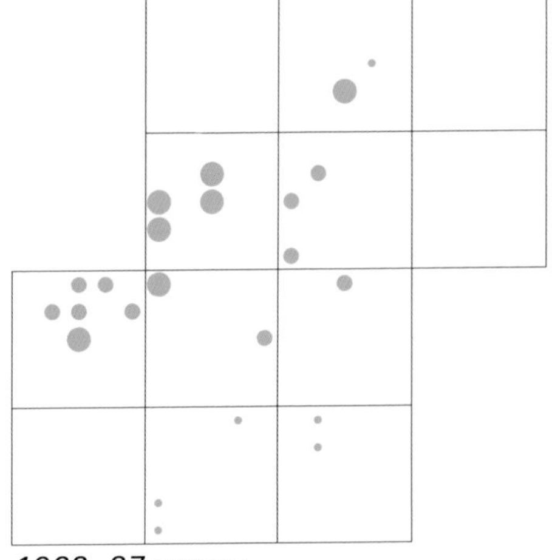

1983–87 survey

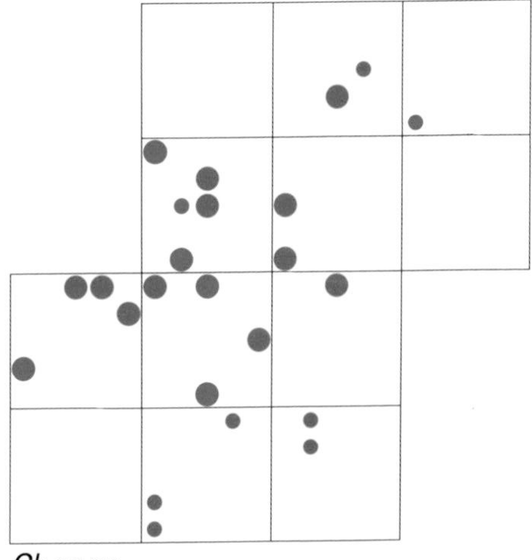

Change

Yellow Wagtail

Motacilla flava

RICHARD TYLER

Migrant breeder

UK conservation status: AMBER

Long-term UK trend (1970–2006): -69%

Recent population trend (1994–2007): -47%

Numbers of tetrads with breeding records:

	2003–07		1983–87
	13 squares	12 squares	12 squares
Possible	**18**	18	19
Probable	**18**	18	8
Confirmed	**10**	9	9
Totals	**46**	**45**	**36**

Yellow Wagtails are traditionally associated with lowland meadows, but in this survey, as in that of the 1980s, they showed a distinct liking for areas planted with legumes, even on higher ground. They were also found associated with crops of wheat and oilseed rape. Location of the Yellow Wagtail is straightforward thanks to its distinctive call, and although its song is modest, it does openly carry food to young, allowing breeding to be confirmed. Family parties in typical breeding habitat also provided evidence of successful breeding.

At first glance, the number of records and the percentage of confirmed breeding tetrads suggest that there had been very little alteration in the area occupied. However, the change map illustrates a definite westward shift in its location. In the 1980s it was primarily found in the northeast of the region, in two distinct areas: between Chipping Campden and Shipston in the Stour Valley, and in the Evenlode Valley. In this survey, both of these populations had largely disappeared, but a new area had been colonized–the higher arable farmland between Northleach and Andoversford on the dip slope. The small population in the Avon Vale remained, although the maps probably overestimate the number of occupied sites here, because the crop rotation cycle caused the birds to move from one tetrad to another, depending on which fields were planted with legumes.

It was believed that there had been a population increase during the first Atlas survey period; this time, anecdotal comments were that the population was in decline. It may be that there was a peak in numbers in the early to mid-1990s.

Species sponsored by Patrick Buxton

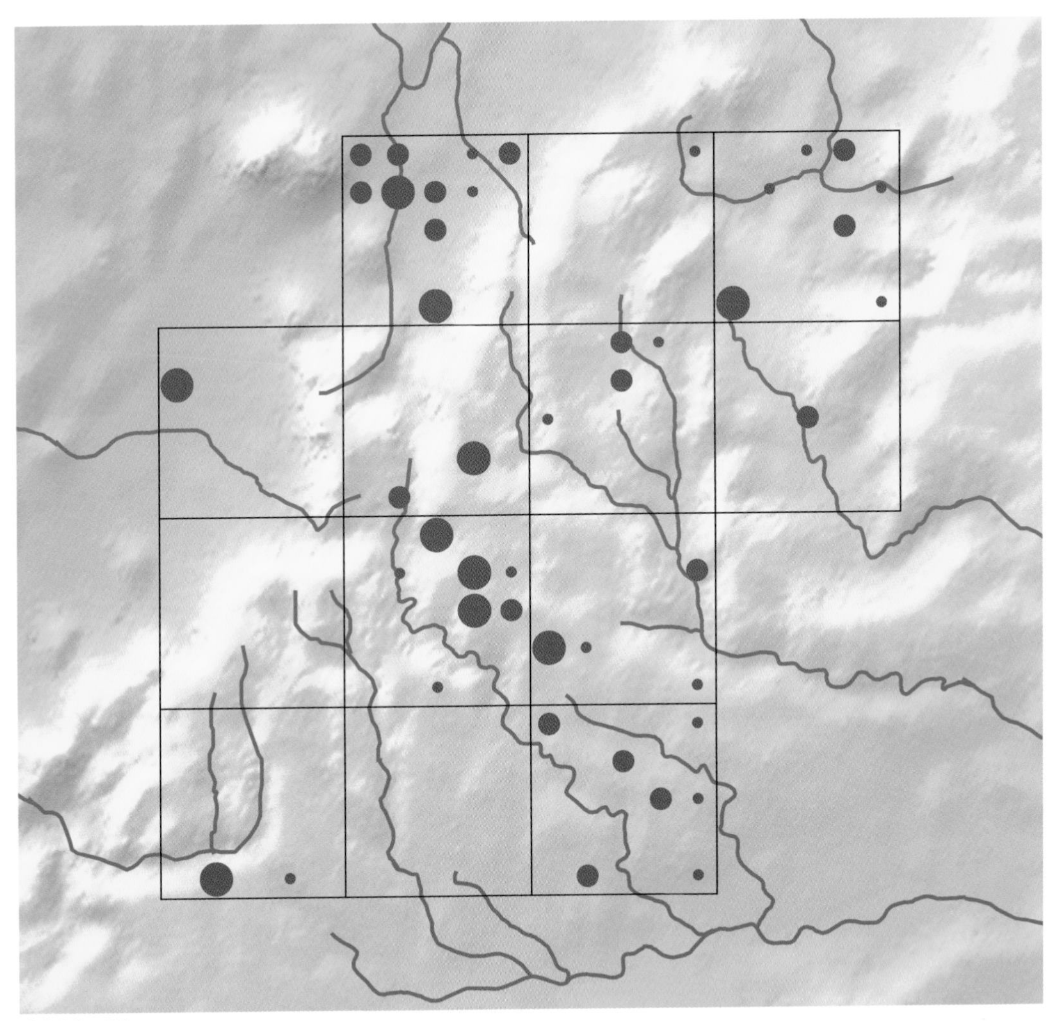

2003–07 survey

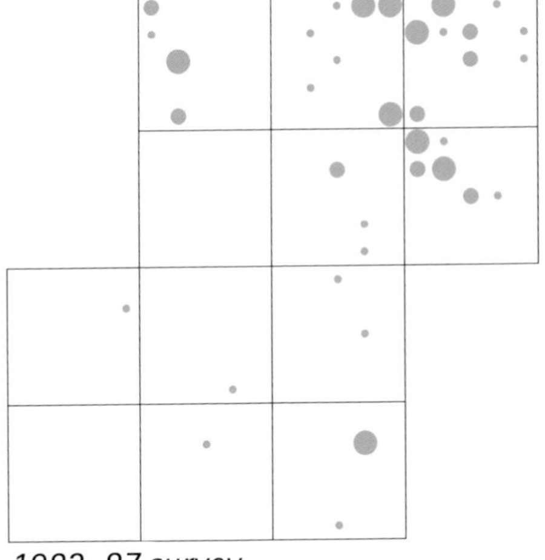

1983–87 survey

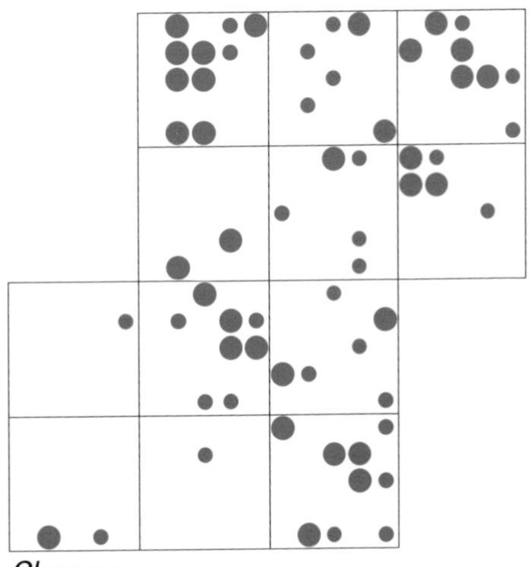

Change

Grey Wagtail

Motacilla cinerea

The loud song and confiding nature of Grey Wagtails made location of breeding birds straightforward, and confirmation of breeding was also relatively easy through observation of food carrying, with about 45% of records being in the highest category.

Grey Wagtails traditionally breed close to shallow fast-flowing streams, and this type of habitat provided the majority of records in this survey. They were also frequently found alongside other small shallow waterbodies, very minor tributaries and even at sewage works. Most widely distributed along the Rivers Frome, Churn and Coln as well as the upper reaches of the Windrush, there were however also a few records from the Evenlode Valley, which is a much slower-moving river system. In the survey in the 1980s Grey Wagtails were absent from this area and it was thought that such sluggish waters were not suited to the species.

Overall, there seems to have been a marked expansion in the distribution of the species, with 130% more occupied tetrads, which is similar to that seen with the Pied Wagtail (although from a smaller baseline). The increase in the percentage of confirmed breeding records (150%) is also very similar to that seen with Pied Wagtail, as is the change map, which suggests that this growth has been fairly uniformly distributed across the region. Although the greater survey effort may have contributed to this, there seems little doubt that there has been a real expansion in range.

Resident breeder

UK conservation status: AMBER

Long-term UK trend (1975–2006): -23%

Recent population trend (1994–2007): +26%

Numbers of tetrads with breeding records:

	2003–07		1983–87
	13 squares	12 squares	12 squares
Possible	**35**	34	16
Probable	**26**	26	12
Confirmed	**50**	43	17
Totals	111	103	45

ROB BROOKES

Species sponsored by Mark Farmer

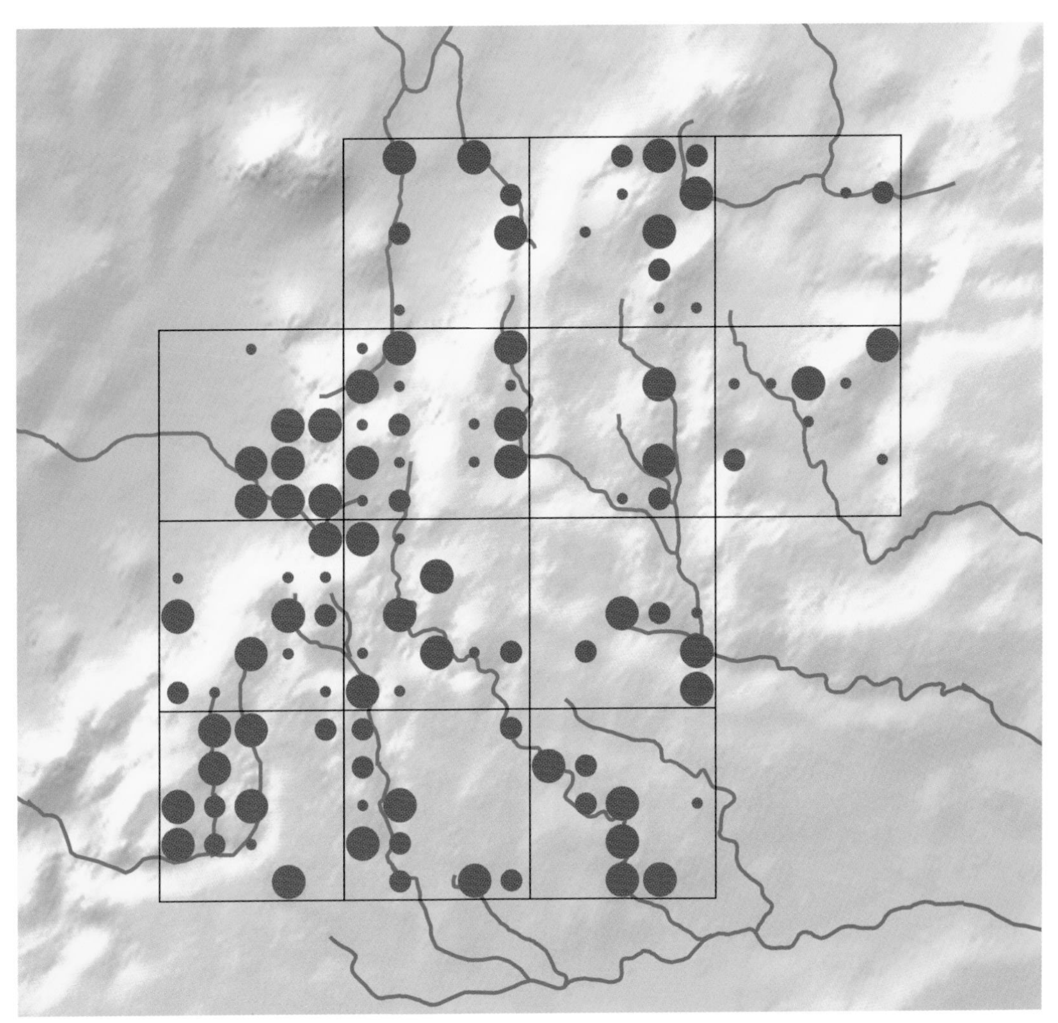

2003–07 survey

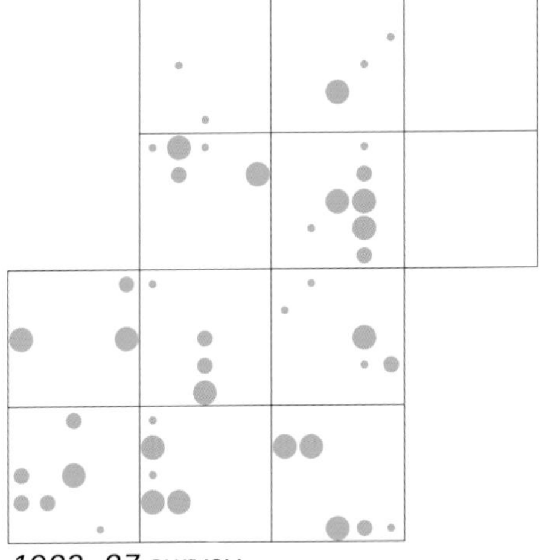

1983–87 survey

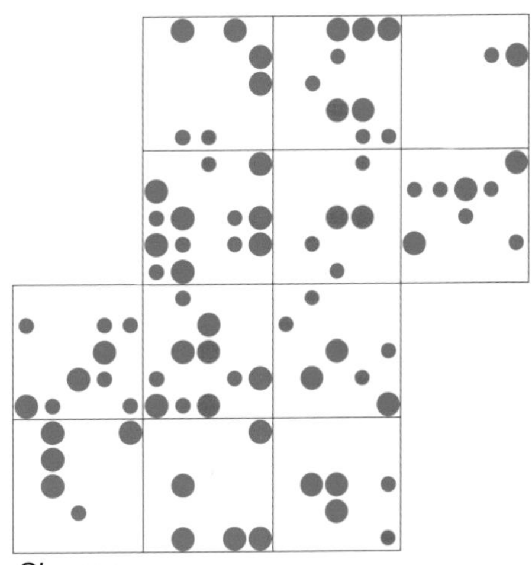

Change

Pied Wagtail

Motacilla alba yarrellii

ROB BROOKES

Resident breeder

UK conservation status: GREEN

Long-term UK trend (1970–2006): +45%

Recent population trend (1994–2007): +15%

Numbers of tetrads with breeding records:

	2003–07		1983–87
	13 squares	12 squares	12 squares
Possible	**46**	38	39
Probable	**93**	84	34
Confirmed	**142**	138	37
Totals	**281**	**260**	**110**

Although Pied Wagtails are not noted for prolonged bouts of singing, their habit of calling from roofs, and their distinctive flight and flight call, made location easy. Confirmation of breeding was also straightforward with adults regularly being seen carrying large quantities of food to nestlings and fledglings.

One of the pleasant surprises from the Atlas surveying was the realization of how widespread Pied Wagtails had become. In the 1980s, it was described as being 'widespread but nowhere common'. The situation was not noticeably different in the mid-1990s when, as a result of a small unpublished survey of the species undertaken by the NCOS, it was still thought of as more a winter visitor than a common breeding resident. In the latest survey, it seemed that most farmyards, stables, industrial/workshop sites, water treatment works and similar habitats supported breeding wagtails, with several adults often being seen–suggesting more than one breeding pair in some locations and often several breeding sites within one tetrad.

The change map indicates a real and marked expansion in the species' distribution density, rather than one due mainly to the greater survey effort–with about 135% more occupied tetrads this time. The increase has occurred throughout the survey area, with some areas being well populated where previously there were only two or three isolated records. In addition, there was an increase in confirmed breeding records (from about 35% to about 55% of occupied tetrads), which suggests that the Pied Wagtail is doing well.

The reason for the increase does not appear to be habitat-related–the survey area has always provided what appear to be ideal conditions, with many farms and stables.

Species sponsored by Charlotte and Jessica Nash

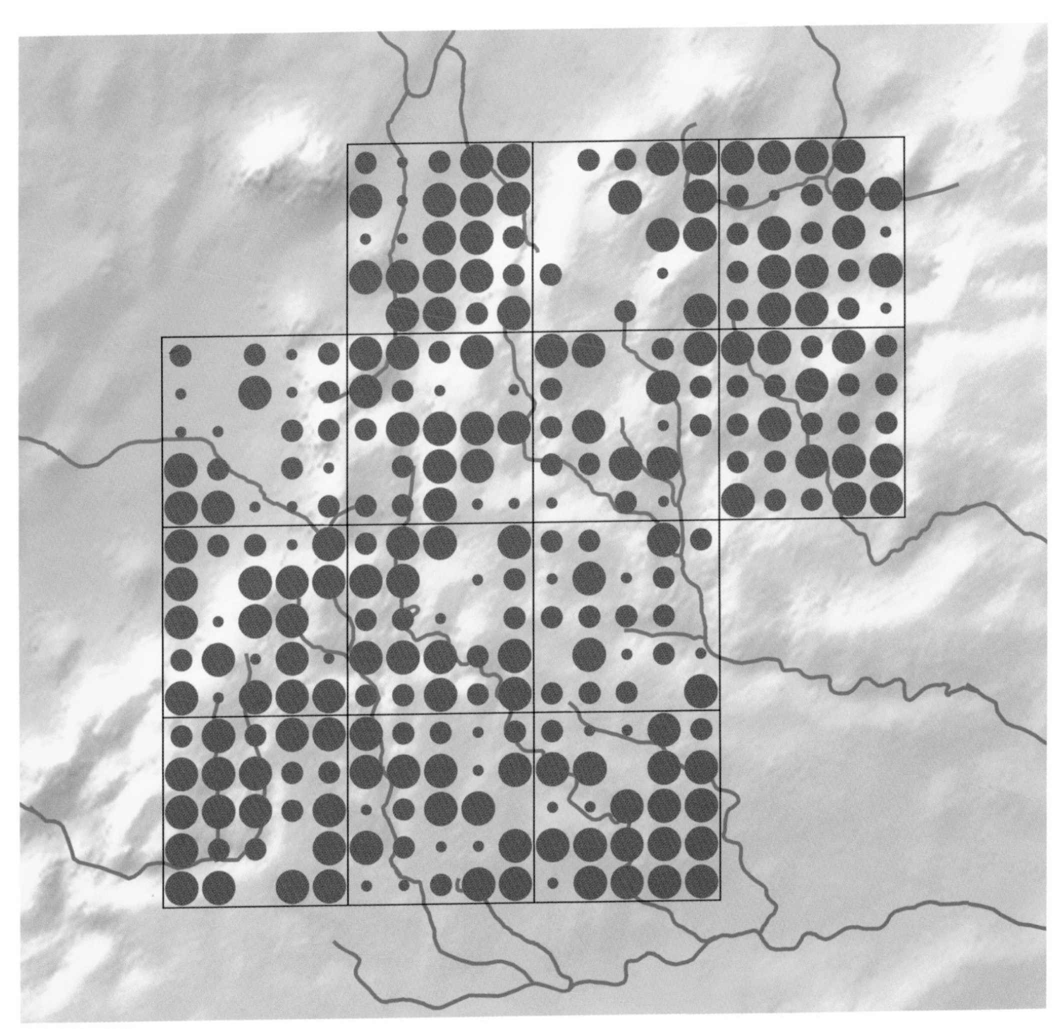

2003–07 survey

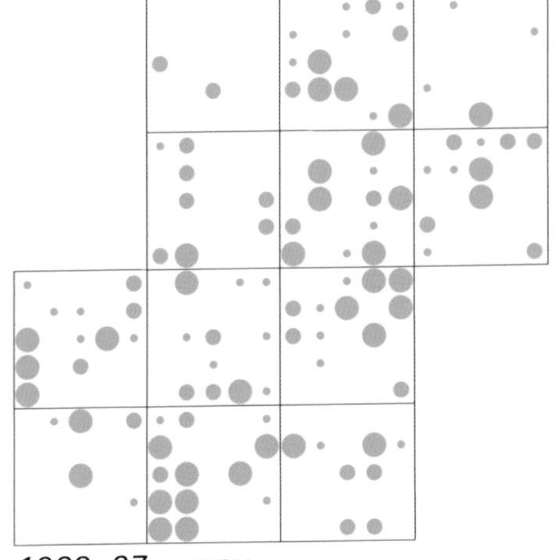

1983–87 survey

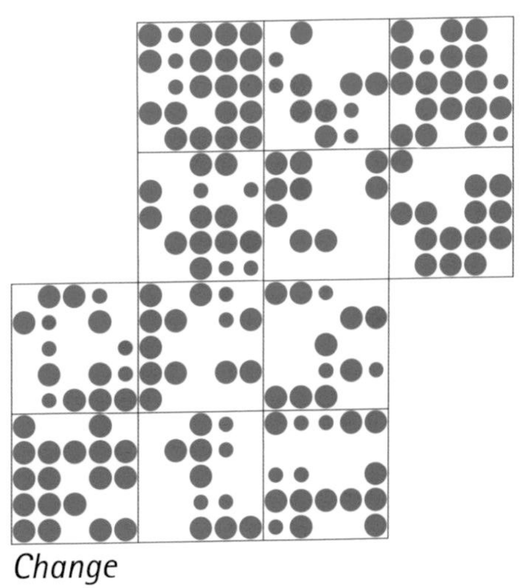

Change

Dipper (White-throated Dipper)

Cinclus cinclus

RICHARD TYLER

Resident breeder

UK conservation status: GREEN

Long-term UK trend (1970–2006): -11%

Recent population trend (1994–2007): -12%

Numbers of tetrads with breeding records:

	2003–07		1983–87
	13 squares	12 squares	12 squares
Possible	3	3	14
Probable	4	3	4
Confirmed	11	11	17
Totals	18	17	35

As Dippers are restricted to a very specialized habitat, surveying likely areas was relatively easy. Singing and territorial behaviour starts early in the year and, once located, the sites could be revisited in the following months for evidence of adults carrying food, or newly fledged young. Most occupied territories yielded confirmed breeding records.

Dippers prefer clear, shallow, fast-flowing streams in which to feed and beside which to nest, and there are many such habitats in the survey area, including several tributaries of the River Thames. Although the streams are still generally clear and fast-flowing, the distribution of Dippers has noticeably contracted since the survey of the 1980s, and there has been a 50% decline in the number of occupied tetrads. Chemical changes (such as increased acidification or nitrification) have been suggested as possible causes for this, through a decline in insect food.

In the 1980s survey, the species was primarily found on the Coln, Churn and Windrush, with one regular site on the River Frome, and it was absent from the Evenlode, Stour and Isbourne. In this survey, there were virtually no records from the Coln, and the numbers from the Windrush river complex were drastically reduced. There were still a number of sites on the Churn, including some new sites towards its source, and the species also bred along the upper reaches of the Isbourne and on the Frome. Overall, the most easterly sites have been lost.

In the 1980s survey, it was believed that the survey area held the most easterly breeding pairs of Dippers in southern England; this appears to have been confirmed by the new survey. Given the national contraction in distribution, the Cotswolds still clearly provides an important marker for the species' fortunes.

Species sponsored by Alexander Nice

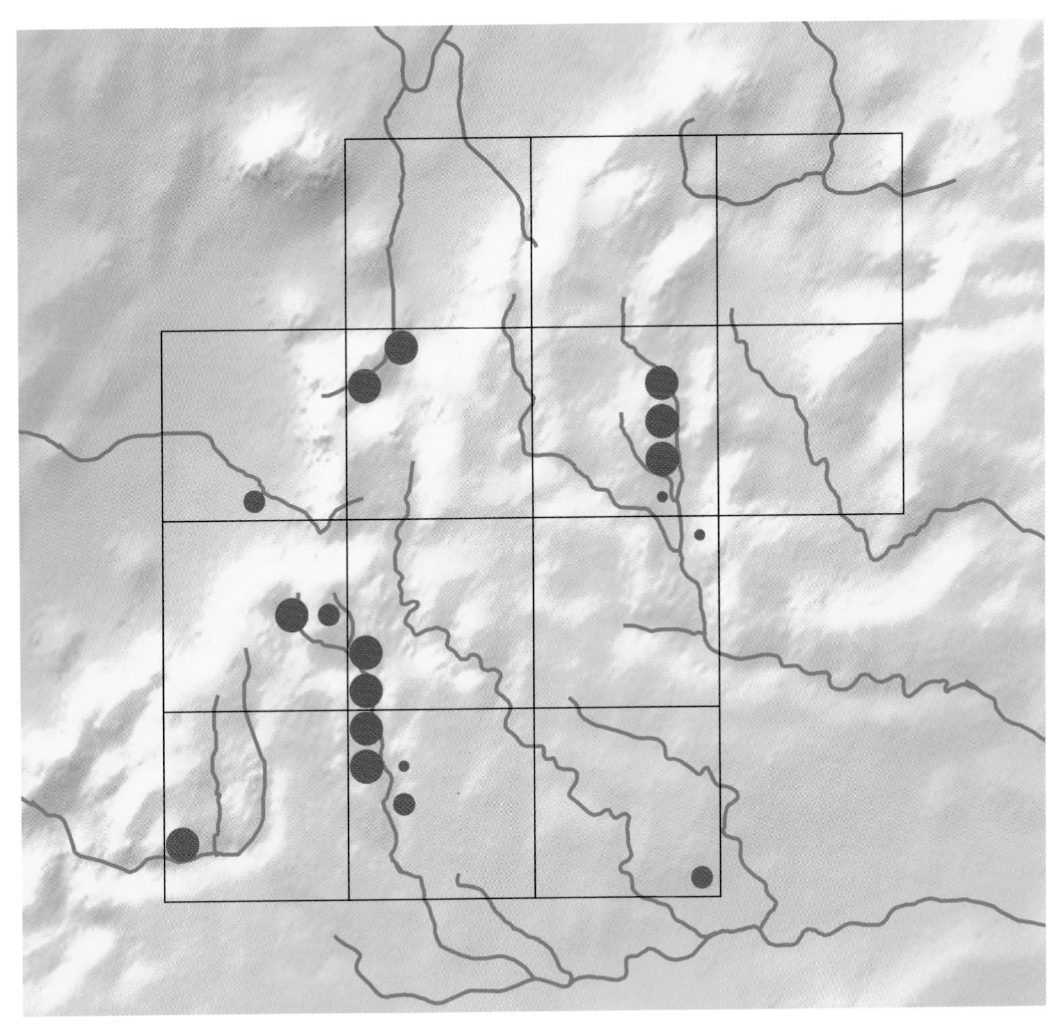

2003–07 survey

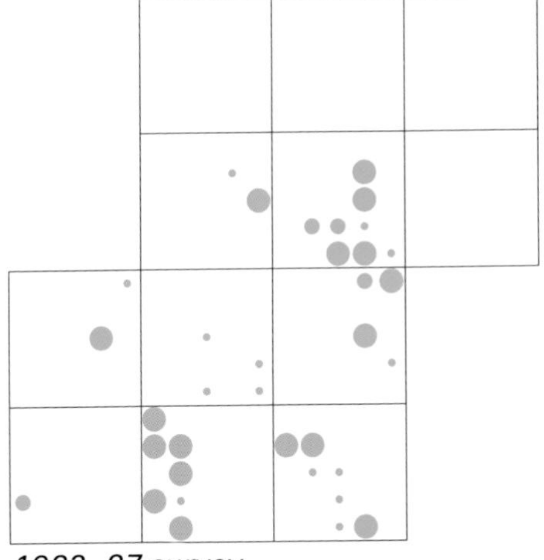

1983–87 survey

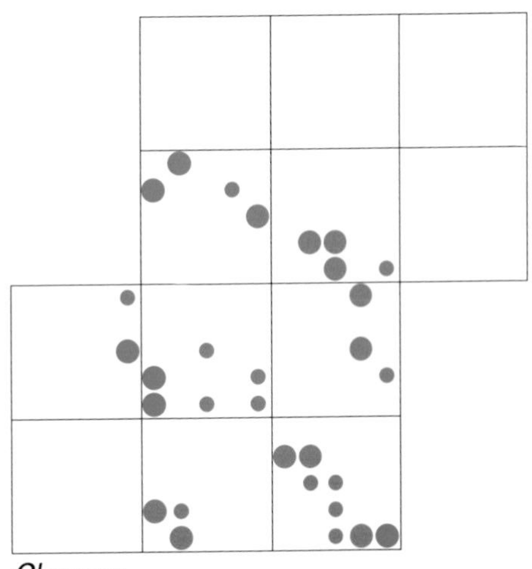

Change

Wren (Winter Wren)

Troglodytes troglodytes

RICHARD TYLER

Resident breeder

UK conservation status: GREEN

Long-term UK trend (1970–2006): +70%

Recent population trend (1994–2007): +25%

Numbers of tetrads with breeding records:

	2003–07		1983–87
	13 squares	12 squares	12 squares
Possible	1	1	5
Probable	165	151	229
Confirmed	159	148	51
Totals	**325**	**300**	**285**

A very familiar bird throughout the region, the Wren was found in a wide range of habitats, including gardens, hedgerows, large and small woods, stone walls, river banks and quarries. Its loud song is familiar to all observers, and territorial birds were very easy to locate. Confirmation of breeding was also regularly achieved by observing birds carrying food (albeit often in small quantities) or faecal sacs, or repeatedly visiting the same site in a hedgerow or bush to the accompaniment of the high-pitched calls of nestlings.

The Wren was one of seven species that were located in all 325 tetrads, and breeding activity was noted in all but one of these. It is a species that has been prone in the past to significant population declines following hard winters, the 1962/63 and 1981/82 winters being the most notable recent examples. However, a rapid recovery in numbers was reported after a cold winter in 1984/85, and since the 1983–87 survey, the Cotswolds have had a series of mild winters and the population has not had to cope with this type of pressure.

Greater survey effort probably accounts for the filling in of tetrads in which Wrens were not found in the 1980s survey, the comment made in the first Atlas being that it probably bred in every tetrad. However, the trebling in the number of confirmed breeding records does suggest that the number of breeding pairs may have increased (*ie* population density has increased), even if the distribution is virtually unchanged.

Species sponsored by Dursley Birdwatching and Preservation Society

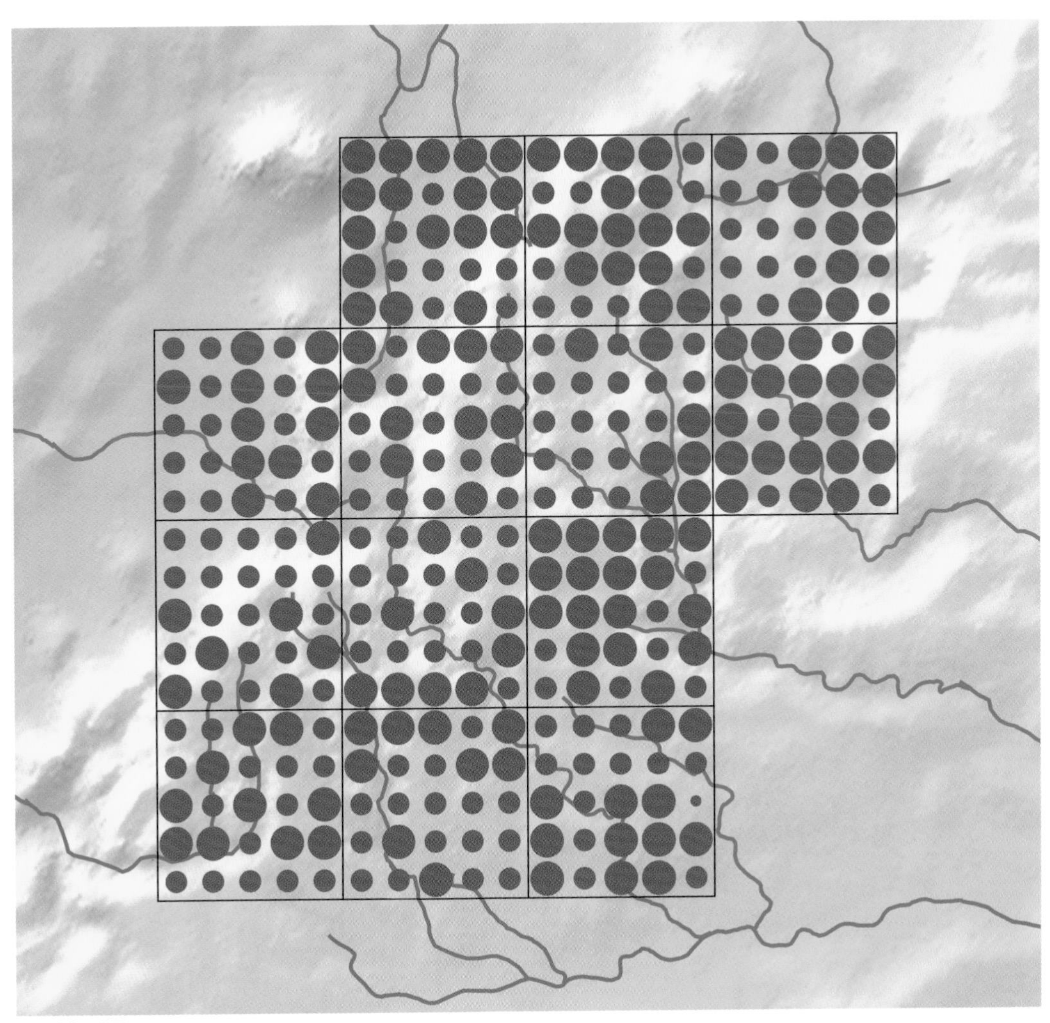

2003–07 survey

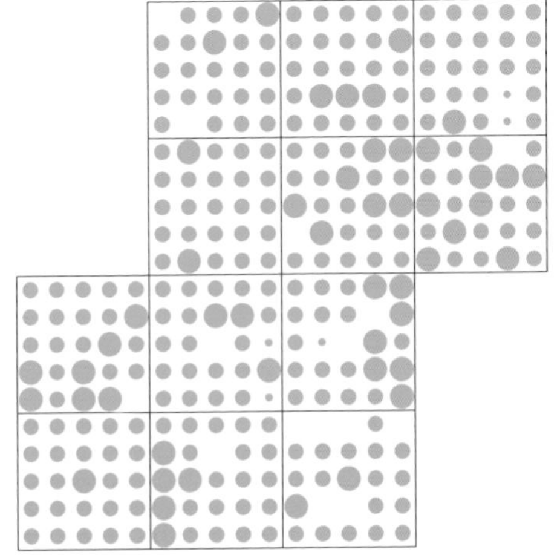

1983–87 survey

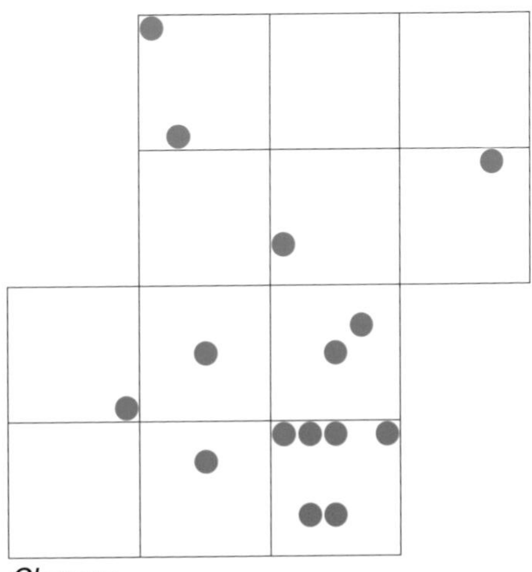

Change

Dunnock (Hedge Accentor)

Prunella modularis

The Dunnock is a very familiar species and is a regular breeder in parks and gardens, as well as in rural hedgerows, scrub and, to a lesser extent, woodland. Its song is distinctive and well known, and it has a prolonged song period from late winter through into summer, which makes location and recording of probable breeding activity easy. It could sometimes be seen carrying insect food for its young, allowing confirmation of breeding, although this was not as regular an occurrence as might be expected. Breeding confirmation was also obtained by seeing young fledglings begging for food.

Dunnocks were recorded in all 325 tetrads. In the 1980s survey they were not located in 10% of the 300 tetrads. The filling in of the apparently unoccupied tetrads was almost certainly due to the greater survey effort. Interestingly, in the 1980s survey, although the Dunnock was considered to be very common, it was believed to have declined during the survey period. The general perception with observers at the start of the later survey was that the species had continued to decline slightly in the intervening years. Although a distribution survey would not be expected to provide evidence to support this, Dunnocks may be less common than most of the other species for which universal distribution was found, especially away from parks and gardens. In the 1980s, its relative scarcity in woodland was noted, and this probably continues to be the case.

Resident breeder

UK conservation status: AMBER

Long-term UK trend (1970–2006): -25%

Recent population trend (1994–2007): +25%

Numbers of tetrads with breeding records:

	2003–07		1983–87
	13 squares	12 squares	12 squares
Possible	**3**	3	23
Probable	**192**	182	186
Confirmed	**130**	115	61
Totals	**325**	**300**	**270**

ROB BROOKES

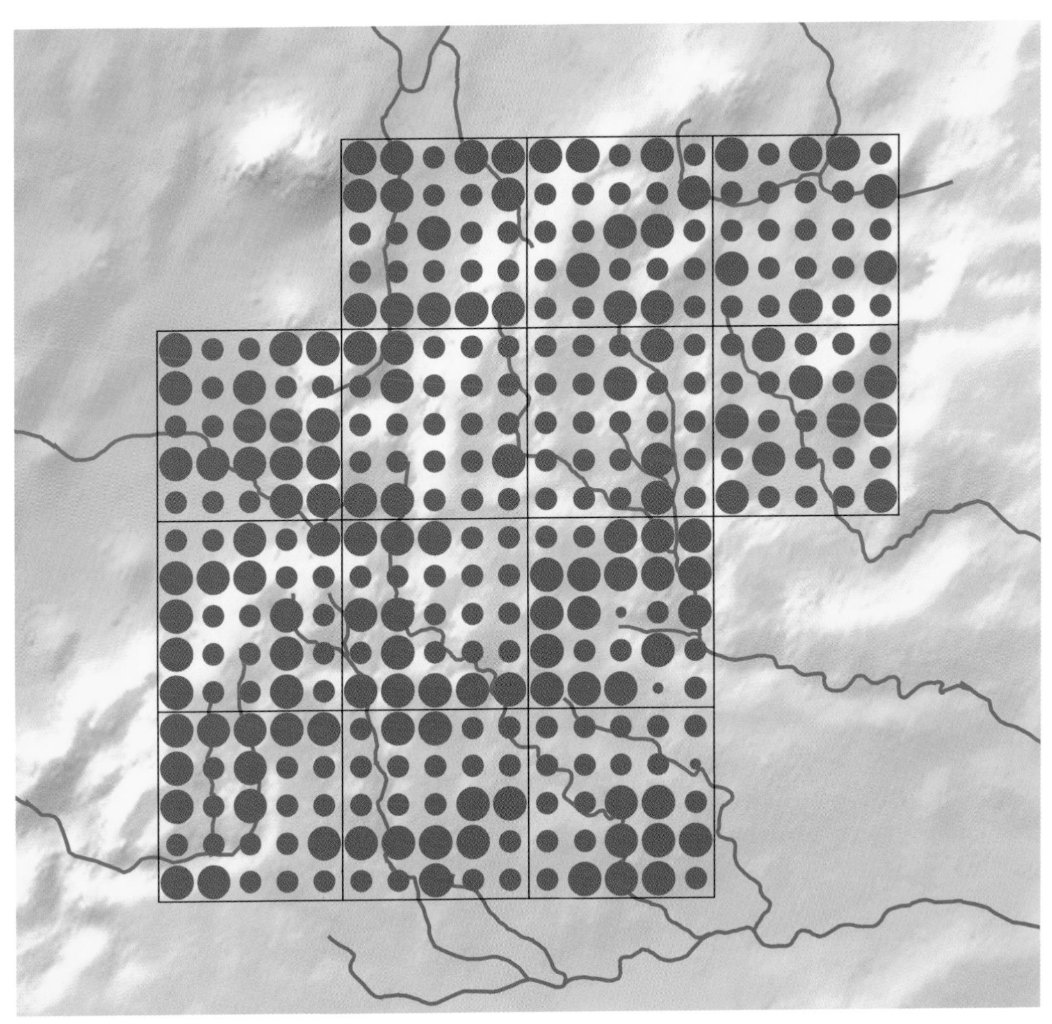

2003–07 survey

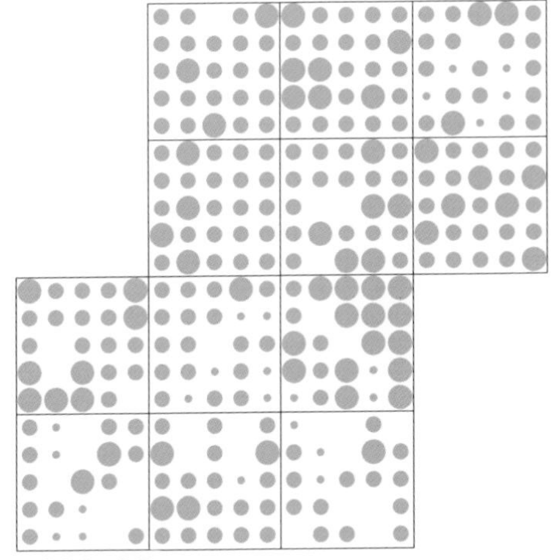

1983–87 survey

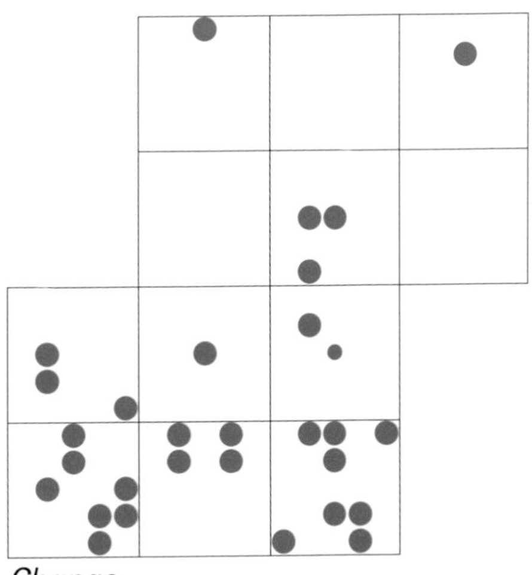

Change

Robin (European Robin)

Erithacus rubecula

ROB BROOKES

Resident breeder

UK conservation status: **GREEN**

Long-term UK trend (1970–2006): +45%

Recent population trend (1994–2007): +21%

Numbers of tetrads with breeding records:

	2003–07		1983–87
	13 squares	12 squares	12 squares
Possible	**0**	0	6
Probable	**120**	115	198
Confirmed	**205**	185	84
Totals	**325**	**300**	**288**

The Robin and the Blackbird have the distinction of being the only two species for which breeding behaviour, rather than mere presence, was noted in every tetrad. As in the 1983–87 survey, Robins were both very common and widespread throughout the area, occupying a variety of habitats, from dense woodland, through open glades and hedgerows to suburban gardens. Even where suitable habitat is limited, on the high tops of the wolds for example, there is likely to be an isolated group of trees or a few bushes around a derelict farm building that will hold a pair.

Robins are extremely familiar birds and there is no doubt that they do actually breed in every tetrad. Their prolonged song period and their highly territorial behaviour made location of birds in the probable breeding category very easy; having many pairs of birds in each tetrad also helped. Their confiding nature, openly carrying food to their young and regularly raising two broods, made confirmation of breeding relatively easy, so nearly two-thirds of the records were of confirmed breeding. This is more than double the results of the 1980s survey which, although undoubtedly due in part to the greater survey effort, may also hint at an increase in population density. If this has occurred, the succession of mild winters in the two decades since the first survey may have been partly responsible.

Species sponsored anonymously

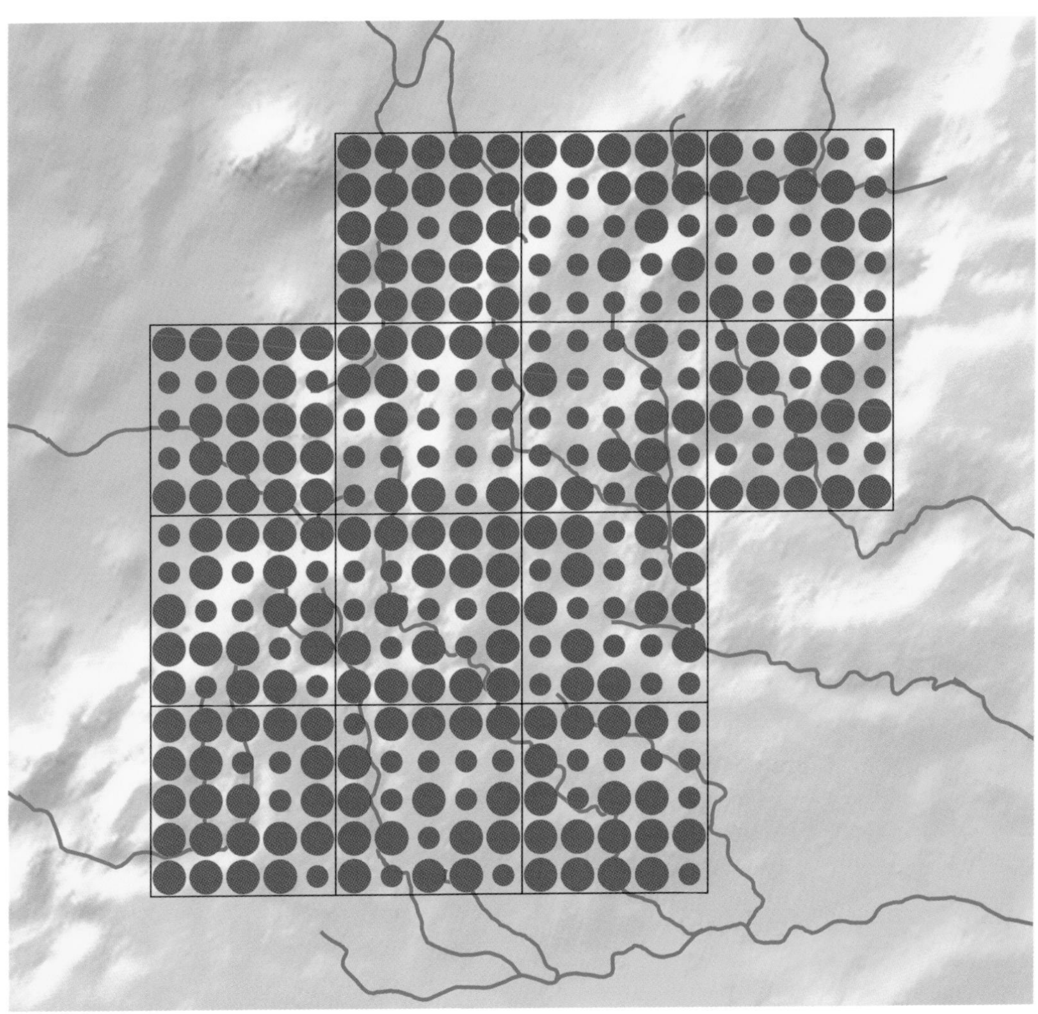
2003–07 survey

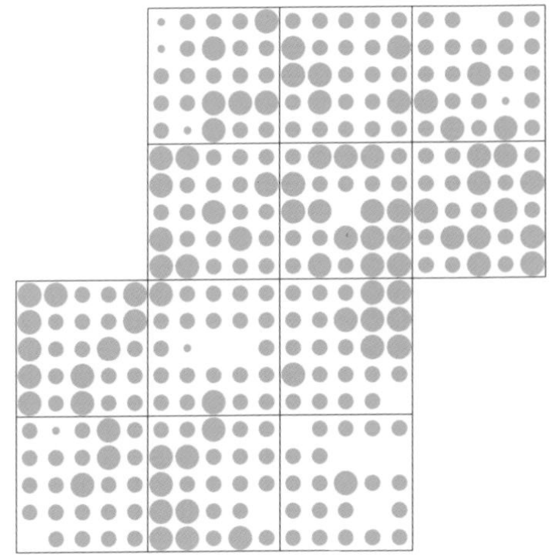
1983–87 survey

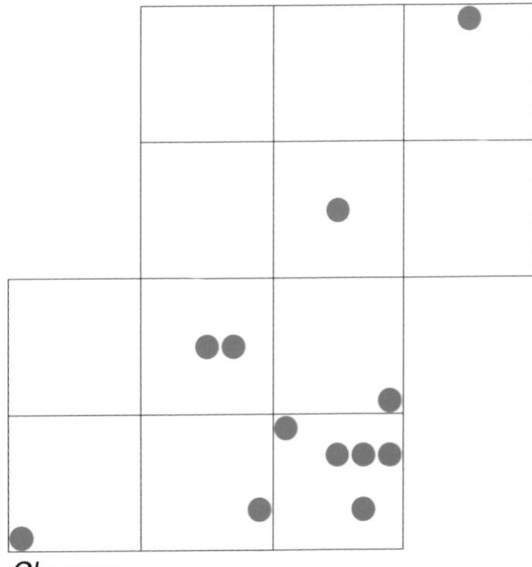
Change

Nightingale (Common Nightingale)

Luscinia megarhynchos

RICHARD TYLER

Migrant breeder

UK conservation status: AMBER

Population trend not available

Numbers of tetrads with breeding records:

	2003–07		1983–87
	13 squares	12 squares	12 squares
Possible	0	0	0
Probable	7	7	28
Confirmed	0	0	0
Totals	7	7	28

Although the Nightingale is very shy and skulking, nesting in dense shrub layers, the male's song is the most vibrant of all British birds, and can be heard from mid- to late April until the beginning of June. Given the level of observer effort in this survey, it is therefore very unlikely that any genuinely breeding birds were missed. It is a bird that sings on passage, however, and there were a handful of other one-off hearings of birds passing through. Interestingly, in many cases these birds stopped off at previous breeding sites, giving hope that they might recolonize the area if there is a general national expansion in range. Confirmation of breeding is very difficult and was not achieved in any of the few remaining sites.

In the first survey in the 1980s, there were believed to be 28 tetrads occupied by Nightingales (although again there was no confirmation of breeding). Even allowing for the uncertainty as to whether some of the birds in the first survey were breeding birds or passage migrants, it is clear that there was a major decline in numbers in the years between the two surveys. In 1991, a small survey by the NCOS revealed a population of approximately 10 singing males in the areas to the east of Wormington village (where breeding was confirmed in the mid-1990s), but these were unoccupied after 2003. The other major strongholds of the Stroud Valleys and Cirencester Park rapidly became unoccupied in the early 1990s.

The few records that were obtained in this survey came mainly from the first three years, in their traditional stronghold in the Avon Vale area adjoining the Cotswold scarp south of Broadway, where their preferred habitat of deciduous woods and thickets with shrubby undergrowth was available. The last of these birds was heard in June 2005 (the third year of the survey). It is possible that Nightingales have now ceased to breed in the recording area.

Species sponsored by Tim Hutton

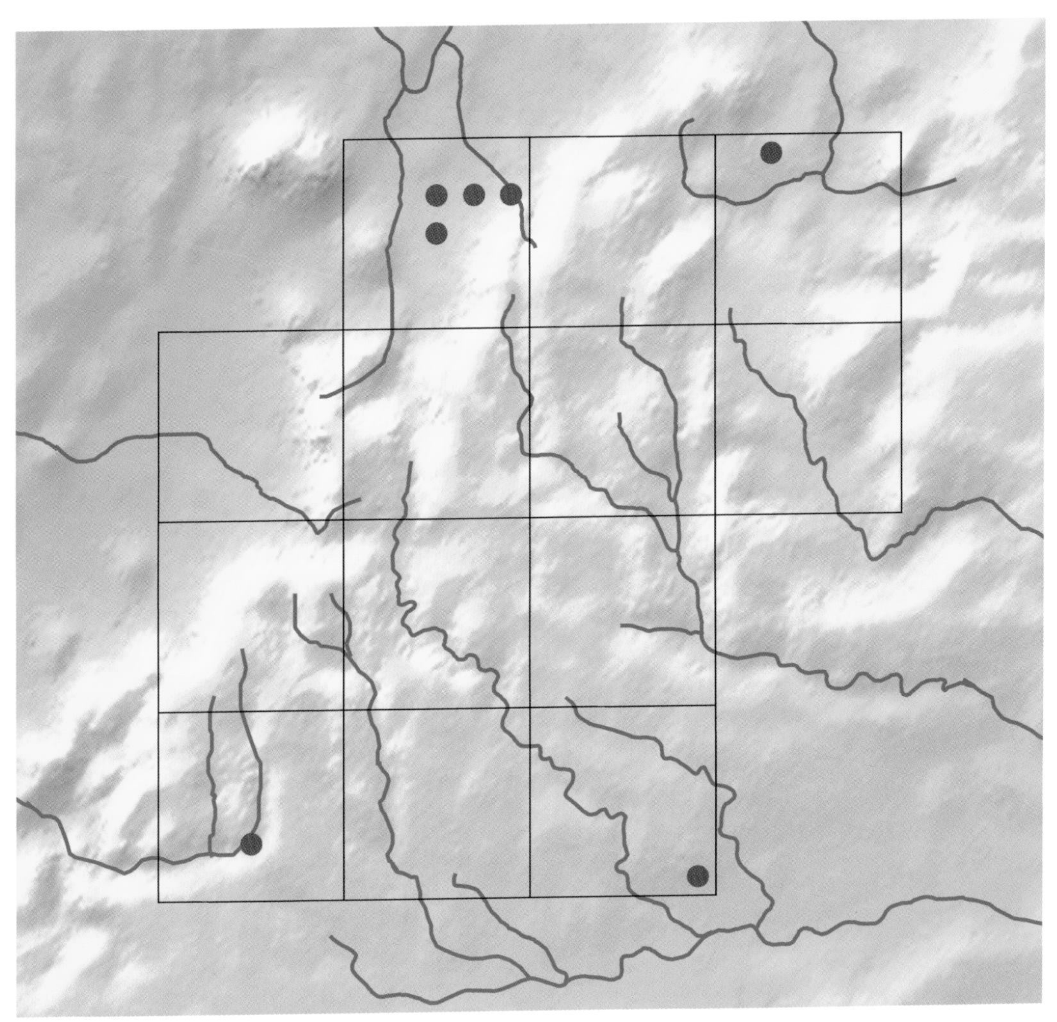

2003–07 survey

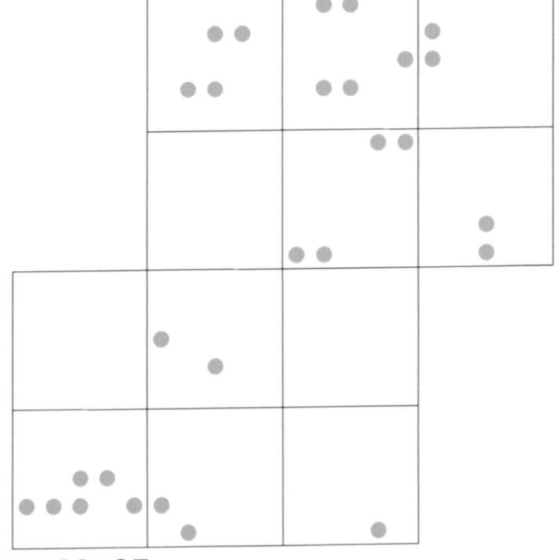

1983–87 survey

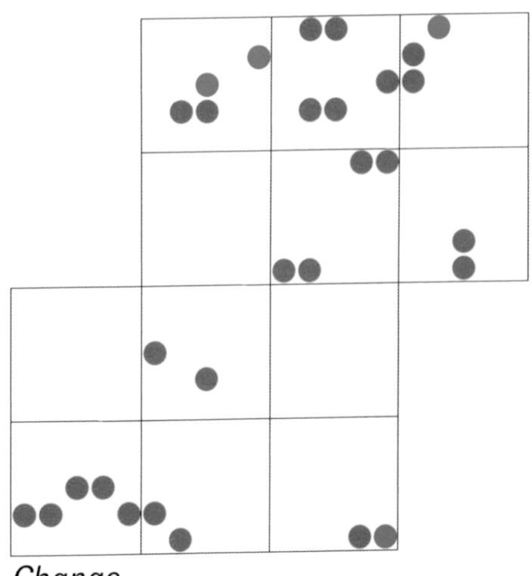

Change

Redstart (Common Redstart)

Phoenicurus phoenicurus

The Redstart is a species of some significance to the NCOS, a sizeable hitherto unrecognized population having been discovered during the survey in the 1980s. This is important because the Cotswolds are on the easterly edge of its range in the southern Midlands although some were still present near Tring in the early 1990s. Optimum habitat appears to be open grazed woodland, which is particularly prevalent on the steep-sided scarp, and sheep-rearing country with its stone walls and scattered bushes.

Although their song is quite distinctive, Redstarts can be overlooked by an observer unfamiliar with the species, and they can sometimes be difficult to see when singing in the canopy of a large tree. On the other hand, adults carrying food to young were often seen, allowing breeding to be readily confirmed.

Since the 1980s survey the number of occupied tetrads has decreased and the change map shows that the species has retreated from its northeastern fringes in the Stour Valley area (it had disappeared as a breeding bird from the nearby area surveyed by the Banbury Ornithological Society in the late 1990s). It is still relatively widely, if thinly, distributed along the scarp and High Wold from Broadway south to Miserden and North Cerney. A significant proportion of its distribution is within the 10 km square SO92, which was not surveyed in the 1980s. It is believed that there has been a moderate decline in numbers to accompany the slight contraction in range, and that the distribution change has been partially masked by more thorough surveying and knowledge of its preferred habitats. In the first Atlas the total population was 'estimated at well over 50 pairs'. Given that some tetrads hold several pairs, it seems likely that, despite the perceived decline, there are still more than 50 pairs in the area.

Migrant breeder

UK conservation status: AMBER

Long-term UK trend (1970–2006): +18%

Recent population trend (1994–2007): +23%

Numbers of tetrads with breeding records:

	2003–07		1983–87
	13 squares	12 squares	12 squares
Possible	**4**	3	16
Probable	**40**	37	28
Confirmed	**14**	12	18
Totals	**58**	**52**	**62**

RICHARD TYLER

Species sponsored by Keith White

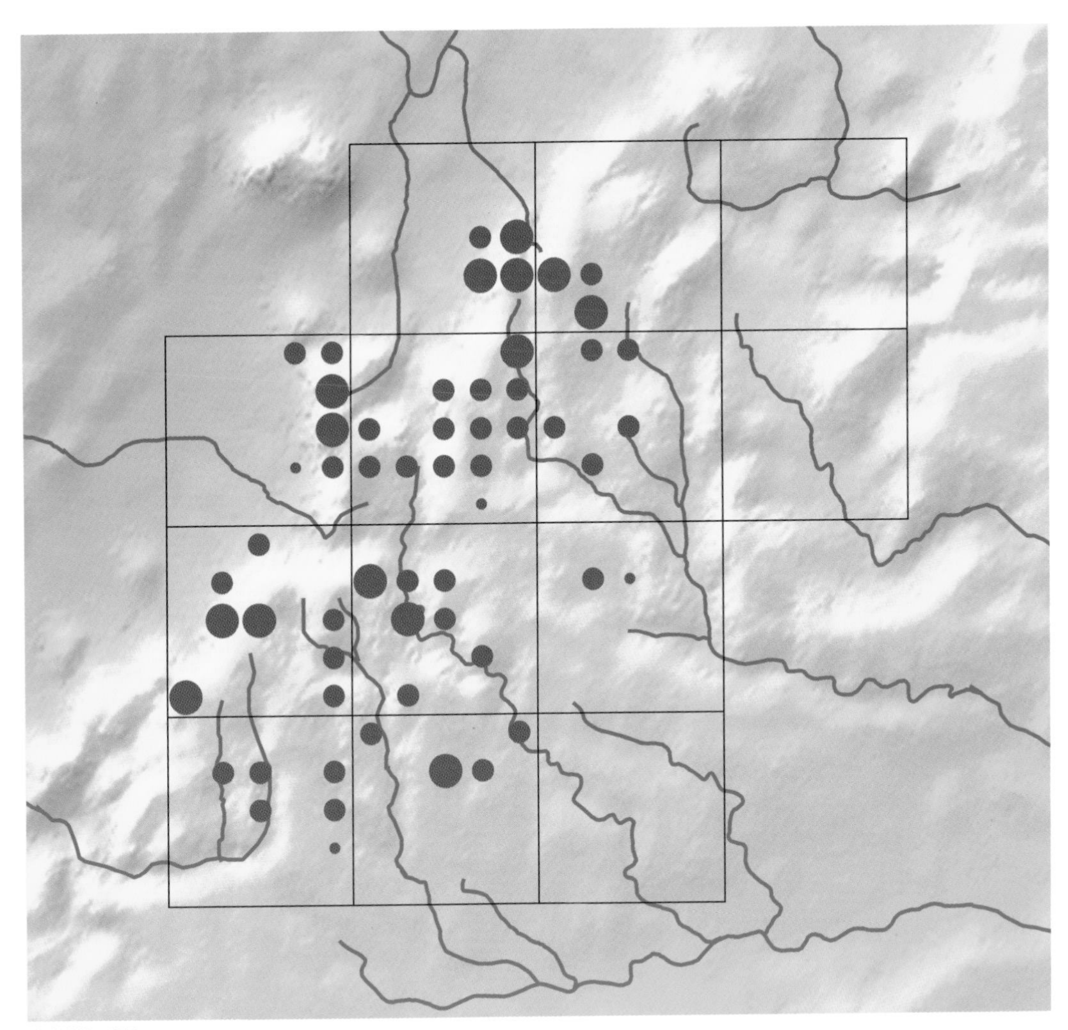
2003–07 survey

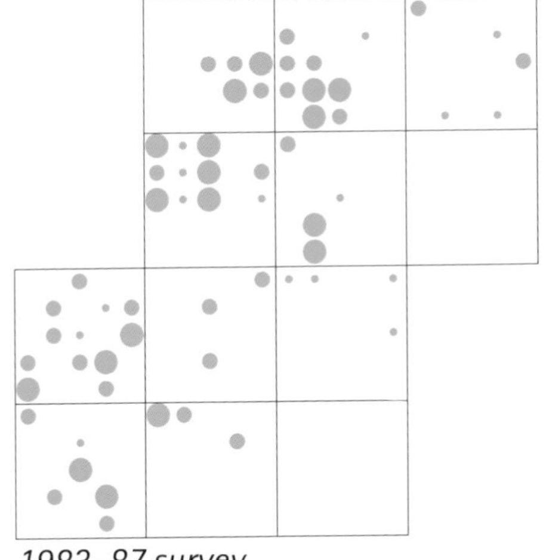
1983–87 survey

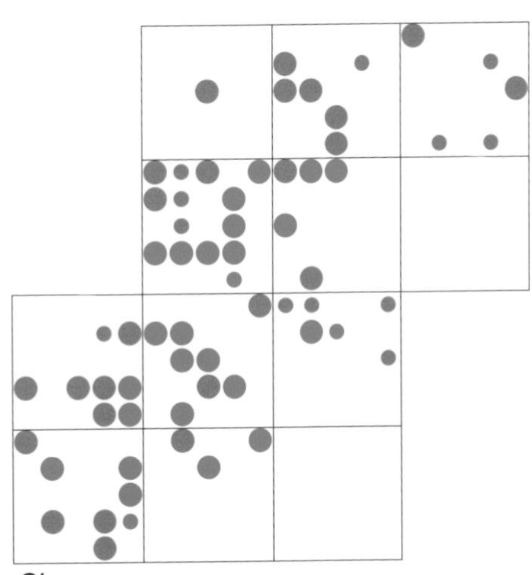
Change

Stonechat
Saxicola torquatus

Stonechats perch prominently on the tops of bushes, are very vocal, and as a result are difficult to overlook. This makes location of territorial birds easy, despite the fact that their song was infrequently heard. Confirmation of breeding is also relatively straightforward through observation of adults carrying food, and young fledglings.

The Stonechat has a very restricted distribution in the survey area, breeding being confirmed in only two tetrads. The only reliable site for Stonechats is on Cleeve Common, although breeding activity was noted in one survey year east of Sherborne. Cleeve Common provides classic inland Stonechat habitat–upland, often windswept, and with sizable patches of gorse amongst rough pasture. In the previous Atlas it was commented that this colony was 'the only one known in Gloucestershire east of the River Severn', and although breeding has been confirmed occasionally at other sites, notably in the Coberley area to the southeast of Cheltenham, it is still probably the only regular breeding site in the area.

DAVE PEARCE

Resident breeder

UK conservation status: AMBER

Long-term UK trend: Not available

Recent population trend (1994–2007): +278%

Numbers of tetrads with breeding records:

	2003–07		1983–87
	13 squares	12 squares	12 squares
Possible	2	2	0
Probable	3	2	0
Confirmed	2	1	1
Totals	7	5	1

Following a major decline in the 1980s, Stonechat numbers across the country have benefited from the succession of mild winters and there is evidence that the species is beginning to occupy a larger part of Cleeve Common. A significant part of this is in the 10 km square SO92, which was not in the recording area in the 1980s survey, with the result that the change map is seriously incomplete. However, there are other places in the recording area with patches of suitable habitat, and it will be interesting to see if these become colonized in the next 20 years.

Species sponsored by Iain and Jill Main

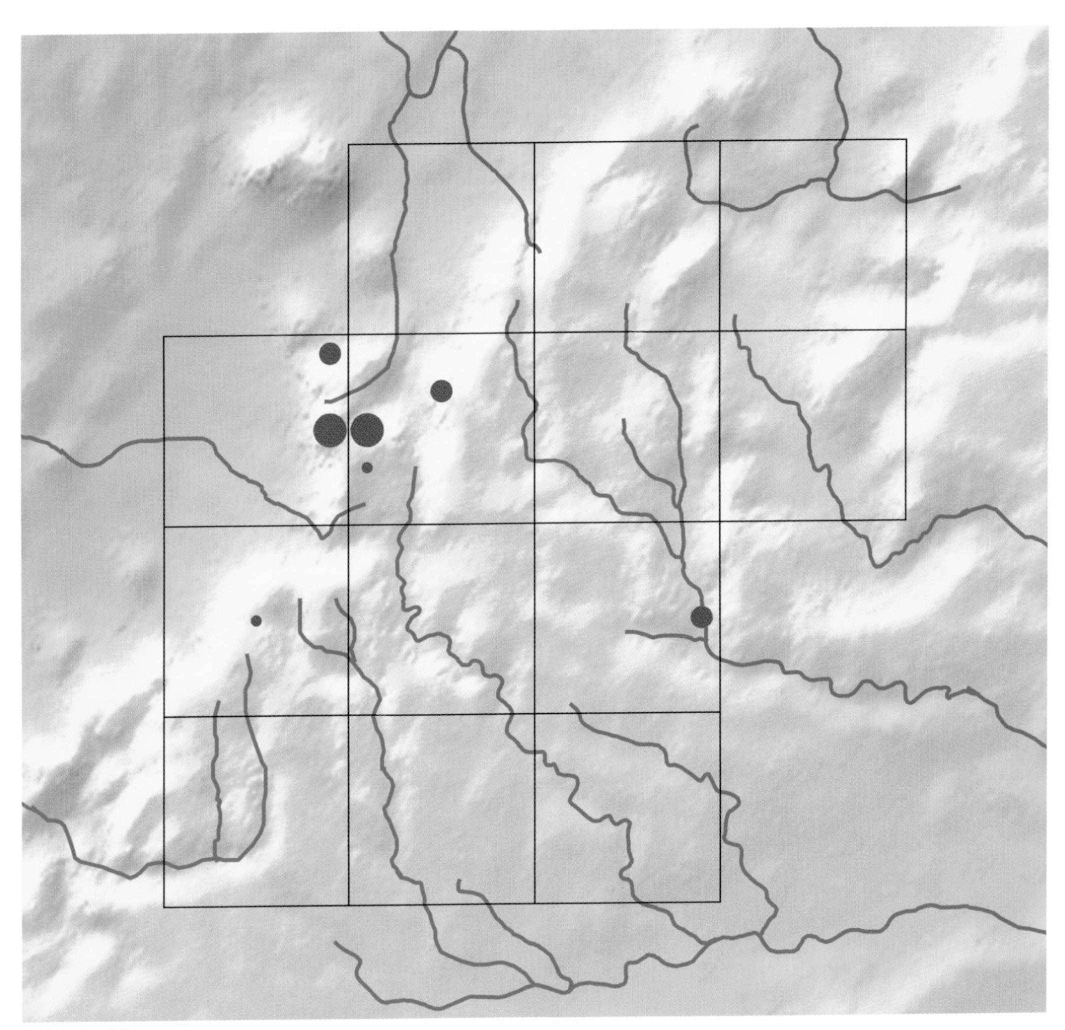

2003–07 survey

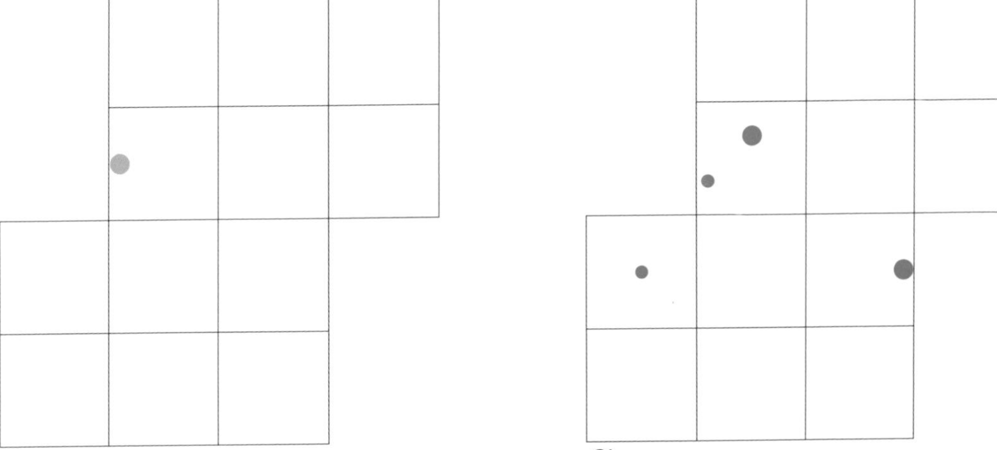

1983–87 survey

Change

Wheatear (Northern Wheatear)

Oenanthe oenanthe

Wheatears are regular spring and autumn passage migrants over much of the survey area, particularly Cleeve Common, Leckhampton Hill and other upland areas. They are found quite often in open farmland, close-cropped pasture and bare earth fields at all altitudes. Migration occurs from March through to early June and this regular passage can give the appearance of pairs holding territory, when in fact it is a series of pairs and individuals staying for only a few days. The return passage builds up rapidly in early August.

Cleeve Common would seem to be the most likely area in which breeding might occur, and it has been observed there occasionally. In the 1980s survey, breeding was confirmed in the Taddington area, south of Snowshill on the High Wold, with adults seen carrying food late in the season. During this survey, the local opinion was that Wheatears were again nesting in the dry stone walls near the same village, but this could not be confirmed despite extensive observation.

Migrant breeder

UK conservation status: **GREEN**

Long-term UK trend: Not available

Recent population trend (1994–2007): +13%

Numbers of tetrads with breeding records:

	2003–07		1983–87
	13 squares	12 squares	12 squares
Possible	**11**	10	3
Probable	**0**	0	5
Confirmed	**0**	0	1
Totals	**11**	**10**	**9**

In an unrelated observation, a very young bird with partially unfeathered wings was found on a hedgerow in farmland near Gotherington, in the Vale north of Cheltenham. While this bird might have been from a local nest, possibly on Cleeve Common, it is impossible to be sure.

RICHARD TYLER

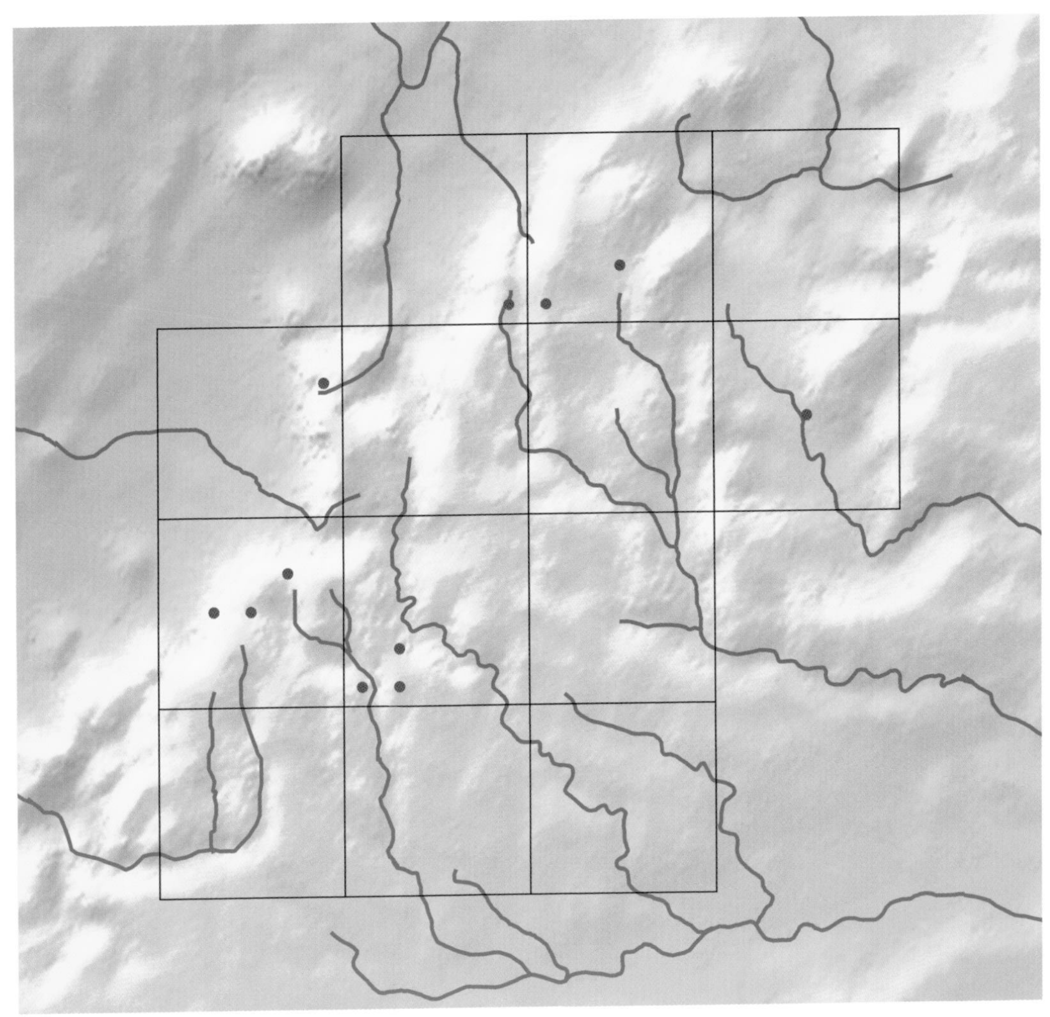

2003–07 survey

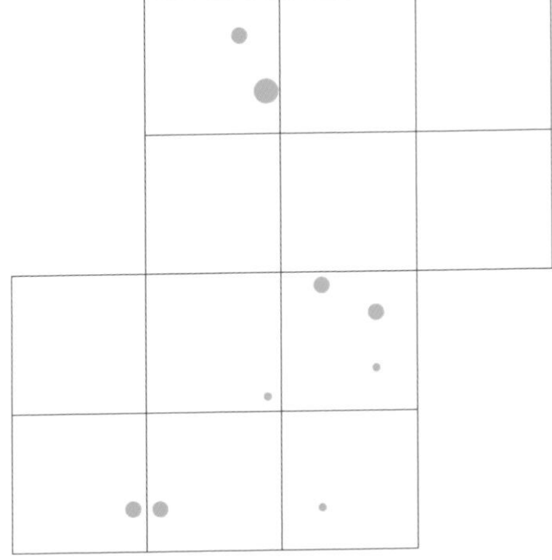

1983–87 survey

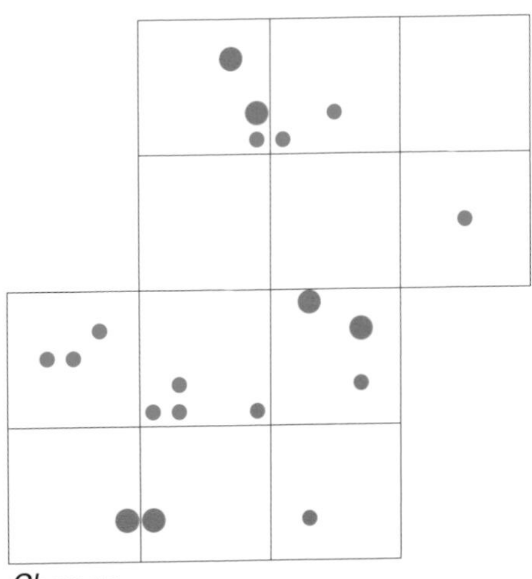

Change

Blackbird (Common Blackbird)

Turdus merula

RICHARD TYLER

Resident breeder

UK conservation status: **GREEN**

Long-term UK trend (1970–2006): -16%

Recent population trend (1994–2007): +24%

Numbers of tetrads with breeding records:

	2003–07		1983–87
	13 squares	12 squares	12 squares
Possible	**0**	0	5
Probable	**70**	67	106
Confirmed	**255**	233	186
Totals	**325**	300	297

Blackbirds are familiar even to non-birdwatchers from their melodious song and their presence in gardens. Although their numbers have in fact decreased nationally since the 1970s, there is no doubt that they are still very common in our survey area. Feeding largely on ground invertebrates, they do not move far from cover when searching for food and, as they do not have any specific habitat requirements that limit their occurrence, they can be found in any situation that offers a small bush or a hedge in which to build a nest.

Since Blackbirds have a protracted breeding season—they were observed regularly nesting as early as mid-March—and as they are multiple-brooded, confirmation of breeding was relatively simple, through observation of adults carrying food and by seeing young fledglings. This, combined with the high density of birds, meant that breeding activity was noted in all tetrads and confirmed in nearly 80% of them, compared with 63% in the previous survey.

A distribution survey such as this is unable detect any population changes in such a widespread bird, but there was no evidence that there had been any major alteration in numbers since the survey in the 1980s.

Species sponsored by Jill Lewis

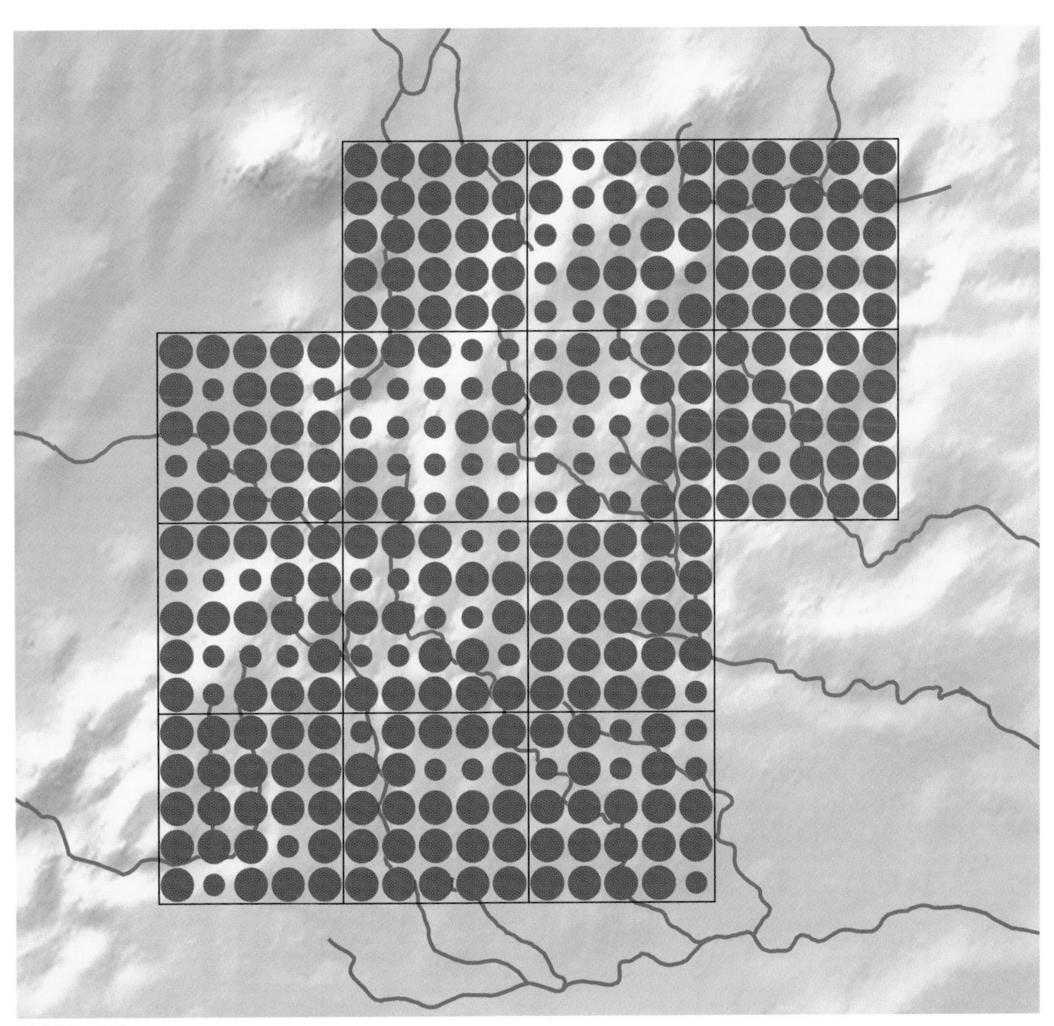

2003–07 survey

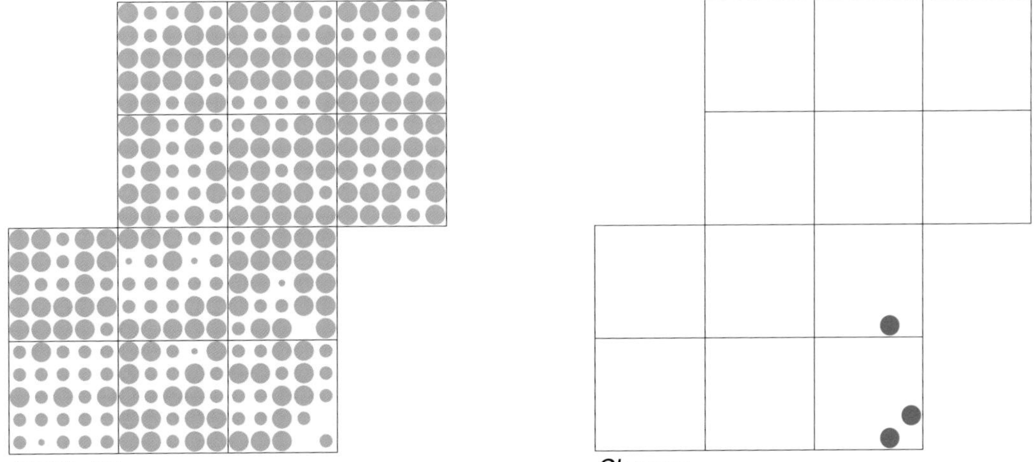

1983–87 survey

Change

Song Thrush
Turdus philomelos

The loud, repetitive and persistent song of the Song Thrush made location of territorial birds very easy. It also tends to sing at dawn and/or dusk over a long period of time, and is not a species in the habit of holding territory for a short while before moving on. Although known to eat snails, it feeds its young mainly on worms. This allowed confirmation of breeding in over 40% of occupied tetrads.

Nationally, there has been concern over the status of the Song Thrush for some years, especially in farmland and gardens. Together with woodland, these constitute the major part of the habitat of the Cotswolds. However, both the survey results and the general feeling of observers is that the species is holding up quite well in the survey area. Comments made in the first Atlas are probably still largely true: '... in summer, it is most commonly found in and around towns and villages, and far less often in open farmland. However, it does occur in good numbers in woodland ...'.

The data indicate that breeding Song Thrushes were recorded in 98% of tetrads, which is almost a 20% increase on the result from the 1980s survey; this may be entirely due to the greater survey effort. However, it is significantly less common than the Blackbird, as it was in the 1980s. Its perceived decline, in the eyes of the casual observer, may be partially due to its rather unobtrusive nature when not actively singing. Given the ease and confidence with which territorial birds could be located, it is likely that the data from the new survey give a true indication of its distribution, and a reliable reflection of its ubiquity as a breeding bird in the recording area. However, the survey does not provide any data about the population density of the species.

Resident breeder

Gloucestershire BAP species

UK conservation status: **RED**

Long-term UK trend (1970–2006): -50%

Recent population trend (1994–2007): +18%

Numbers of tetrads with breeding records:

	2003–07		1983–87
	13 squares	12 squares	12 squares
Possible	**2**	2	18
Probable	**185**	176	156
Confirmed	**131**	116	73
Totals	**318**	**294**	**247**

RICHARD TYLER

Species sponsored by Ian Cox

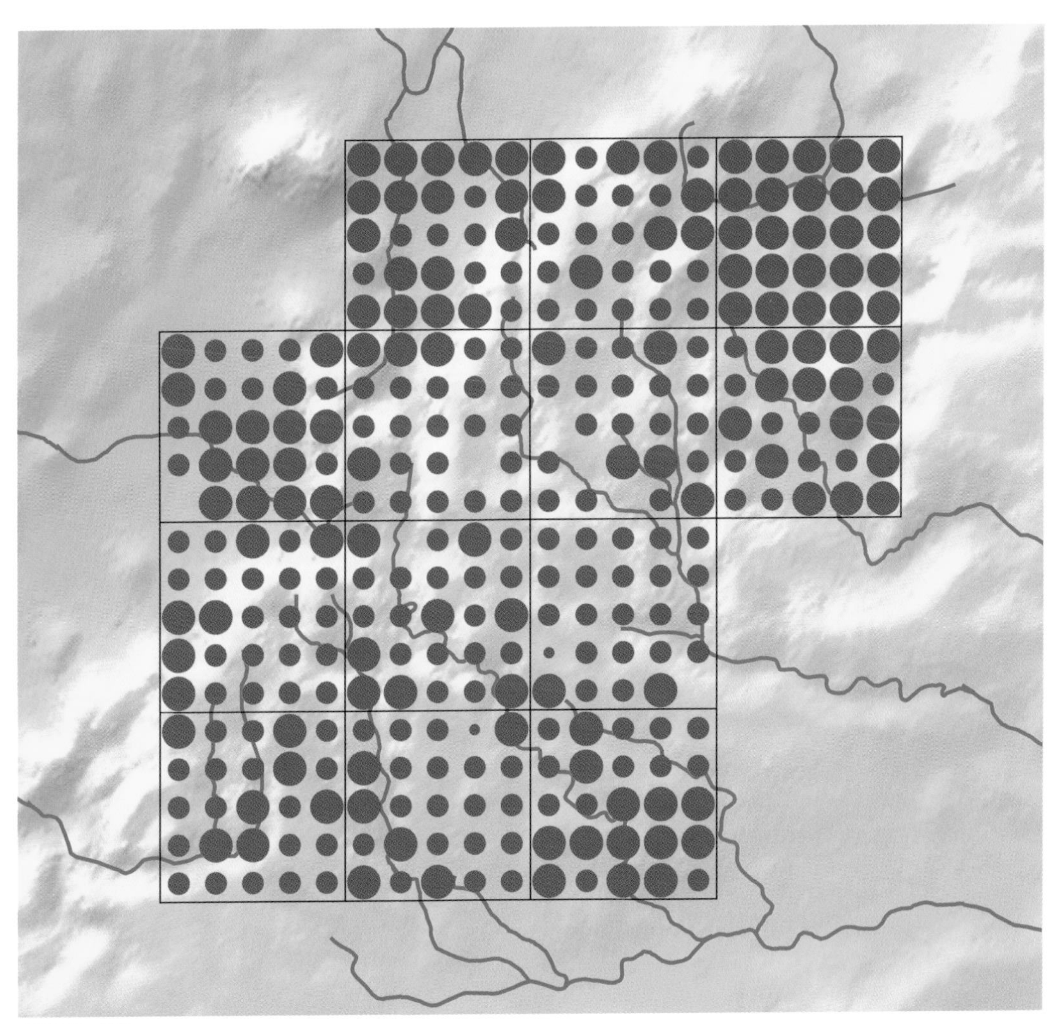

2003–07 survey

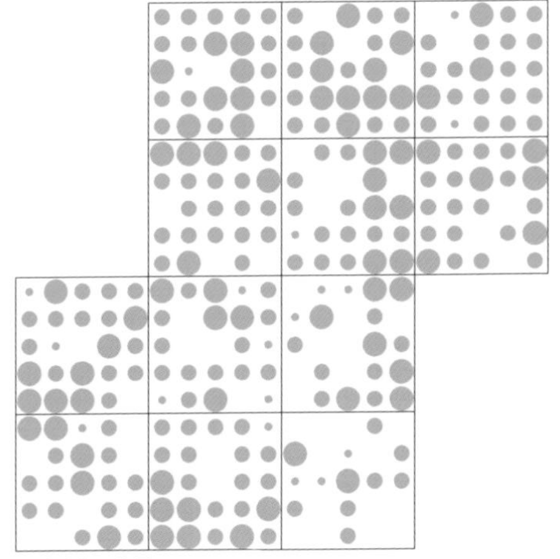

1983–87 survey

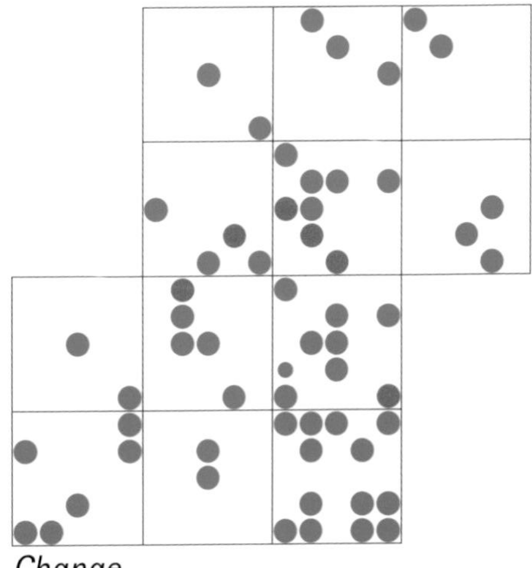

Change

Mistle Thrush

Turdus viscivorus

RICHARD TYLER

Resident breeder

UK conservation status: AMBER

Long-term UK trend (1970–2006): -43%

Recent population trend (1994–2007): -12%

Numbers of tetrads with breeding records:

	2003–07		1983–87
	13 squares	12 squares	12 squares
Possible	**30**	28	35
Probable	**142**	132	101
Confirmed	**124**	113	92
Totals	**296**	**273**	**228**

The loud and far-carrying song of the Mistle Thrush can generally be heard from very early spring until late May, most often before midday. Mistle Thrushes usually sing from a high perch, such as the top of an isolated tree, so locating breeding territories is fairly straightforward. Their nests, usually in the fork of a tree, are relatively easy to find, and adults openly forage for their young (often carrying food some distance). Furthermore Mistle Thrushes noisily and aggressively defend their nest sites. This all makes confirmation of breeding a fairly simple matter. However, care had to be taken not to assume that family parties with well-developed juveniles originated in the area in which they were found, as they are known to roam a considerable distance from their nest.

Mistle Thrushes have a preference for more open country with tall trees and specifically habitat near to meadows and pasture for feeding. Although the wide distribution of Mistle Thrushes is similar to that of the Song Thrush, they are much less numerous in most areas. As with other species however, this wide distribution indicates the existence of at least small patches of suitable habitat throughout much of the survey area. In the first Atlas it was commented that they were most common in the north and west of the survey area, and the 1980s distribution map shows only sparse distribution in the Thames tributaries area and the southeastern part of the dip slope, where large arable areas predominate. The change map suggests that they have moved into this area, possibly as a result of an increase in the number of horse paddocks, but that there has been a slight decrease in the Stour Valley. However, the apparent gains may be largely due to increased survey effort.

Species sponsored anonymously

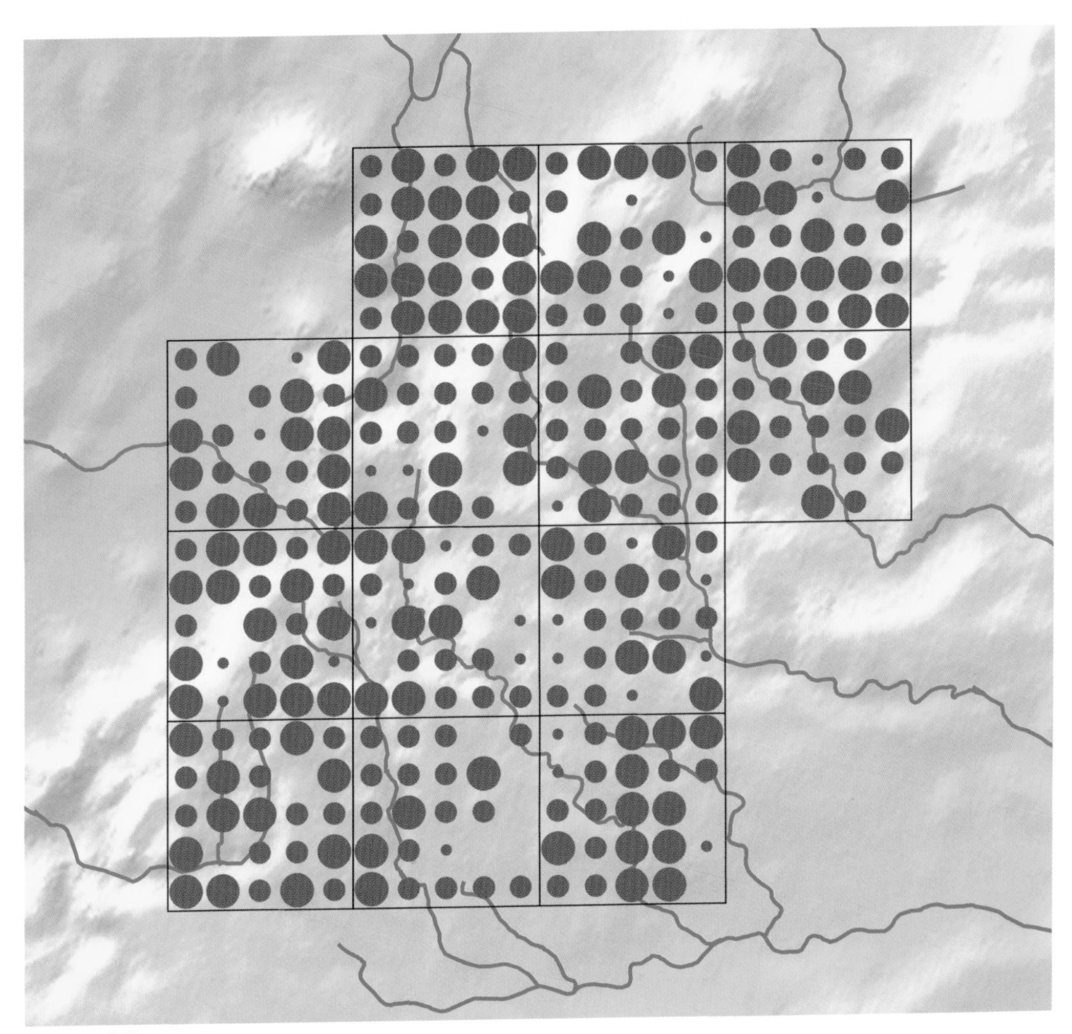

2003–07 survey

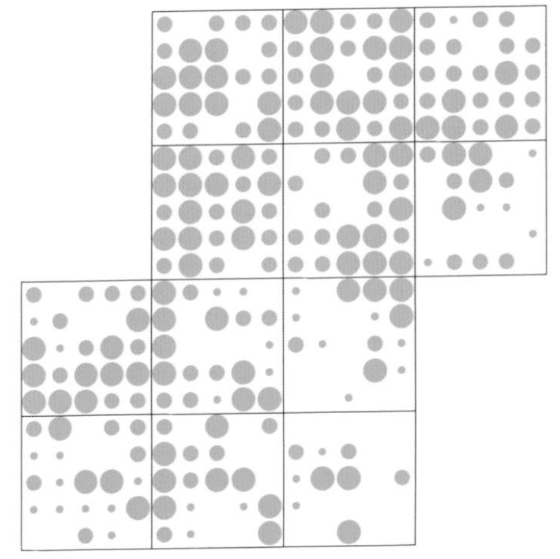

1983–87 survey

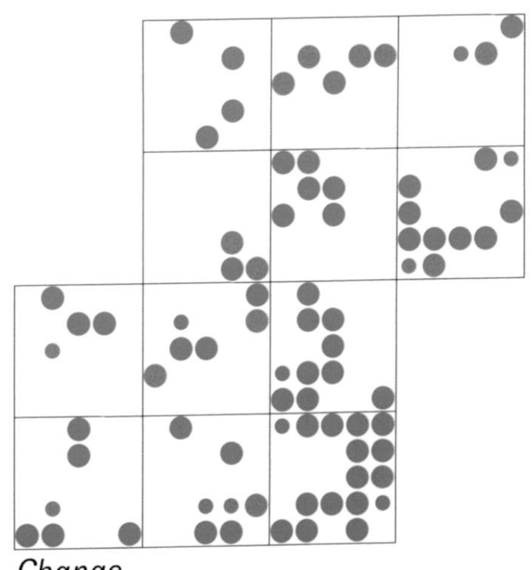

Change

Grasshopper Warbler (Common Grasshopper Warbler)

Locustella naevia

RICHARD TYLER

Migrant breeder

UK conservation status: RED

Long-term UK trend: Not available

Recent population trend (1994–2007): +68%

Numbers of tetrads with breeding records:

	2003–07		1983–87
	13 squares	12 squares	12 squares
Possible	0	0	1
Probable	14	11	18
Confirmed	0	0	0
Totals	14	11	19

Although in some districts Grasshopper Warblers can be found in tangled vegetation near water, in the Cotswolds they are more often found in dry situations amongst gorse and brambles, typically on or near the slopes of Cleeve Common. The distribution map clearly shows that the breeding population is almost entirely restricted to this area, part of which falls within the 10 km square SO92 which was not covered in the first survey. This contrasts with the more scattered distribution recorded in the 1980s, although the first Atlas did comment that 'the distribution map may give an inflated impression of the true breeding status of this species. Approximately half the records refer to single males singing, probably while on passage'. In this survey, a specific effort was made not to record passage birds; however, the lack of many outliers on the map suggests that there has probably been a contraction in the distribution of the breeding population. Grasshopper Warblers were certainly known to hold territory in some of the roadside verges in the Avon Vale in the early 1990s, and these fragmentary populations have now disappeared.

Given the intensity of coverage in this survey and the persistent reeling song of the bird, it seems unlikely that many birds genuinely holding territory would have been missed, despite the fact that the reeling is inaudible to some people. Thus, it is likely that the current distribution map is a fair reflection of its current status. As was the case in the first Atlas, the secretive nature of the bird and the overall rarity of actual sightings meant that breeding was nowhere confirmed.

Species sponsored by Dave Pearce

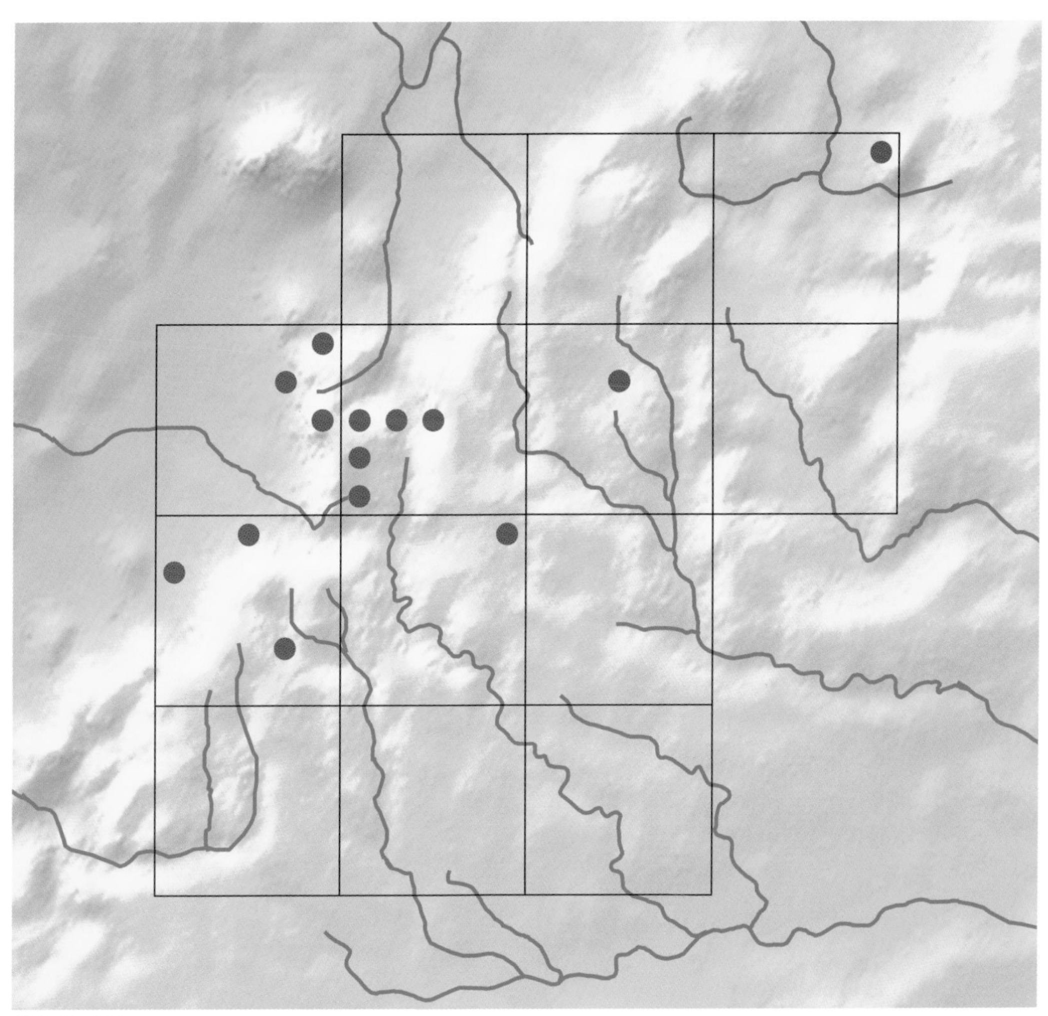
2003–07 survey

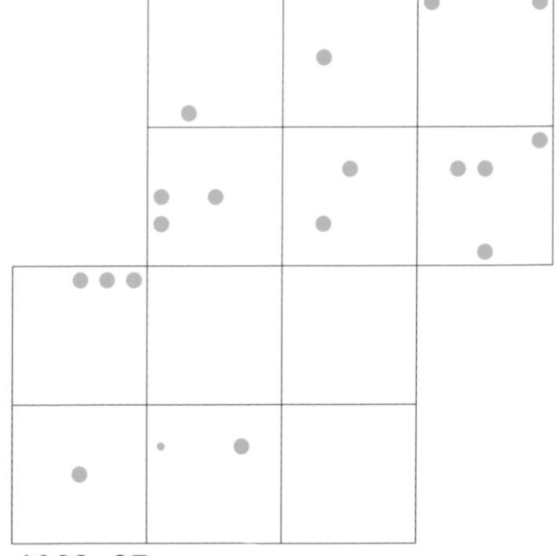
1983–87 survey

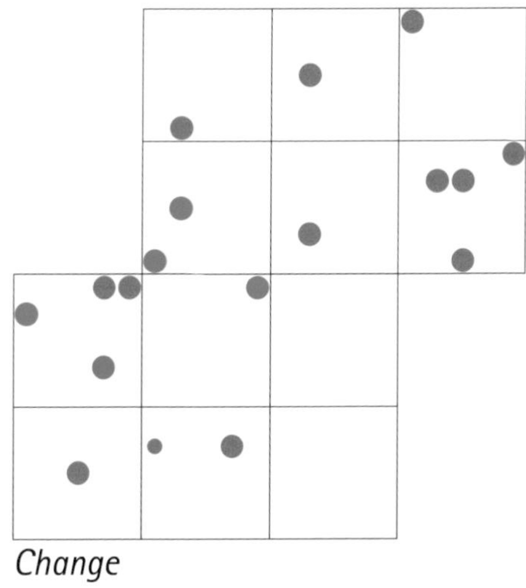
Change

Sedge Warbler

Acrocephalus schoenobaenus

RICHARD TYLER

Migrant breeder

UK conservation status: GREEN

Long-term UK trend (1970–2006): -21%

Recent population trend (1994–2007): +8%

Numbers of tetrads with breeding records:

	2003–07		1983–87
	13 squares	12 squares	12 squares
Possible	2	2	1
Probable	14	13	16
Confirmed	3	3	0
Totals	19	**18**	**17**

The song of the Sedge Warbler is both penetrating and distinctive and, given its protracted song period from early April to mid-July, and the relative ease of pinpointing its location, it seems unlikely that many birds genuinely holding territory were missed. Significant caution had to be observed, however, to exclude singing passage birds which were sometimes found in hedgerows in arable areas for a day or two. Confirmation of breeding is more difficult, as the young are usually fed in thick vegetation.

Although Sedge Warblers were found in their typical habitat of lush waterside vegetation around Bourton and Fairford Pits as well as along the valleys of the Windrush, Evenlode, Stour and Isbourne river systems, it is interesting to note that not all suitable locations appeared to be occupied. Sedge Warblers were also found sporadically in arable situations, most notably in or near oilseed rape fields–indeed one of the few confirmed breeding records, in the Avon Vale, was in such a location.

The change map suggests that, although many of the actual tetrads in which the birds were found in the two surveys were different, the number of occupied tetrads was much the same, and the general areas in which they are found have not changed significantly. It was suggested in the first Atlas that, although confirmation of breeding was difficult, breeding probably did take place in most of the major areas. It seems likely that the same was true during this survey.

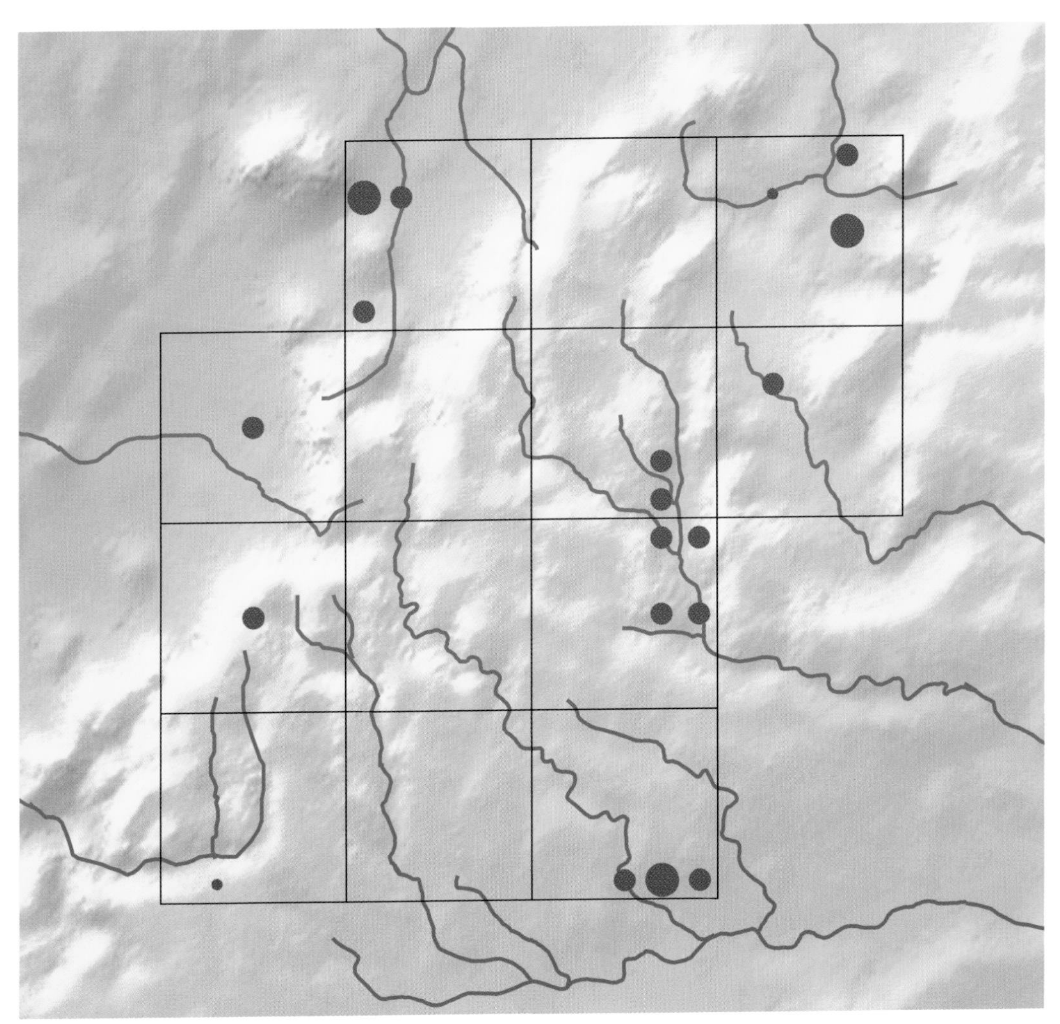
2003–07 survey

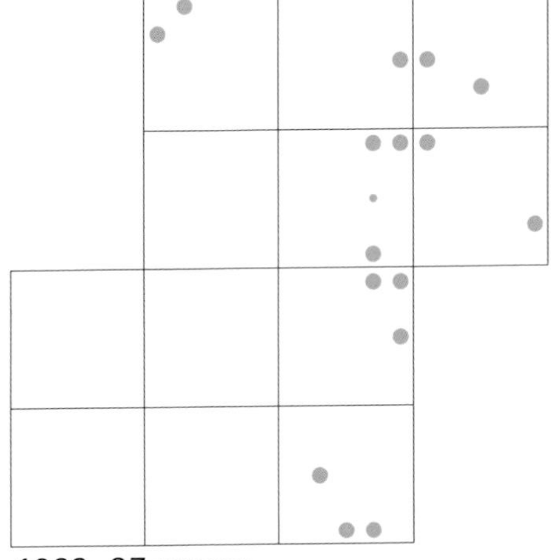
1983–87 survey

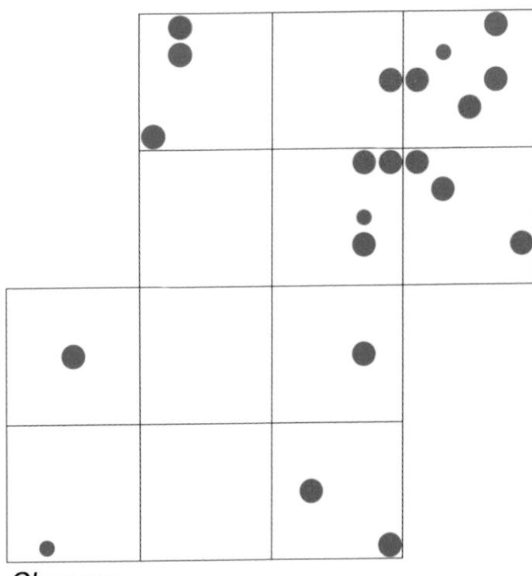
Change

Reed Warbler (Eurasian Reed Warbler)

Acrocephalus scirpaceus

Reed Warblers are much more restrictive in their requirements than Sedge Warblers, needing extensive tracts of the Common Reed *Phragmites*; although they do not necessarily build their nests within the reeds, they do so within close proximity. The Reed Warbler's song is easily recognizable and it tends to sing throughout the breeding period, making location of birds easy, although confirmation of breeding is more difficult.

In the first Atlas survey, Reed Warblers were found only in two separate locations—Fairford Pits and the northern section of Bourton Pits—and it was believed that this represented the true breeding status of the species at that time. Interestingly, there were no breeding records of Reed Warblers during the 2003–07 survey period at either the northern or southern section of Bourton Pits, although they have held territory there many times in the intervening years. However, the new distribution map shows a scattering of records from around the survey area, illustrating a marked expansion in the species' distribution.

Pockets of reed beds have become more numerous and extensive since the 1980s, and Reed Warblers seem to take advantage of them fairly rapidly. They were found, among other places, at Dowdeswell Reservoir and along the River Stour. Nationally there has been a significant expansion in the range of Reed Warblers in the last few decades, and there appears to have been a moderate influx into the study area. It may be that the species is set to become significantly more established in the Cotswolds in the coming years, given sufficient habitat. More study would be required to confirm that some of the records are not of transient passage birds—singing birds were sometimes found in non-breeding habitat in late May.

Migrant breeder

UK conservation status: GREEN

Long-term UK trend (1970–2006): +135%

Recent population trend (1994–2007): +26%

Numbers of tetrads with breeding records:

	2003–07		1983–87
	13 squares	12 squares	12 squares
Possible	**0**	0	0
Probable	**12**	12	3
Confirmed	**2**	2	0
Totals	**14**	**14**	**3**

RICHARD TYLER

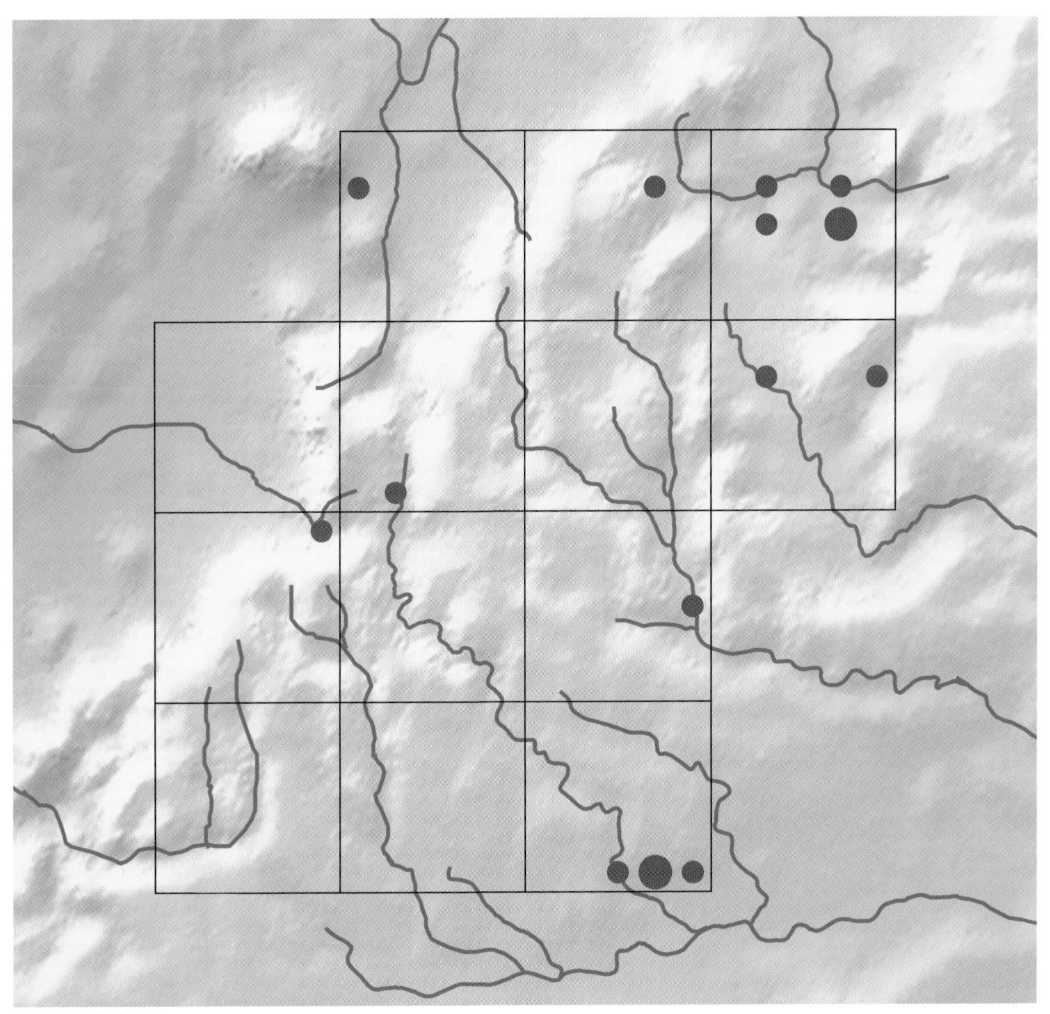

2003–07 survey

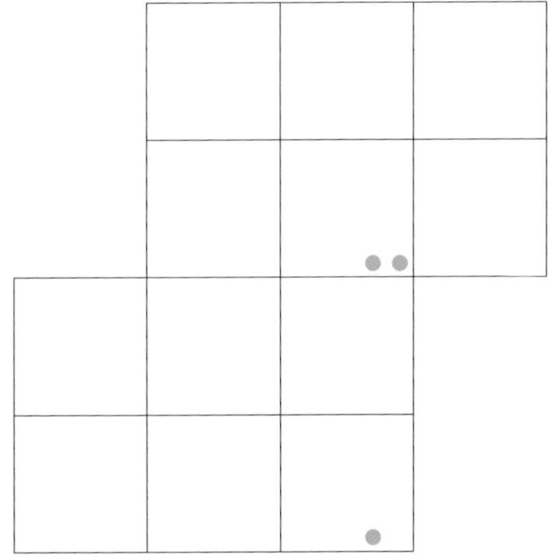

1983–87 survey

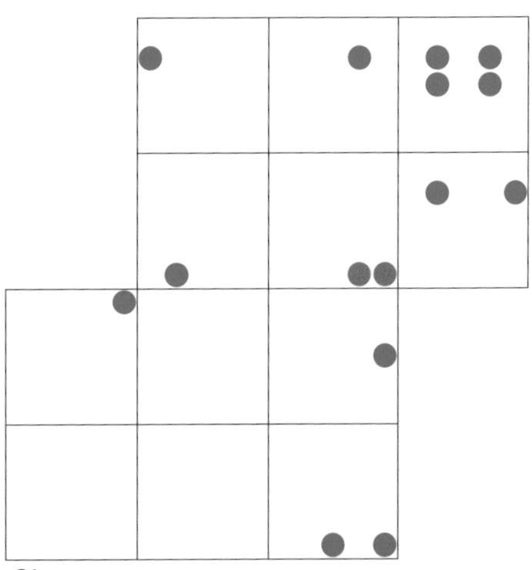

Change

Blackcap

Sylvia atricapilla

RICHARD TYLER

Migrant breeder

UK conservation status: GREEN

Long-term UK trend (1970–2006): +147%

Recent population trend (1994–2007): +62%

Numbers of tetrads with breeding records:

	2003–07		1983–87
	13 squares	12 squares	12 squares
Possible	**0**	0	1
Probable	**255**	237	208
Confirmed	**60**	55	30
Totals	**315**	**292**	**239**

Along with the Chiffchaff, the Blackcap is one of the most widely distributed warblers in the survey area, and probably the most numerous. It favours well-established woodland or groups of trees with a good shrub layer and despite it often remaining well concealed, it is easy to identify from a brief view. Its song is also loud and familiar, and location of birds holding territory was fairly easy. However, confirmation of breeding was difficult; although it openly carries food to young, this was not regularly observed.

The distribution map shows the Blackcap to be widespread, being absent only from parts of the open arable Severn Vale and the rolling cereal fields in the Thames tributaries region. The change map suggests a pronounced expansion in range. Although the recorded increase (22%) in the number of occupied tetrads is within the limits that might be attributable to the greater survey effort, it is probable that there has been an increase in population numbers since the 1980s. The relative distribution compared with that of the Garden Warbler has changed in the years between the two surveys. In the report of the 1980s survey, although it was recognized that the Blackcap was the more numerous and more widely distributed, the similarity in distribution warranted comment. The current situation is that, unlike the Garden Warbler, the Blackcap is almost ubiquitous. General observations over the 20 years since the 1980s survey have suggested that Blackcaps are very common, whereas Garden Warblers are quite uncommon.

Species sponsored by John and Viv Phillips

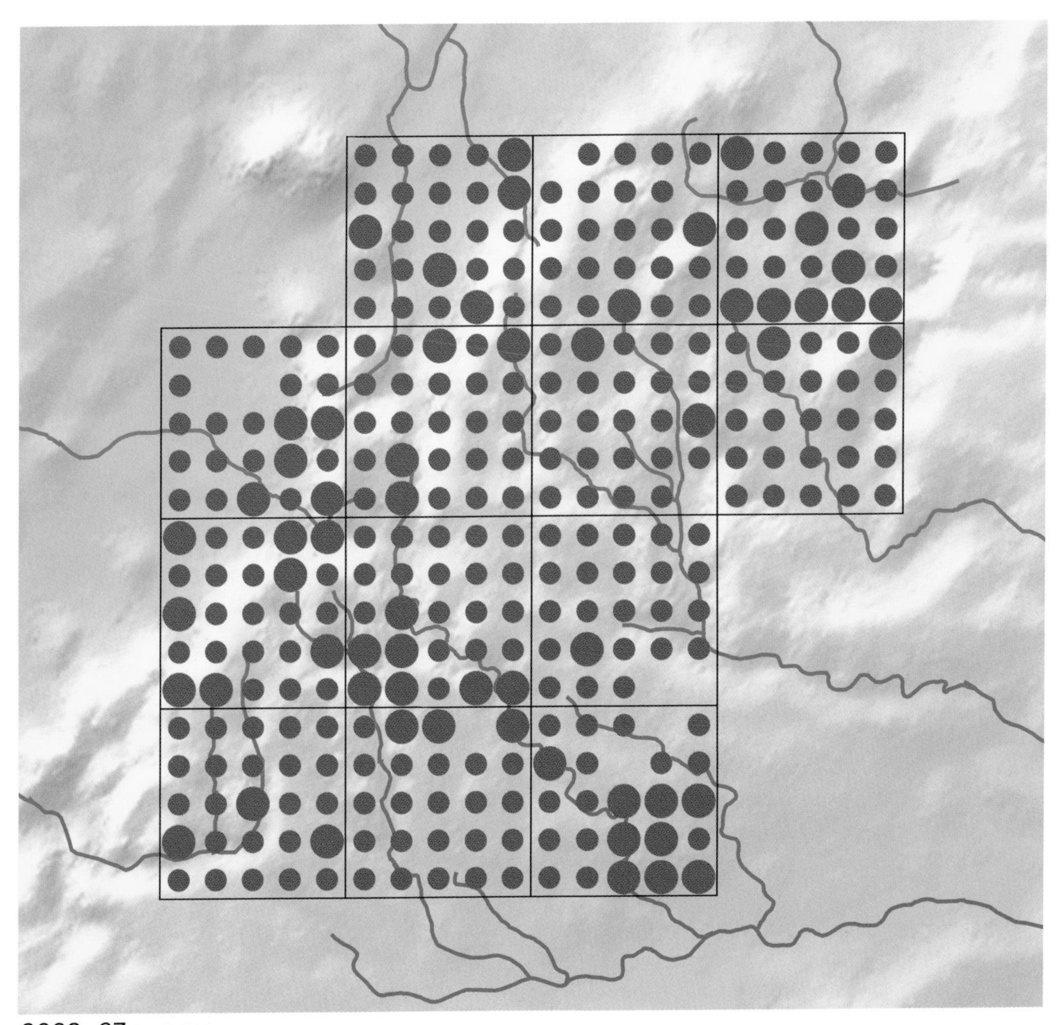

2003–07 survey

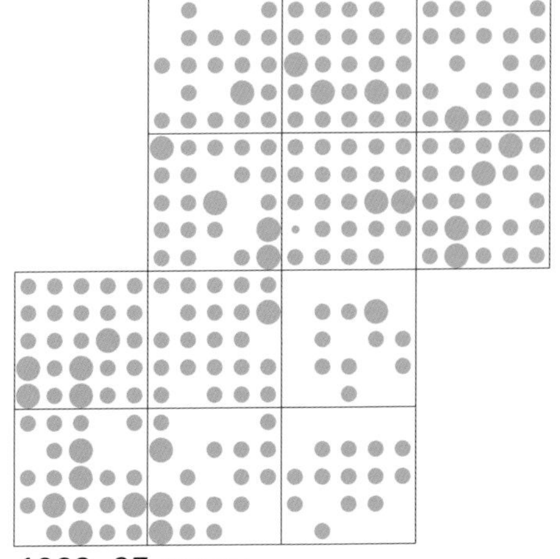

1983–87 survey

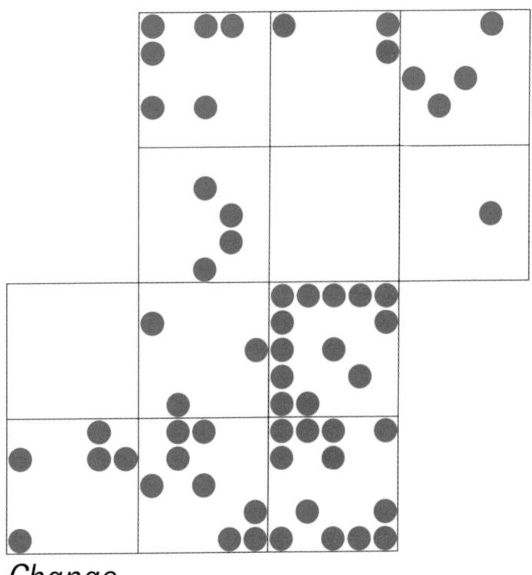

Change

Garden Warbler

Sylvia borin

RICHARD TYLER

Migrant breeder

UK conservation status: **GREEN**

Long-term UK trend (1970–2006): -1%

Recent population trend (1994–2007): -5%

Numbers of tetrads with breeding records:

	2003–07		1983–87
	13 squares	12 squares	12 squares
Possible	**3**	3	4
Probable	**124**	114	147
Confirmed	**8**	8	17
Totals	**135**	**125**	**168**

In the survey area the inappropriately named Garden Warbler is predominantly a bird of scrub, bushes, new plantations and woodland edges. Its song is similar to that of the Blackcap, but less well known, and it may be that it has been underrecorded as a result.

Comparison of the results of this survey with those from 1983–87 indicates a 25% reduction in number of occupied tetrads in the 12-square area, with most of the loss appearing to be in the east, in the Evenlode Valley and the northern parts of the High Wold. Bearing in mind the greater survey effort, the real change may be significantly larger, which is interesting as the first Atlas stated: 'It is believed that the population has increased during the survey period.' It seems that this trend was reversed in the intervening years, though the species is prone to year-to-year variations in numbers. It is still well represented along the scarp from Broadway southwest to Cheltenham and one particular area that appears to have been newly colonized is the gravel-pit complex near Fairford, which has become more extensive since the 1980s survey.

The presumed overall decrease in numbers of breeding pairs is supported by the fact that there were only eight tetrads in which breeding was confirmed, compared with 17 in the first survey. This is despite the general increase in survey effort, and the particular emphasis that was put on confirming breeding. Although breeding is difficult to prove for this species, the very small number of confirmed breeding records accord with the general feeling amongst observers that Garden Warblers are rather thinly distributed in most areas.

Species sponsored by Richard and Pat Gunn

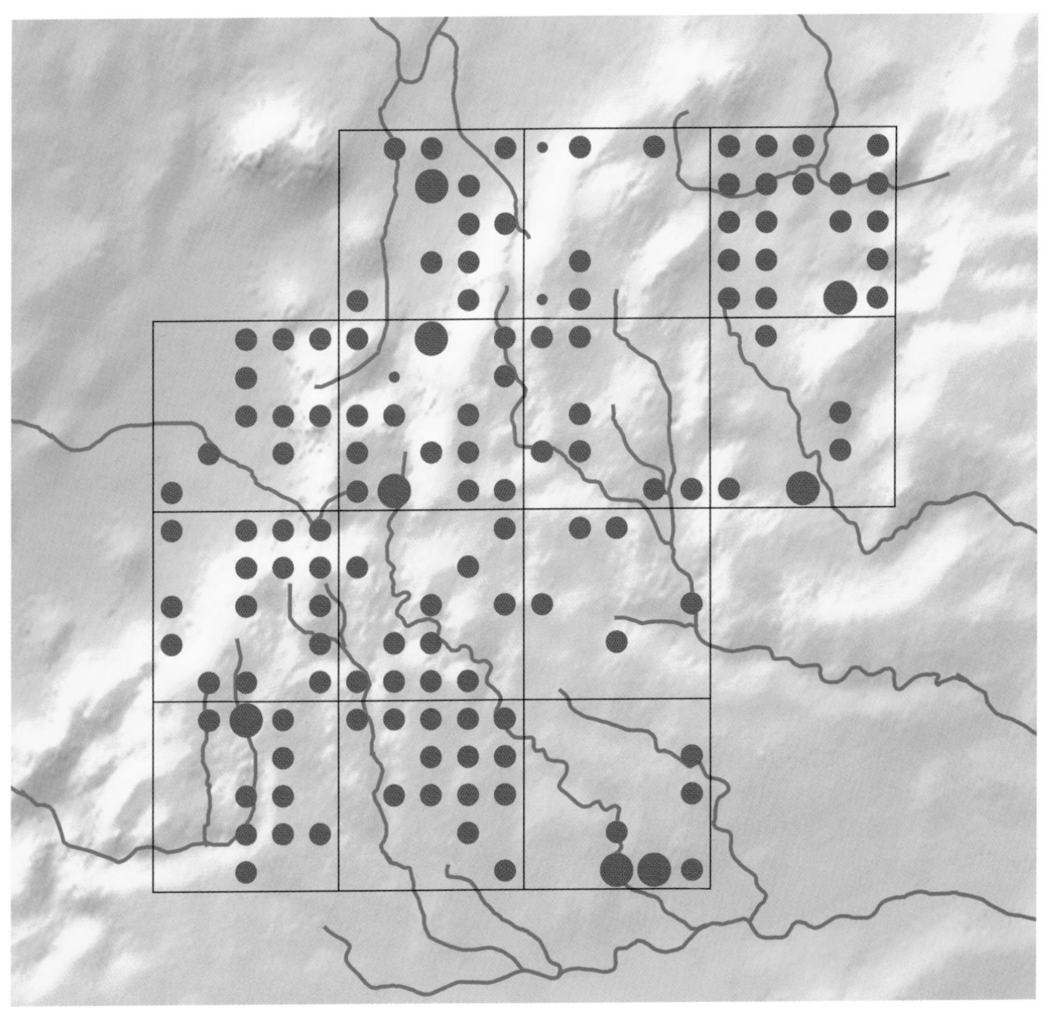

2003–07 survey

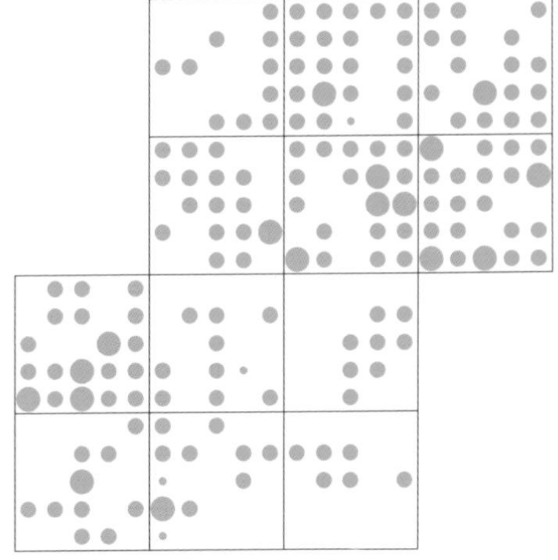

1983–87 survey

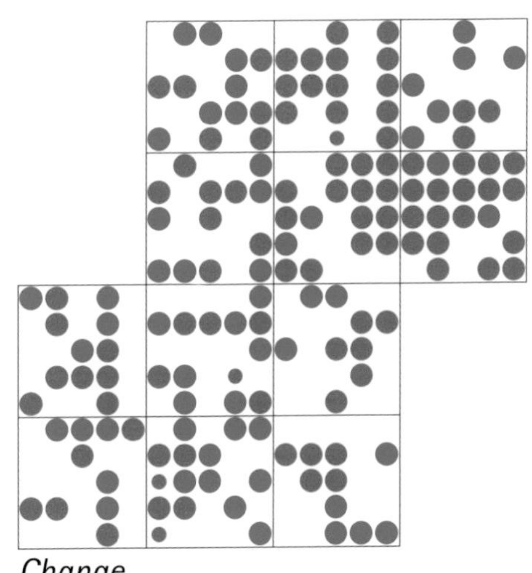

Change

Lesser Whitethroat

Sylvia curruca

DAVE MORGAN

Migrant breeder

UK conservation status: GREEN

Long-term UK trend (1970–2006): +8%

Recent population trend (1994–2007): -12%

Numbers of tetrads with breeding records:

	2003–07		1983–87
	13 squares	12 squares	12 squares
Possible	**6**	6	7
Probable	**128**	115	82
Confirmed	**21**	20	11
Totals	**155**	**141**	**100**

Although it is found in a variety of situations, the Lesser Whitethroat's typical habitat preference is for tall hedgerows bordering pasture. Distribution during the first Atlas survey was described as being 'patchy'. It still seems to be that way, with no obvious trend or pattern, although there is a hint of a slight negative correlation with hilliness–suggesting that it favours flatter areas.

More than most species, its distribution maps may be significantly influenced by subtle changes in survey methods. The Lesser Whitethroat has a very short initial song period, about three weeks after it arrives in late April or early May with a second bout of singing often heard later in the summer. Thus, it can be very easy for observers to miss occupied territories when trying to cover several areas in a short space of time. Another significant variable is the fluctuation from year to year in the numbers of Lesser Whitethroats found. Its spring migration is through Israel, crossing into eastern Europe, so that it arrives in the UK from the east, rather than from the south. Consequently weather patterns tend to have a different effect on both the timing of its arrival and the numbers arriving, compared with other small passerines.

It is clearly less widely distributed than the Whitethroat, and it is significantly less numerous. The moderate increase in the recorded number of occupied tetrads is probably not significant, but rather an effect of the greater survey effort. That said, the change map paints a very interesting picture, with unexplained losses from the northern part of the High Wold and the Evenlode Valley, and gains in the Avon Vale and the Thames tributaries area.

Species sponsored by Iain and Jill Main

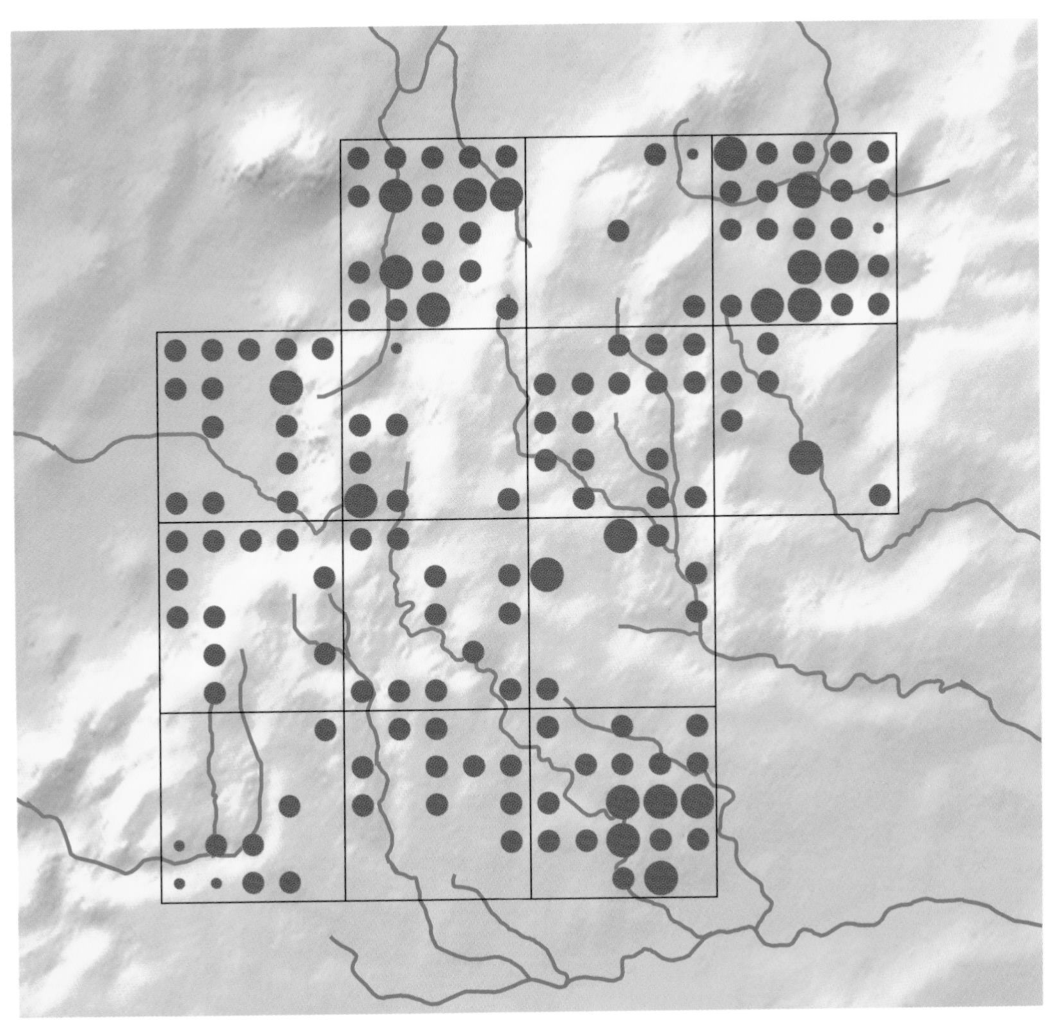
2003–07 survey

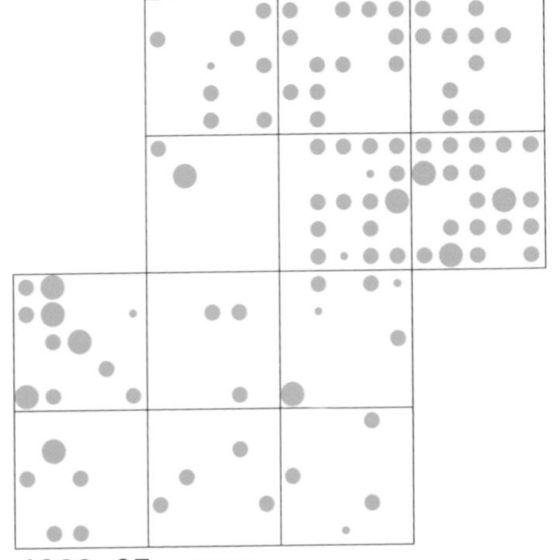
1983–87 survey

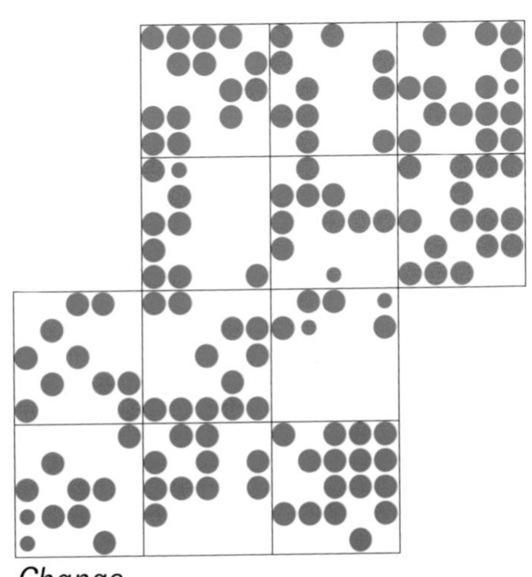
Change

Whitethroat (Common Whitethroat)

Sylvia communis

Whitethroats have a noticeable song and a distinctive song flight, making territories relatively easy to survey. They are also vocally agitated when their nest site is approached, so that, with a little patience, confirmation of breeding is not too difficult at many sites. Indeed, the total number and proportion of confirmed breeding tetrads (29%) was higher than for any other warbler. Although birds singing on passage could potentially inflate the apparent distribution of the species, efforts were made not to record migrants as birds holding territory, and the map is thought to give an accurate picture of the species' status.

A bird well suited to the type of open situations found in much of the Cotswolds, the Whitethroat has taken advantage of the many broad, low hedges that have been created, particularly bordering arable land. It is widely spread throughout the survey area, being recorded in over 90% of the tetrads, and is quite numerous in many locations.

The change map shows that there has been a significant increase in the distribution density of the species since the survey of the 1980s, with the total number of records having increased by about 70%. This accords with the international population crash in the late 1960s and early 1970s attributed to droughts in the Sahel region, whose effects were still being felt during the 1980s survey. Comments in the first Atlas suggested that a recovery in numbers was taking place at that time and the new survey appears to confirm this, with much of the expansion in range having occurred during the late 1980s.

Migrant breeder

UK conservation status: GREEN

Long-term UK trend (1970–2006): -3%

Recent population trend (1994–2007): +31%

Numbers of tetrads with breeding records:

	2003–07		1983–87
	13 squares	12 squares	12 squares
Possible	**3**	3	11
Probable	**207**	191	116
Confirmed	**87**	79	31
Totals	**297**	**273**	**158**

RICHARD TYLER

Species sponsored by Ian Cox

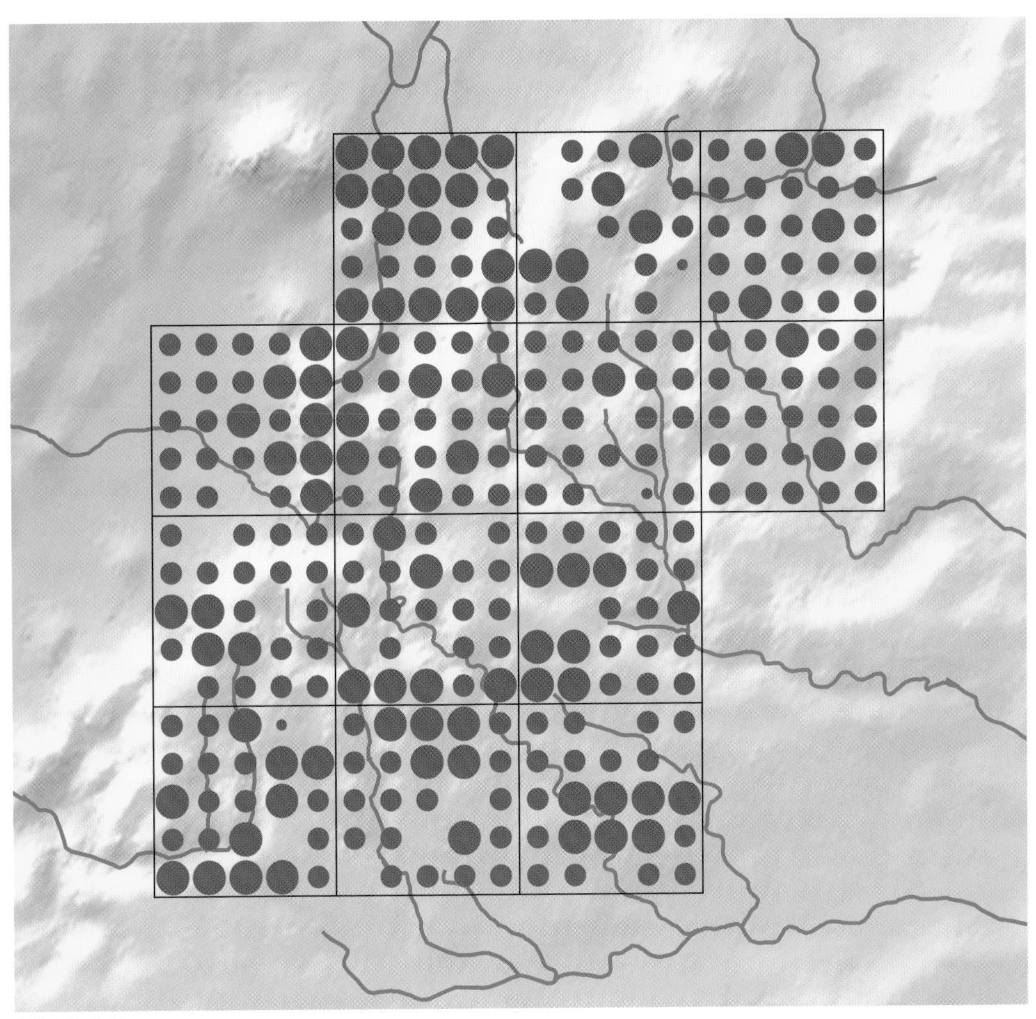

2003–07 survey

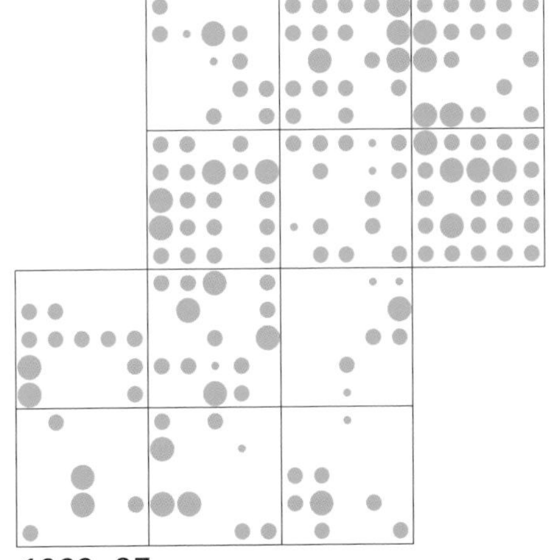

1983–87 survey

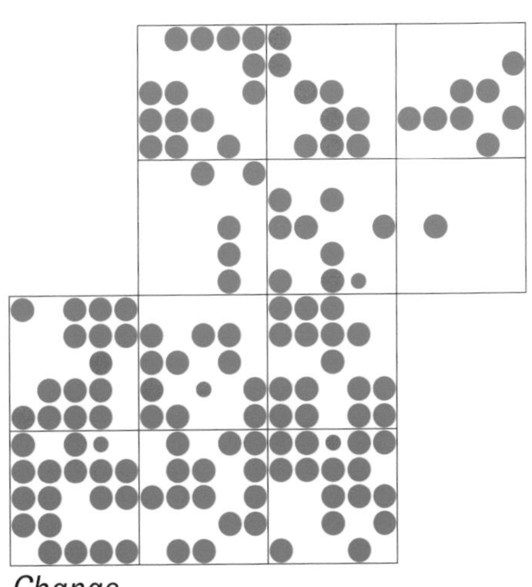

Change

Chiffchaff (Common Chiffchaff)

Phylloscopus collybita

GRAHAM WATSON

Migrant breeder

UK conservation status: GREEN

Long-term UK trend (1970–2006): +24%

Recent population trend (1994–2007): +46%

Numbers of tetrads with breeding records:

	2003–07		1983–87
	13 squares	12 squares	12 squares
Possible	**5**	5	9
Probable	**250**	234	194
Confirmed	**56**	47	20
Totals	311	**286**	**223**

Along with the Blackcap, the Chiffchaff is one of the most numerous and widely distributed warblers in the survey area, and is one of the most abundant breeding species in its favoured habitat of broad-leaved woodland. It was found in most tetrads, its familiar song making territorial birds easy to locate. Although Chiffchaffs do sing while on migration, overrecording of migrants was not considered to have been significant. Breeding was proved in just under one-fifth of occupied tetrads, usually by observation of adults carrying food or by seeing newly fledged young.

The data show an increase in the number of occupied tetrads. Although some of the increase could be attributed to the greater survey effort, it seems unlikely to explain it all—the number of unoccupied tetrads was reduced substantially, from 77 to 14. The general feeling of observers is that Chiffchaffs became significantly more widespread and numerous during the period between the two surveys, a belief supported by the 135% increase in the number of tetrads in which breeding was confirmed. In the first Atlas it was stated: 'The need for taller trees probably accounts for it being rather less abundant than the Willow Warbler'. The relative abundance and distribution of the two species has clearly been reversed; this is discussed in more detail in the Willow Warbler species account.

Species sponsored by Beryl Smith

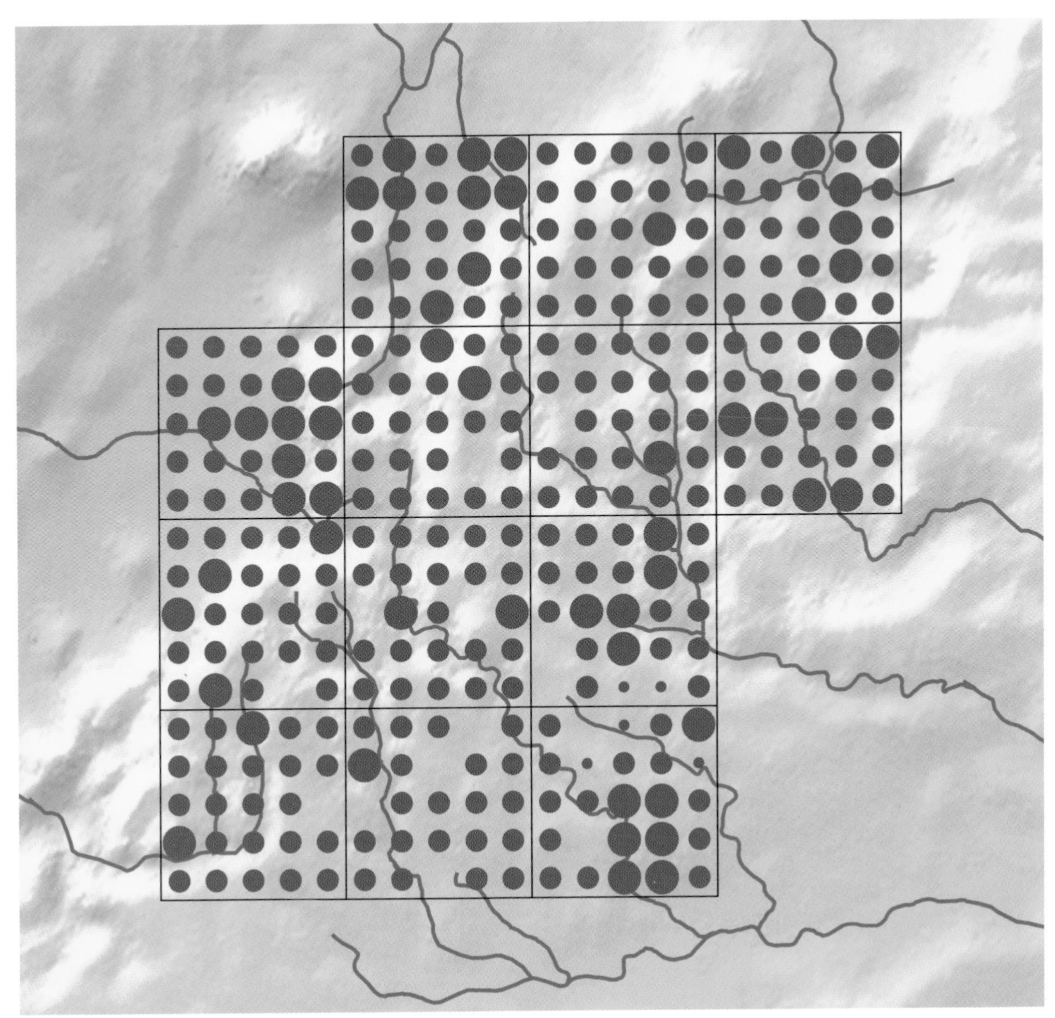

2003–07 survey

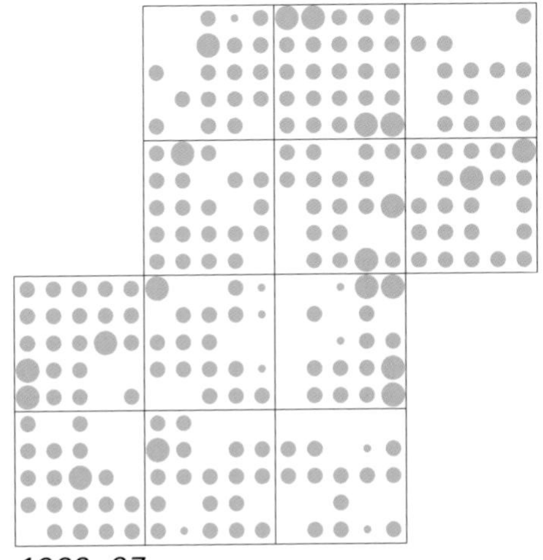

1983–87 survey

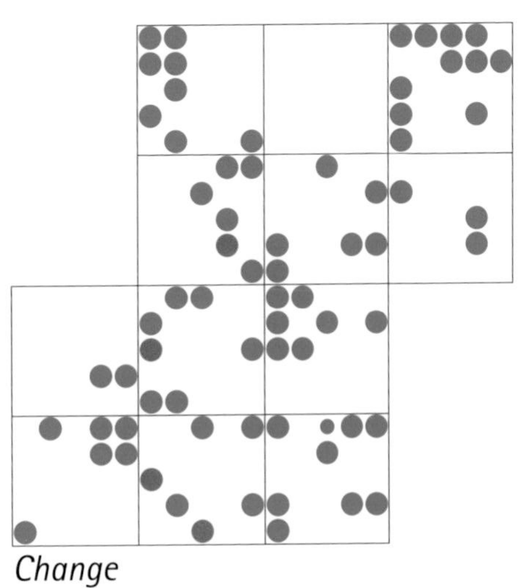

Change

Willow Warbler

Phylloscopus trochilus

The song of the Willow Warbler is both familiar and distinctive, and allowed breeding areas to be easily located. However, they habitually sing on migration and care had to be taken not to record passage birds. Confirmation of breeding was generally achieved by observing adults carrying food to their young.

The relative distributions of the Willow Warbler in the two surveys reflects the decline in their numbers. In the first Atlas it was stated, 'Willow Warblers are very widely and evenly distributed throughout the study area' and, even more tellingly, 'It is likely that Willow Warblers do breed in every tetrad within the study area.' This clearly is no longer the case, with the number of occupied tetrads falling 14% in the 12-square recording area, and the proportion of confirmed breeding tetrads being also less than half that in the first Atlas, despite the greater survey effort. The general feeling amongst observers was that the numbers of singing birds had also markedly declined, with usually only a few pairs in a tetrad with suitable habitat. Whatever the reason for the decline, fewer birds are reaching the area to breed, and they have retreated to their preferred habitat. This is usually bushes or new plantations, mostly found in upland uncultivated areas and often on sloping ground. The change map illustrates a significant thinning of the distribution outside those areas.

RICHARD TYLER

Migrant breeder

UK conservation status: AMBER

Long-term UK trend (1970–2006): -44%

Recent population trend (1994–2007): +1%

Numbers of tetrads with breeding records:

	2003–07		1983–87
	13 squares	12 squares	12 squares
Possible	**7**	6	6
Probable	**224**	211	211
Confirmed	**24**	23	61
Totals	**255**	240	278

Comparing the change maps for Willow Warblers and Chiffchaffs shows many losses and very few gains for Willow Warblers, but increases for Chiffchaffs. In the 1980s Willow Warblers were more widely distributed than Chiffchaffs; the reverse is now true. Similarly, in the 1980s the number of tetrads with confirmed breeding was 3:1 in favour of Willow Warbler; the ratio is now 2:1 in favour of Chiffchaff. This decline in Willow Warbler numbers fits in with the national picture, where it appears to be losing ground in southern Britain, although doing better in the north. For the Chiffchaff, in contrast, both numbers and range seem to have increased.

Species sponsored anonymously

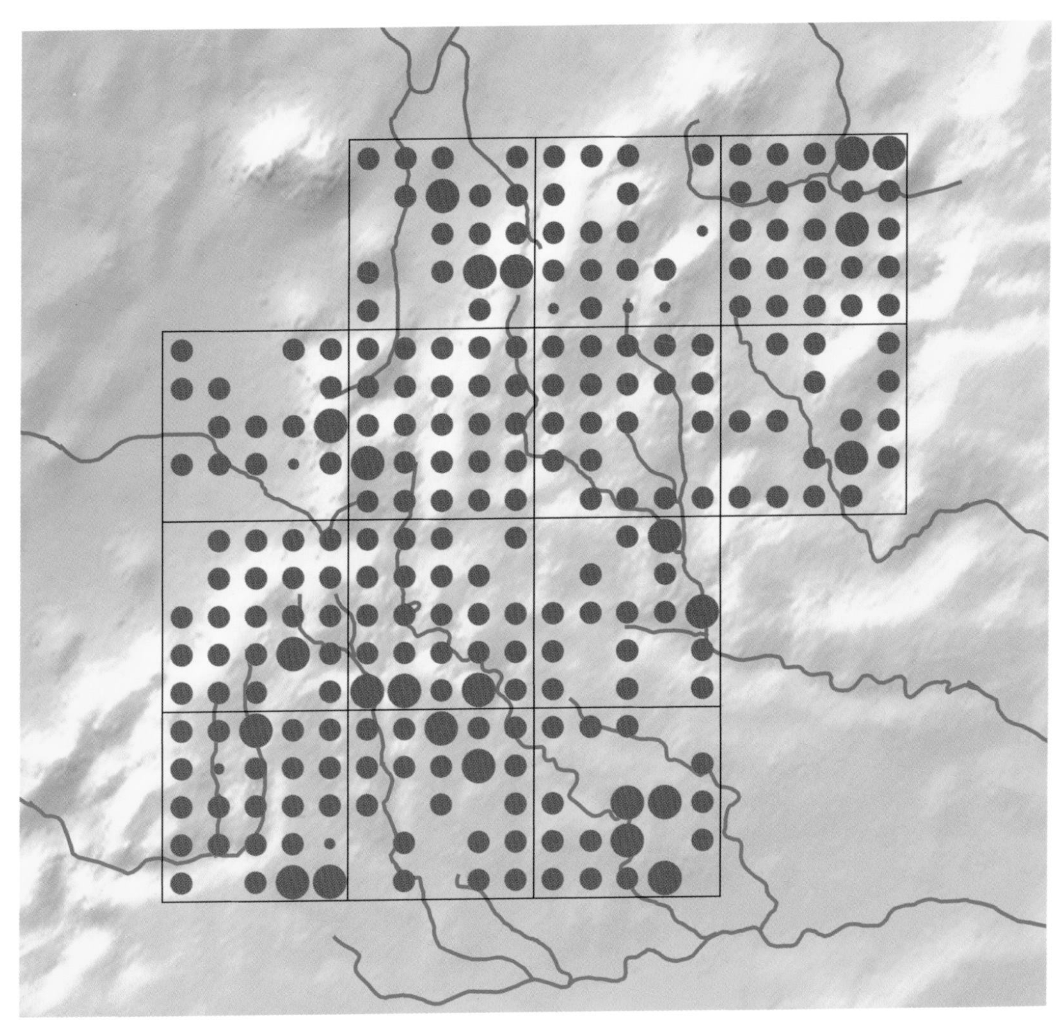

2003–07 survey

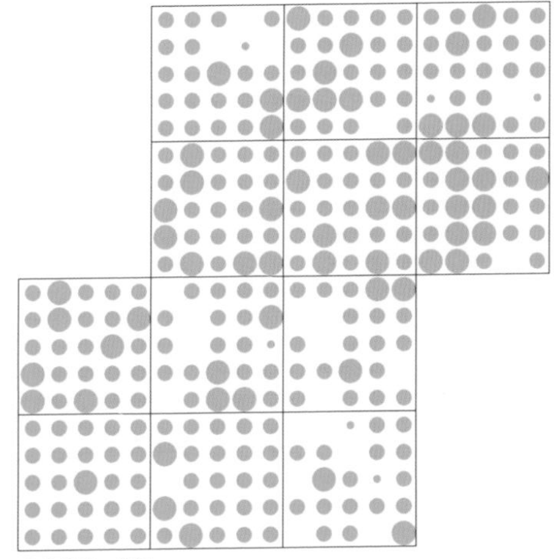

1983–87 survey

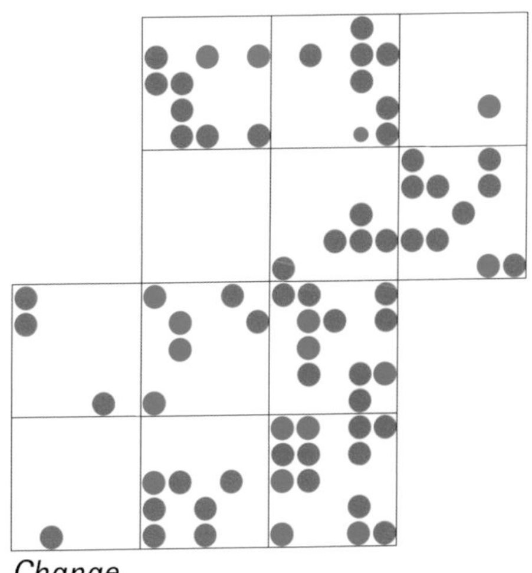

Change

Goldcrest

Regulus regulus

RICHARD TYLER

Resident breeder

UK conservation status: AMBER

Long-term UK trend (1970–2006): -17%

Recent population trend (1994–2007): +50%

Numbers of tetrads with breeding records:

	2003–07		1983–87
	13 squares	12 squares	12 squares
Possible	14	13	8
Probable	223	210	116
Confirmed	53	47	16
Totals	290	270	140

Goldcrests are most easily located by their song, which can be heard throughout much of the year (although breeding typically starts in April). However, because of its high-pitched nature it is inaudible to some birdwatchers, and so the species can be overlooked. Confirmation of breeding is not easy: although adults openly carry food to the young, this is usually difficult to spot in the depths of a conifer tree. In addition, not only are the birds small, but so is the amount of food that they carry.

The smallest breeding bird in the Cotswolds is primarily associated with conifers in the breeding season. Most woods in the recording area contain some, as do churchyards, local parks and gardens, so the species is both widely distributed and quite common in suitable habitat.

Compared with the first survey, the total number of confirmed breeding records nearly trebled and there was also a 90% increase in the number of occupied tetrads. Neither of these outcomes can be attributed entirely to the greater survey effort. There seems little doubt that the distribution has expanded or that the population has increased significantly, the species probably having benefited from a series of mild winters since the late 1980s.

Species sponsored by Siobhan Barker and Christopher Main

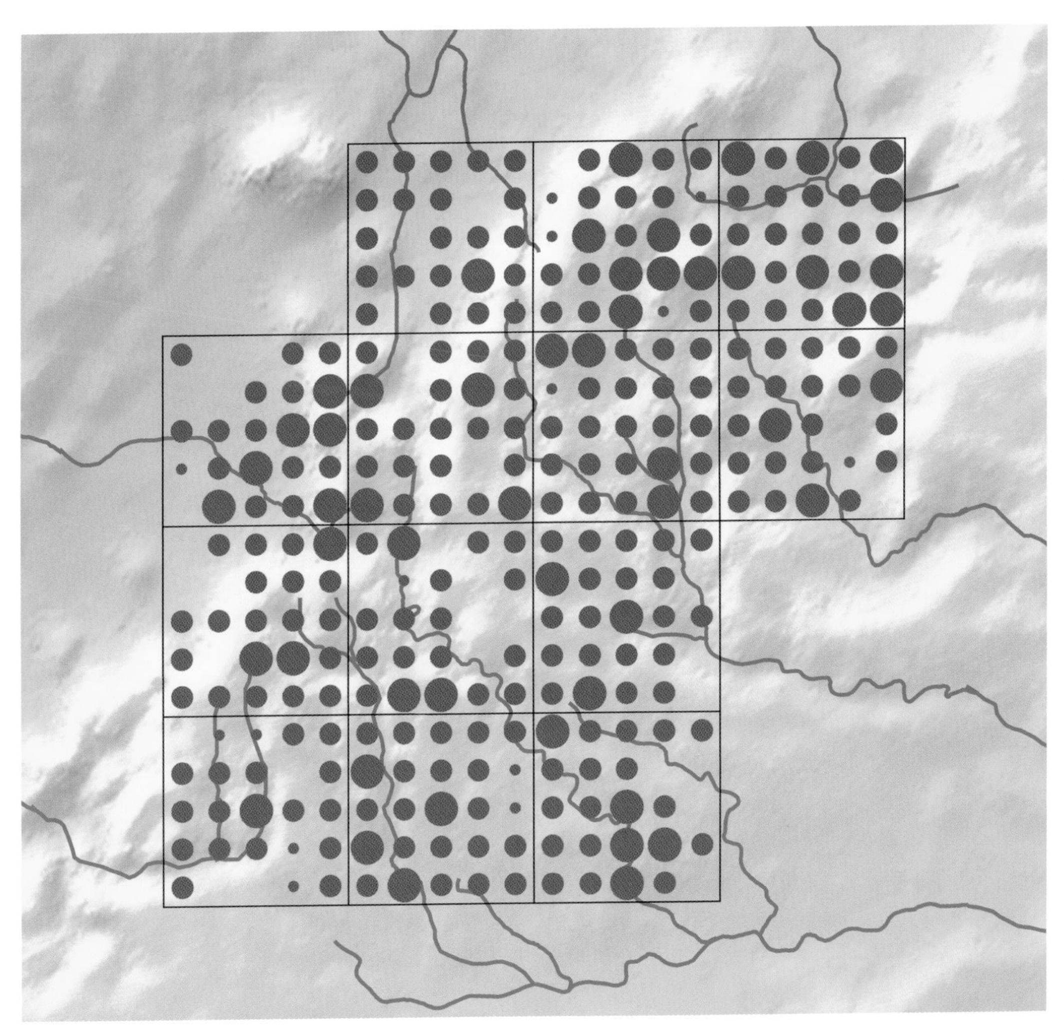
2003–07 survey

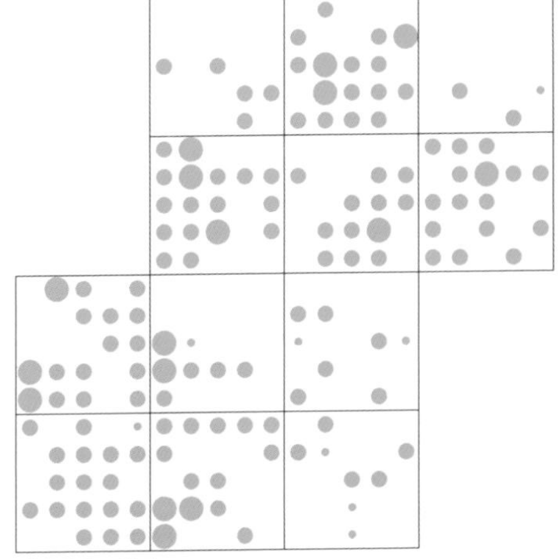
1983–87 survey

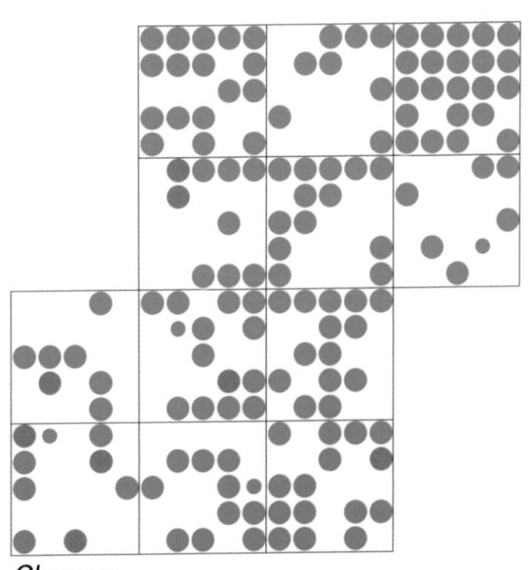
Change

Spotted Flycatcher

Muscicapa striata

RICHARD TYLER

Migrant breeder

Gloucestershire BAP species

UK conservation status: RED

Long-term UK trend (1970–2006): -85%

Recent population trend (1994–2007): -59%

Numbers of tetrads with breeding records:

	2003–07		1983–87
	13 squares	12 squares	12 squares
Possible	**31**	29	34
Probable	**62**	60	57
Confirmed	**55**	53	61
Totals	148	142	152

Spotted Flycatchers are among the last regular migrant breeding birds to reach the Cotswolds, usually around the second week of May. Their typical breeding habitat is open woodland, but they are most familiar in the recording area as village birds. Vocally, they are one of the most nondescript passerine species of the region, and in woodland they typically sing from high in the already developed canopy. This often makes locating them difficult, as the song is quite similar to the calls of, for example, young members of the thrush family. However, in villages they are more visible on fences, telegraph wires, television aerials and rooftops, as well as in garden trees, and it is these areas that provided most records. Breeding was confirmed in nearly 40% of cases, usually from observation of adults carrying food to their young.

The distribution map shows a rather random pattern of occupied tetrads, but this largely reflects the distribution of suitable woodland and villages (see Figures 6 and 7, page 12). There were few records from Vale areas around Cheltenham, or the Stroud Valleys; distribution was also limited in the higher cereal growing areas, as it was in the 1980s survey. Before the survey began, it was felt that there had been a significant decline in Spotted Flycatcher numbers since the 1980s, the previous Atlas having described the population as 'stable'. Although the results suggest that this contraction in range has not been exceptional, the greater survey effort in 2003–07 has probably had a significant effect in masking the change, *ie* it is greater than the change map suggests. The Spotted Flycatcher being a fairly unobtrusive bird, extra survey effort is likely to have resulted in additional views into, for example, private gardens and would be expected to increase the number of records appreciably. The conclusion is that the contraction in range observed certainly reflects a population decline since the 1980s.

Species sponsored by Iain and Jill Main

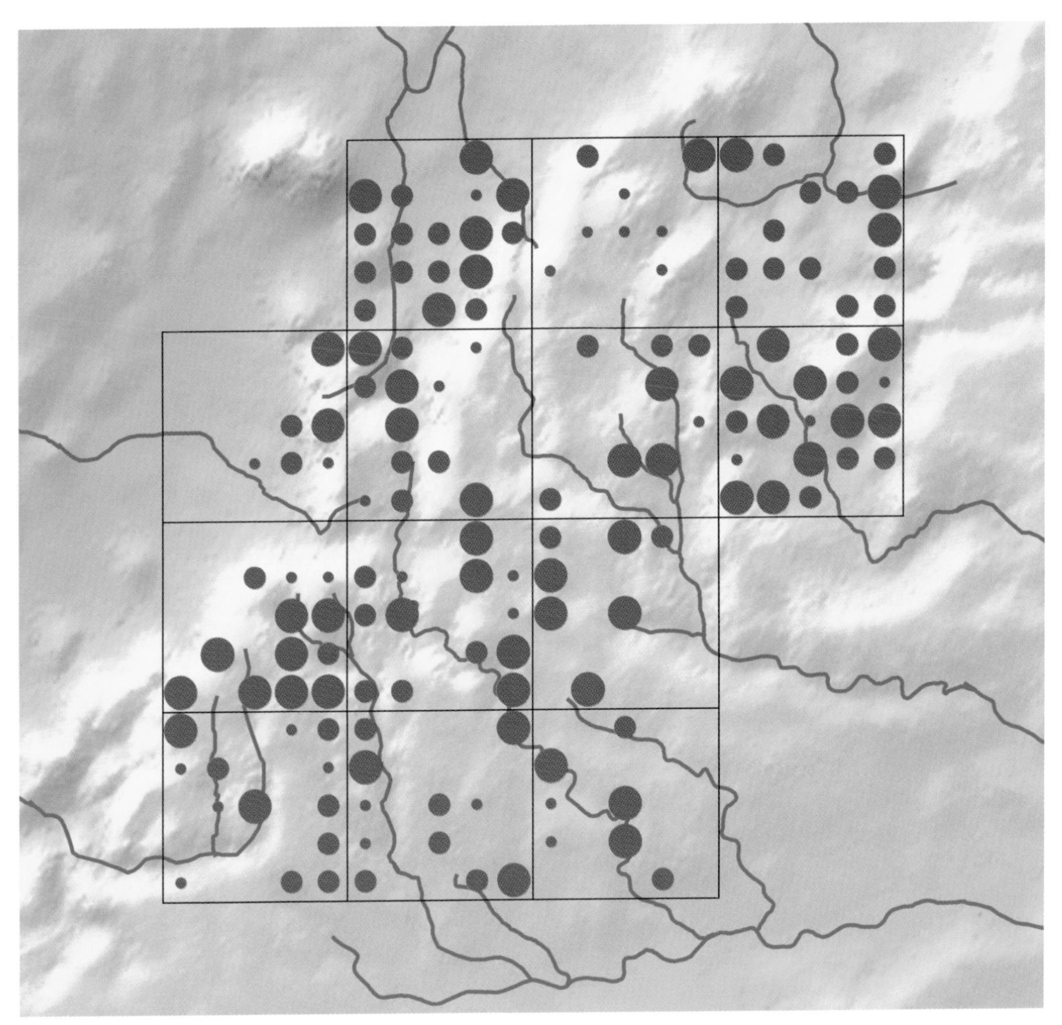

2003–07 survey

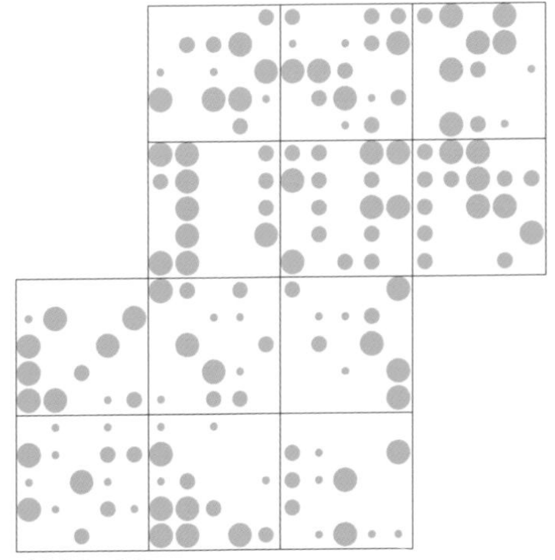

1983–87 survey

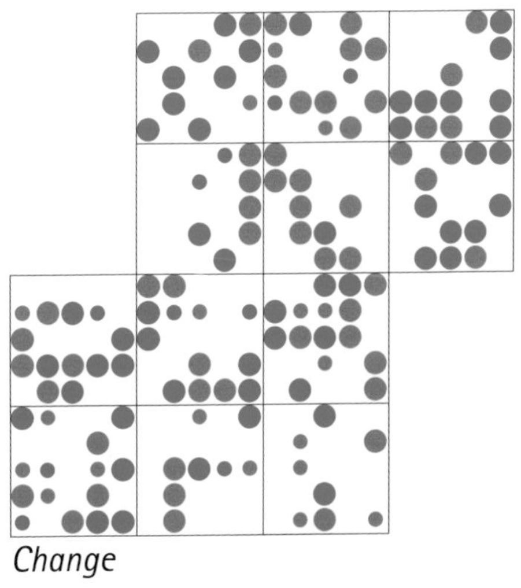

Change

Long-tailed Tit

Aegithalos caudatus

Long-Tailed Tits are among the most familiar and widespread birds of the region, being found in a variety of habitats with bushes and trees. They are usually the first species of the tit group to start breeding. Pairs are easily located by their frequent and distinctive calls in early spring as they prospect for nest sites, and could regularly be seen carrying nesting material. More than any other species, breeding criterion B (nest building) was the primary source of probable breeding records. However, the majority of records were of confirmed breeding, through observation of parents carrying insect food to their nest sites, and of noisy family parties with recently fledged young. Care had to be taken not to record older families, as they tend to roam quite widely soon after fledging.

The species was present in most tetrads as shown by the distribution map, open arable areas being largely responsible for the few absences. This contrasts significantly with the map in the previous Atlas, which shows a much patchier distribution, with major absences in arable areas, especially in the Thames Vale. However, it should be cautioned that the first Atlas commented that the species was 'widely distributed in the study area', and it may be that it was under-recorded during the 1980s survey. That proviso notwithstanding, the change map clearly shows an expansion in distribution. The proportion of confirmed breeding records was much higher in the present survey (60% compared with 35% in the 1980s) which, considering the greater number of total breeding records (a 70% increase), supports the general feeling amongst observers that numbers have increased significantly since the early 1990s. The series of mild winters before and during the new survey may have helped this expansion.

Resident breeder

UK conservation status: GREEN

Long-term UK trend (1970–2006): +49%

Recent population trend (1994–2007): +8%

Numbers of tetrads with breeding records:

	2003–07		1983–87
	13 squares	12 squares	12 squares
Possible	**36**	33	22
Probable	**76**	74	74
Confirmed	**172**	156	58
Totals	284	263	154

RICHARD TYLER

Species sponsored by Wendy Bridgman

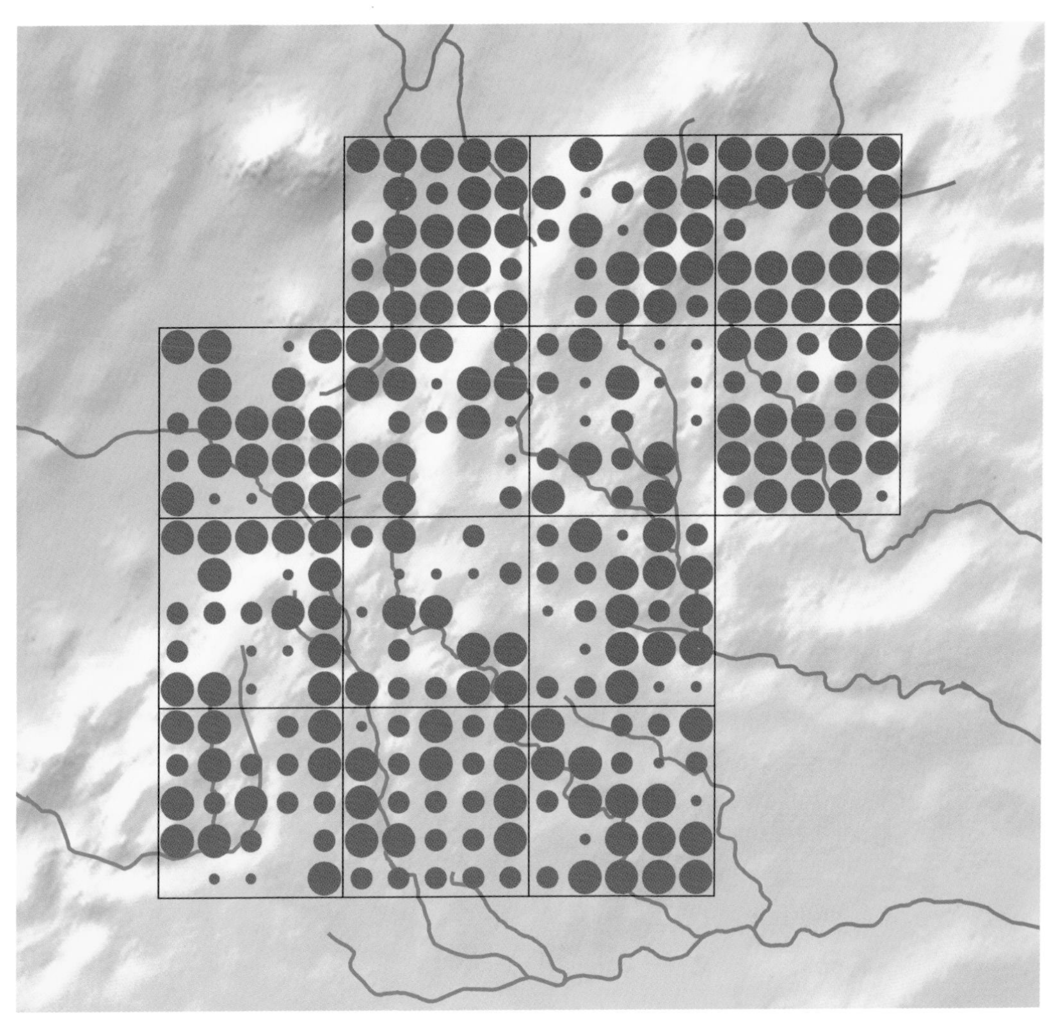

2003–07 survey

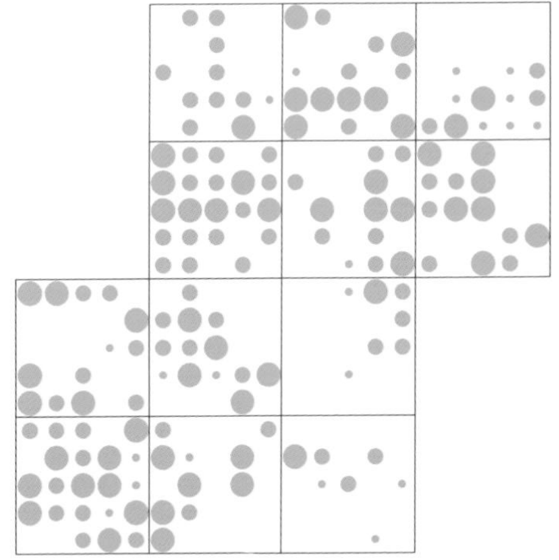

1983–87 survey

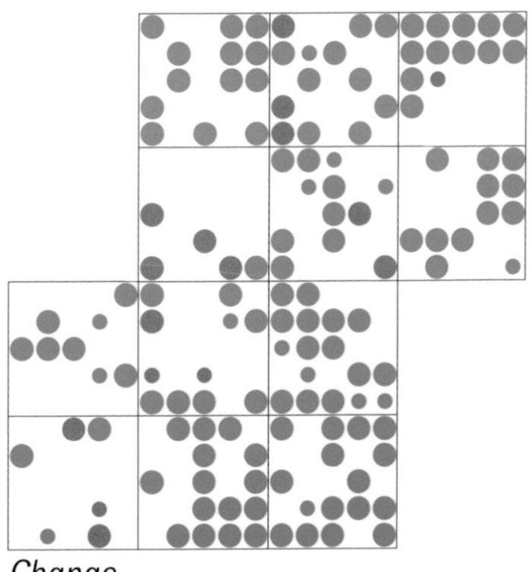

Change

Blue Tit
Cyanistes caeruleus

The Blue Tit is one of the most numerous and widely distributed species in the survey area. It inhabits a wide variety of habitats, being found in almost any location with bushes or trees; deciduous woodland, parks and gardens are particularly favoured. Its natural nest locations are holes in trees and it readily uses nestboxes, but any type of hole will often suffice, including pipes, cracks in masonry on buildings and other man-made artefacts. The number of confirmed breeding records, significantly higher than found in the first survey, is probably due to the increased survey effort. It is also testament not only to the relative ease of confirming breeding by observing adults carrying food and entering nest holes, or newly fledged and begging young, but also to the density of breeding pairs.

Resident breeder

UK conservation status: **GREEN**

Long-term UK trend (1970–2006): +33%

Recent population trend (1994–2007): +14%

Numbers of tetrads with breeding records:

	2003–07		1983–87
	13 squares	12 squares	12 squares
Possible	**3**	2	27
Probable	**37**	36	89
Confirmed	**285**	262	177
Totals	**325**	**300**	**293**

As was found in the results of the first Atlas, the new survey shows that Blue Tits breed throughout the region. The Winter Random Square Survey provides interesting data regarding their abundance in comparison with other tits; data from 1995 to 2007 suggest that they are 50–100% more numerous than Great Tits, and about 10 times more numerous than Coal and Marsh Tits. Although this refers to winter rather than breeding populations, it does suggest that Blue Tits are comfortably the most common tit of the region.

ROB BROOKES

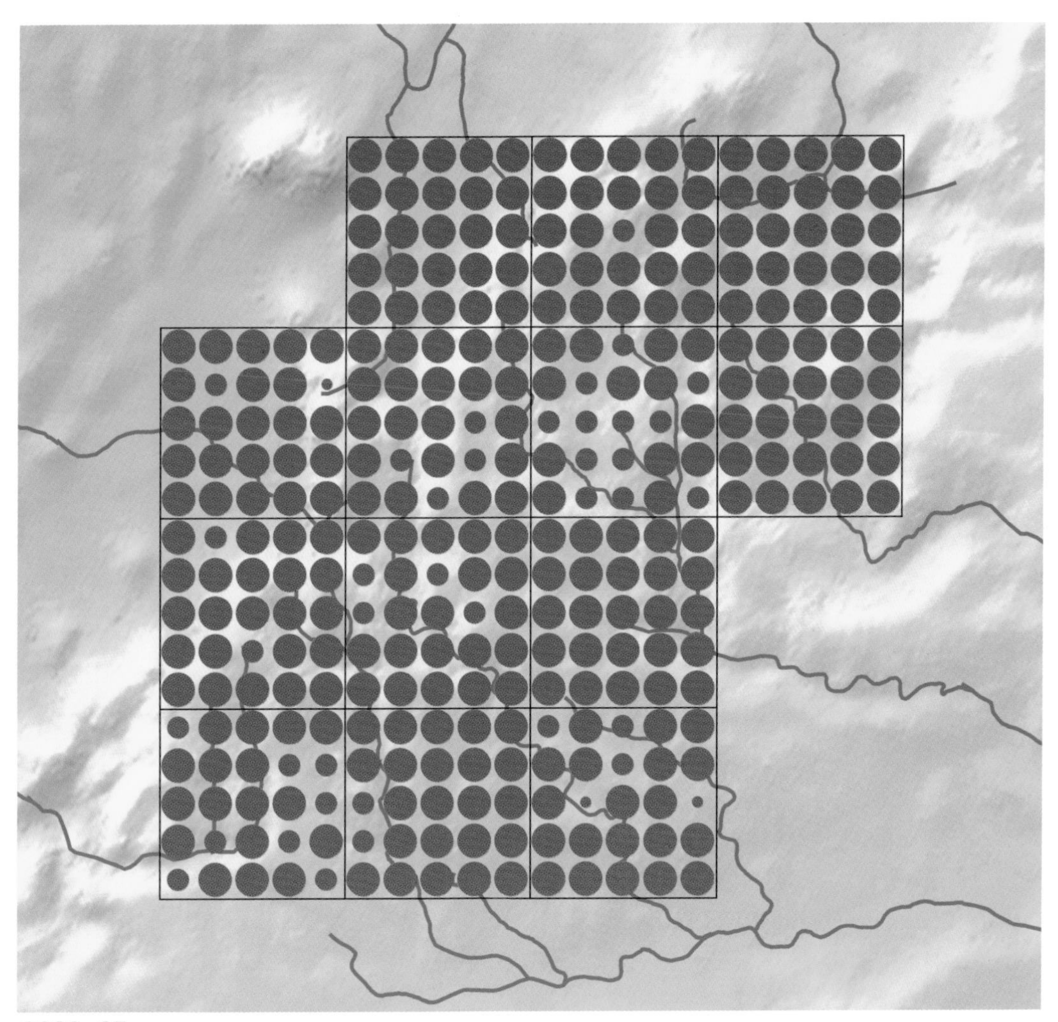
2003–07 survey

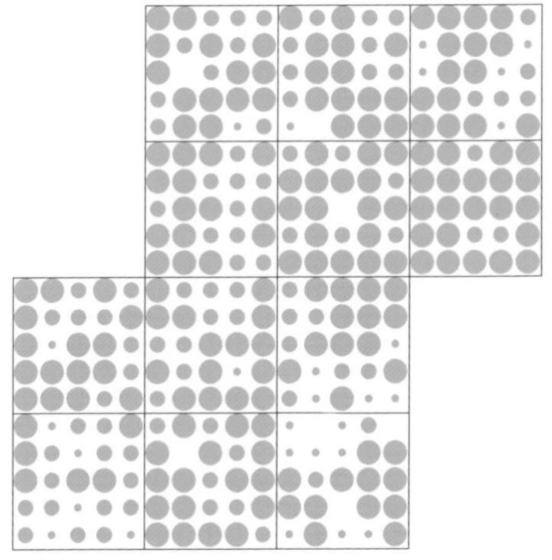
1983–87 survey

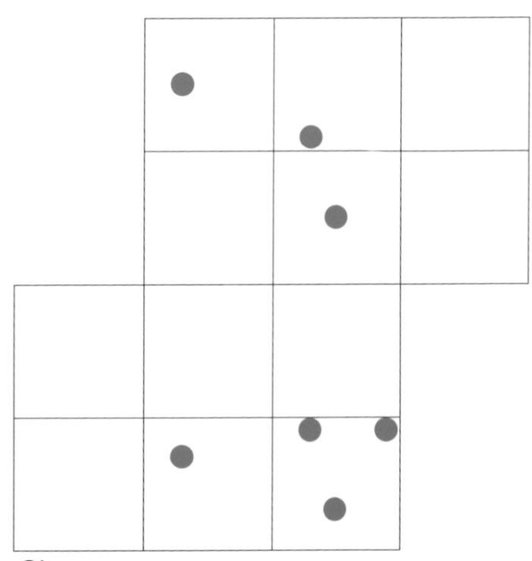
Change

Great Tit

Parus major

ROB BROOKES

Resident breeder

UK conservation status: GREEN

Long-term UK trend (1970–2006): +91%

Recent population trend (1994–2007): +55%

Numbers of tetrads with breeding records:

	2003–07		1983–87
	13 squares	12 squares	12 squares
Possible	**2**	2	25
Probable	**70**	63	118
Confirmed	**251**	233	137
Totals	**323**	**298**	**280**

The loud and persistent calls of Great Tits and their distinctive repertoire of song phrases made the location of breeding birds very simple. Breeding was readily confirmed by observation of adults carrying caterpillars and other invertebrates, often from some distance, to young birds in the nest, as well as seeing newly fledged young.

Great Tits are virtually ubiquitous across the Cotswolds, and were found in a variety of habitats including hedgerows in open arable countryside. Like Blue Tits they were seen to make use of a variety of man-made nesting sites, such as stone walls, nestboxes, buildings and pipes, as well as natural holes in trees. It is likely that some form of suitable nest site is available in all parts of the region and, although they were unrecorded in two tetrads, it would be surprising if at least one pair did not successfully breed in each tetrad every year.

In the 1980s survey, Great Tits were found in only 280 of the 300 tetrads, but it is likely that they were also present throughout the region then, and that the gaps were due to lack of available time and observers. In that survey, about half the records were of confirmed breeding; in the new survey, nearly 80% of records were confirmed. This may reflect the greater survey effort as well as any change in population numbers.

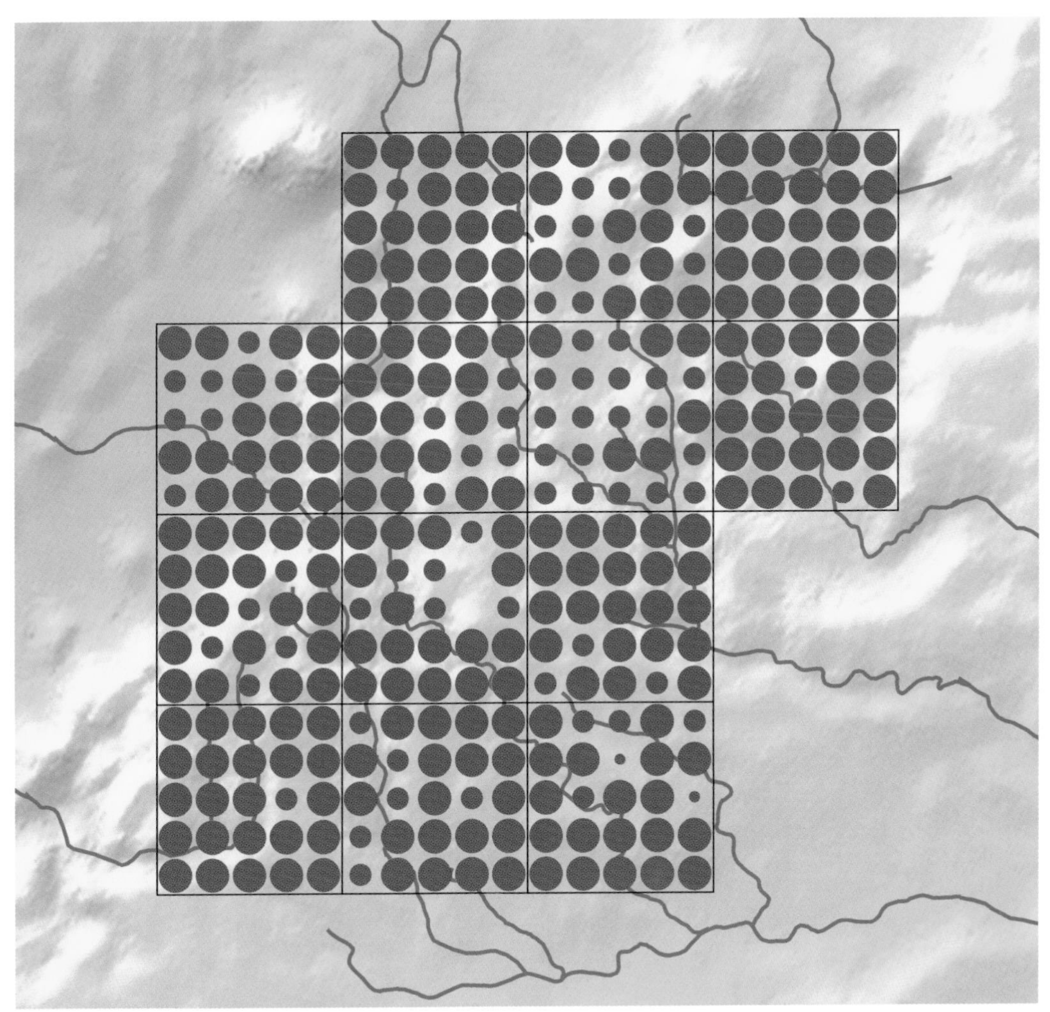

2003–07 survey

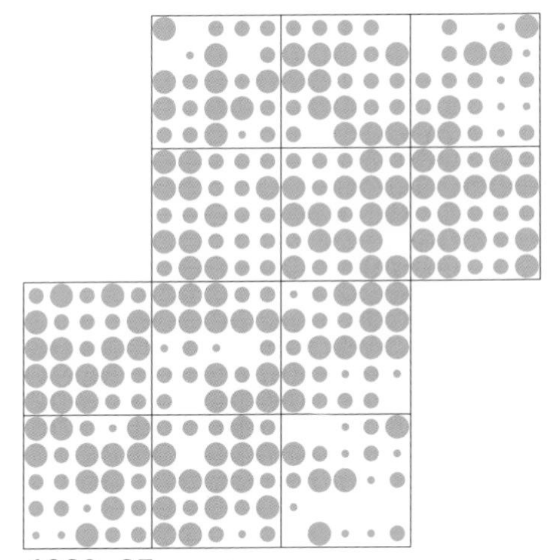

1983–87 survey

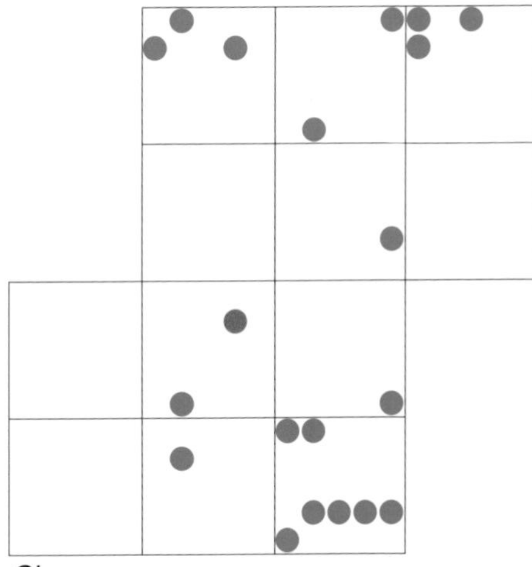

Change

Coal Tit

Periparus ater

The loud and persistent song of the Coal Tit makes the location of territorial birds straightforward. They start to sing in late winter and early spring, and continue singing for longer than either Marsh or Willow Tits. Like other tits they are hole nesters, mainly in trees, but they were frequently found inhabiting holes in old dry stone walls and similar situations in Cotswold villages. Confirmation of breeding was generally by observing either adults carrying food, or newly fledged young with the identifiable yellow tinge to their plumage

Coal Tits are widely distributed throughout the survey area. Although they are generally associated with conifers, such trees can equally be found in mature woodland or a suburban back garden. The distribution map shows fairly uniform coverage, the most noticeable gaps being in the open Severn and Avon Vale farmland in the northwest, and the Evenlode and Stour Valleys in the east.

The change map suggests a significant expansion in range, with an 80% increase in number of occupied tetrads in the original 12-square recording area, and a threefold increase in the number of tetrads in which breeding was confirmed. Comparison with Marsh Tit shows that Coal Tits are now the more widespread species, the reverse of the findings in the 1980s survey. However, Coal Tits are still far less numerous than Great Tits or Blue Tits, although they are probably at least equally plentiful in coniferous woodland.

Resident breeder

UK conservation status: GREEN

Long-term UK trend (1970–2006): +40%

Recent population trend (1994–2007): +19%

Numbers of tetrads with breeding records:

	2003–07		1983–87
	13 squares	12 squares	12 squares
Possible	**15**	14	34
Probable	**142**	136	67
Confirmed	**94**	86	28
Totals	**251**	**236**	**129**

RICHARD TYLER

Species sponsored by Rita Gerry

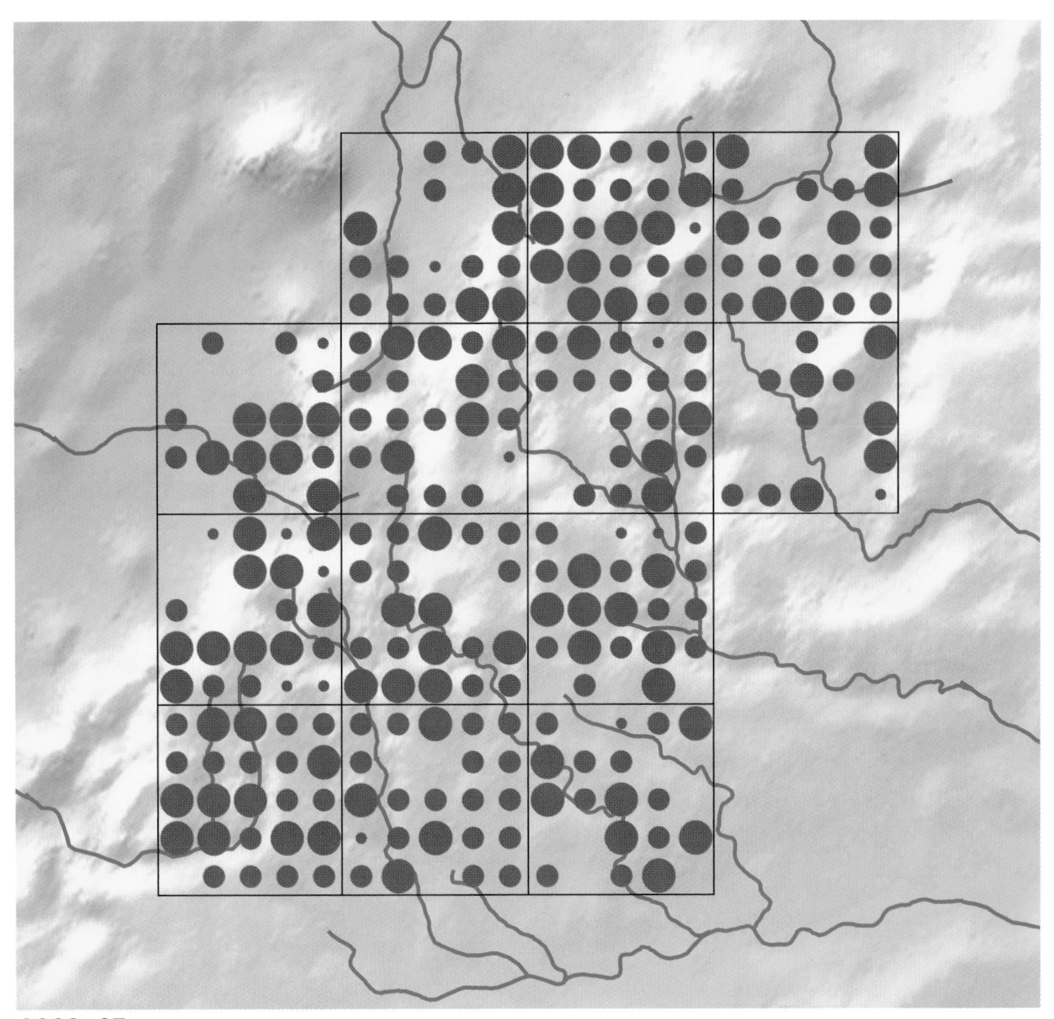
2003–07 survey

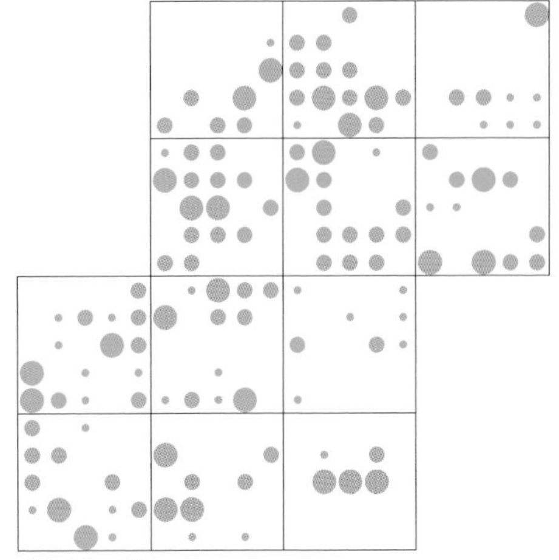
1983–87 survey

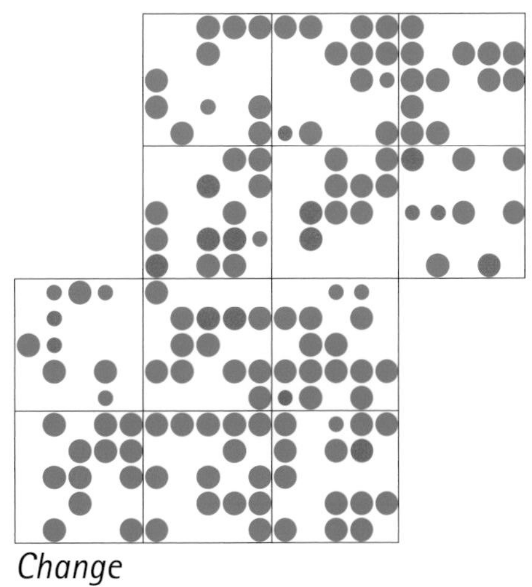
Change

Willow Tit

Poecile montanus

Resident breeder

UK conservation status: **RED**

Long-term UK trend (1970–2006): -88%

Recent population trend (1994–2007): -77%

Numbers of tetrads with breeding records:

	2003–07		1983–87
	13 squares	12 squares	12 squares
Possible	2	2	16
Probable	9	9	26
Confirmed	7	6	6
Totals	18	17	48

Willow Tit was the least common tit species in the area in 1983–87, and this unwanted position has been confirmed by the new survey. The data and the change map clearly show that most of the previously occupied sites have been deserted; it is believed that this reflects a major, almost catastrophic, decline in numbers.

The conclusion from the first Atlas survey was that the species may have been under-recorded, owing to confusion with the very similar Marsh Tit. However, over the past 20 years, the contraction in distribution and decline in numbers of Willow Tits in the Cotswolds has been noted and observers have become much more aware of the vocal differences between the species. Although its song is relatively unfamiliar, it can be a noisy bird with a very characteristic nasal call note which significantly facilitated identification. The favoured locations of the remaining pockets of Willow Tits have also become better known, and so it is likely that most, if not all, of these sites have been found. It now appears to be entirely restricted to a few damp areas with rotting tree stumps in which it can excavate a nest hole, such as around Bourton Pits and Sherborne Brook. The former population in the Winchcombe area (which would still appear to contain suitable habitat) appears to have been completely lost. This decline is very much in line with that found nationally. Confirmed breeding occurred in different years in adjacent 10 km squares (SO91 and SO92) which straddle suitable habitat above Dowdeswell Reservoir and may involve only one pair or territory.

RICHARD TYLER

Species sponsored by Andy Lewis

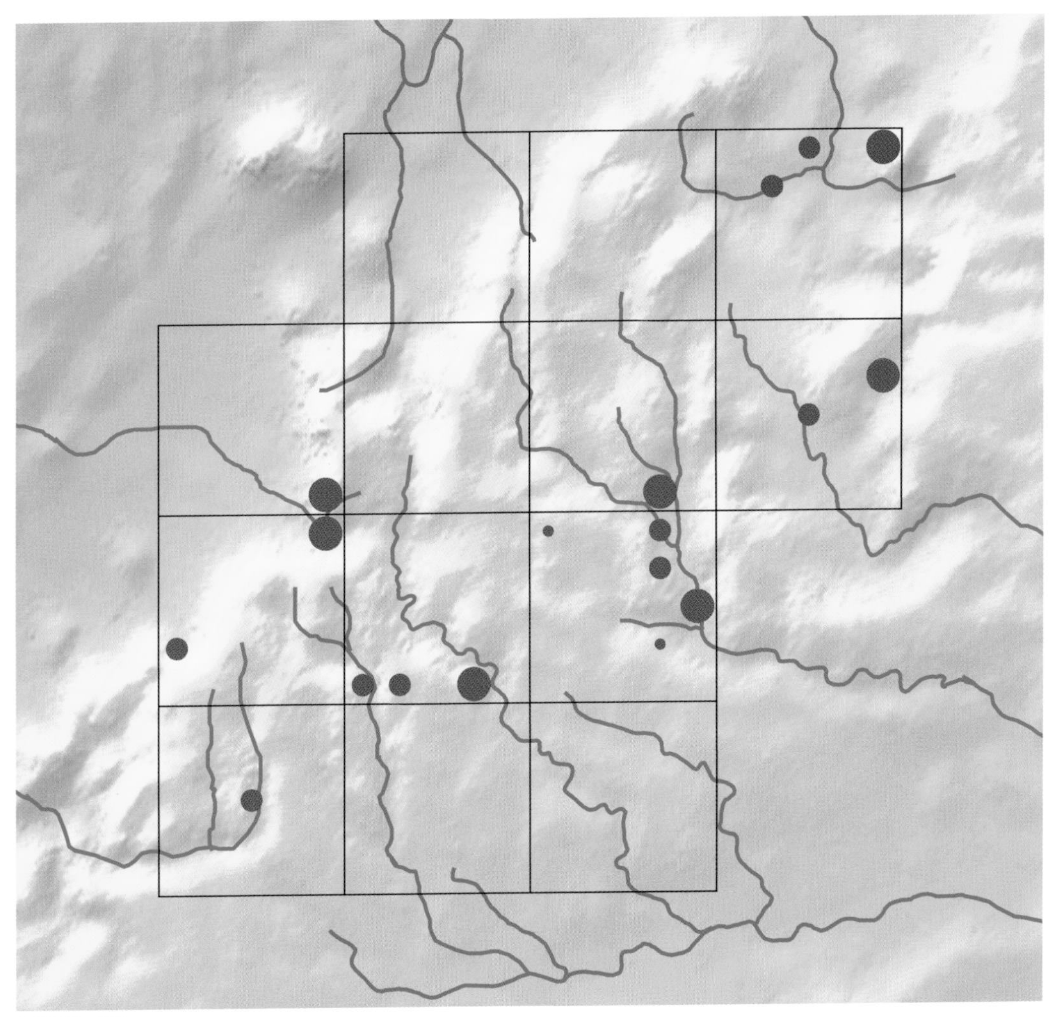

2003–07 survey

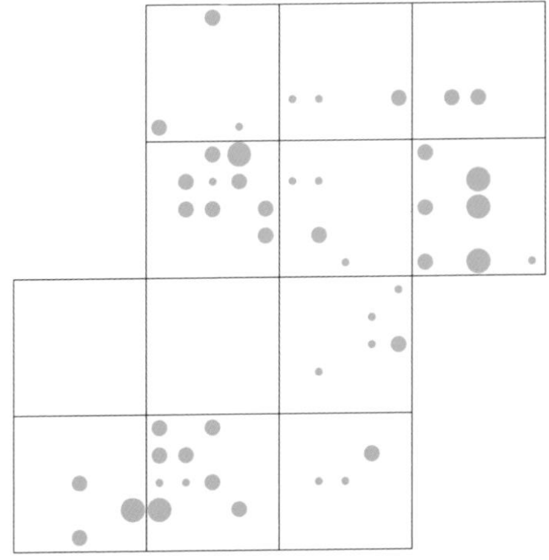

1983–87 survey

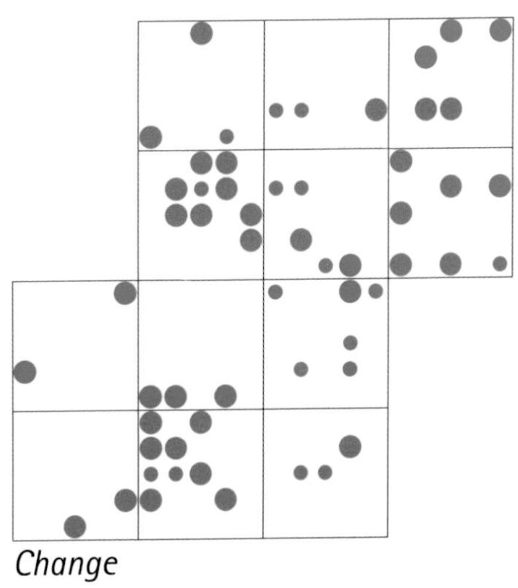

Change

Marsh Tit

Poecile palustris

GRAHAM WATSON

Resident breeder

UK conservation status: RED

Long-term UK trend (1970–2006): -63%

Recent population trend (1994–2007): -6%

Numbers of tetrads with breeding records:

	2003–07		1983–87
	13 squares	12 squares	12 squares
Possible	**38**	38	33
Probable	**93**	90	73
Confirmed	**65**	61	43
Totals	**196**	**189**	**149**

Marsh Tits are typically birds of mature broad-leaved woodland, including small copses. They have a fairly brief song period (March to April), and although their song is quite distinctive, it is heard infrequently. As a result, location of breeding birds was most often by the discovery of pairs of birds from their explosive 'pitchew' call. Breeding was usually confirmed from observation of adults carrying food or from newly fledged young–more confidently in the latter case since Marsh Tits are very sedentary birds, and youngsters almost certainly originated from the tetrad in which they were found.

The distribution map shows Marsh Tits to be fairly uniformly spread. The major unoccupied areas are the Vale in the northwest, much of the Stour Valley in the northeast, and the large arable regions of the Thames tributaries in the southeast. All of these are only sparsely wooded.

As well as their more typical dry woodland habitat, Marsh Tits were also regularly found in wet woodland, such as around Bourton-on-the-Water, which is also a stronghold of the very similar Willow Tit. Whether the demise of the Willow Tit has in any way been caused by competition from Marsh Tits, or whether the latter have merely become more evident in the vacated areas is not clear. It is plausible that confusion between the two species in the 'Willow Tit' habitat previously led to the overlooking of Marsh Tits.

The percentage of confirmed breeding records was similar in the two surveys. Although neither survey provides data on numbers of birds, the feeling of some observers was that numbers had decreased, and that although they are still present in previously occupied areas, the population was lower. This is supported by data from the NCOS Winter Random Square Survey, which suggest a halving of the population in the 10 years from the mid-1990s. The change map indicates no significant redistribution since the previous survey, the increase in number of records being within the limit that could be attributable to the greater survey effort.

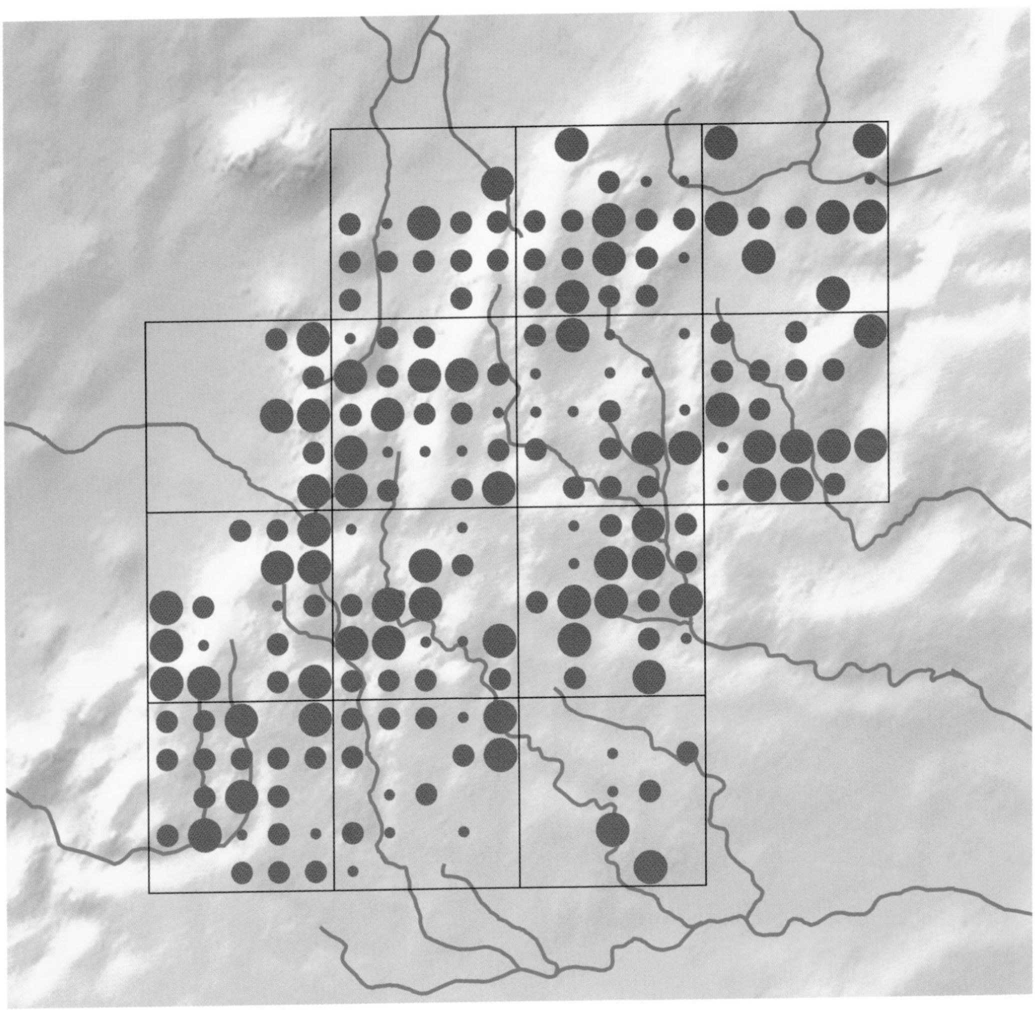

2003–07 survey

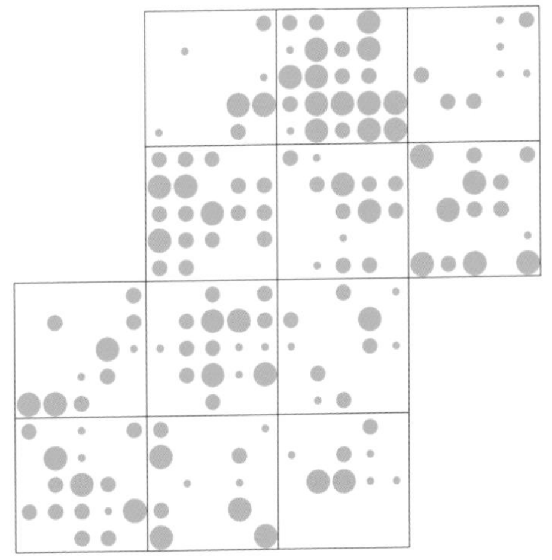

1983–87 survey

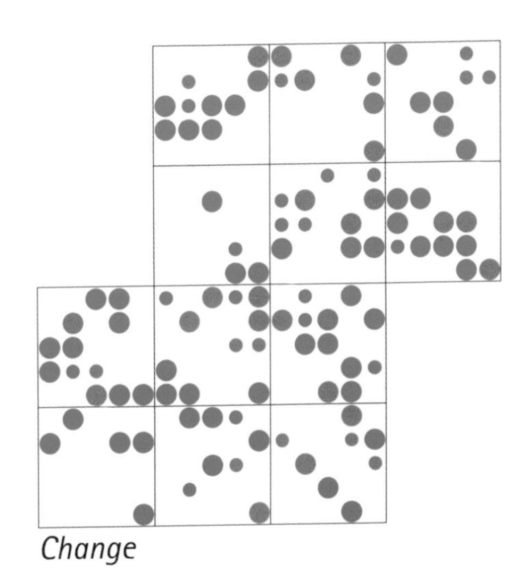

Change

Nuthatch (Wood Nuthatch)

Sitta europaea

RICHARD TYLER

Resident breeder

UK conservation status: GREEN

Long-term UK trend (1970–2006): +175%

Recent population trend (1994–2007): +71%

Numbers of tetrads with breeding records:

	2003–07		1983–87
	13 squares	12 squares	12 squares
Possible	14	14	15
Probable	123	119	64
Confirmed	58	54	25
Totals	**195**	**187**	**104**

Nuthatches set up territory early in the year, and in early spring they are extremely vocal, making their location fairly easy. However, their wide variety of songs is less well known (unlike their loud 'tuit-tuit' contact call). After the end of April they become much quieter, making location and confirmation of successful breeding significantly more difficult. Observation of adults carrying food to nest sites, or observing newly fledged young, were the usual ways to obtain breeding confirmation.

Although they were found in a similar number of tetrads to Treecreepers, the distribution pattern is slightly different as Nuthatches are more tied to mature and overmature broad-leaved trees. In the first survey, the highest distribution density appeared to be in the woodlands of the dip slope to the north and west of Cirencester, and along the scarp from the south of Cheltenham northeast to Broadway. The data show an 80% increase in the number of occupied tetrads (some of this will be due to the greater survey effort) and the change map suggests a spreading out from the previous core area. The availability of suitable breeding sites has probably not changed significantly, and so any expansion in range would involve colonization of previously available, but unoccupied, sites.

Species sponsored by Sarah Rouche

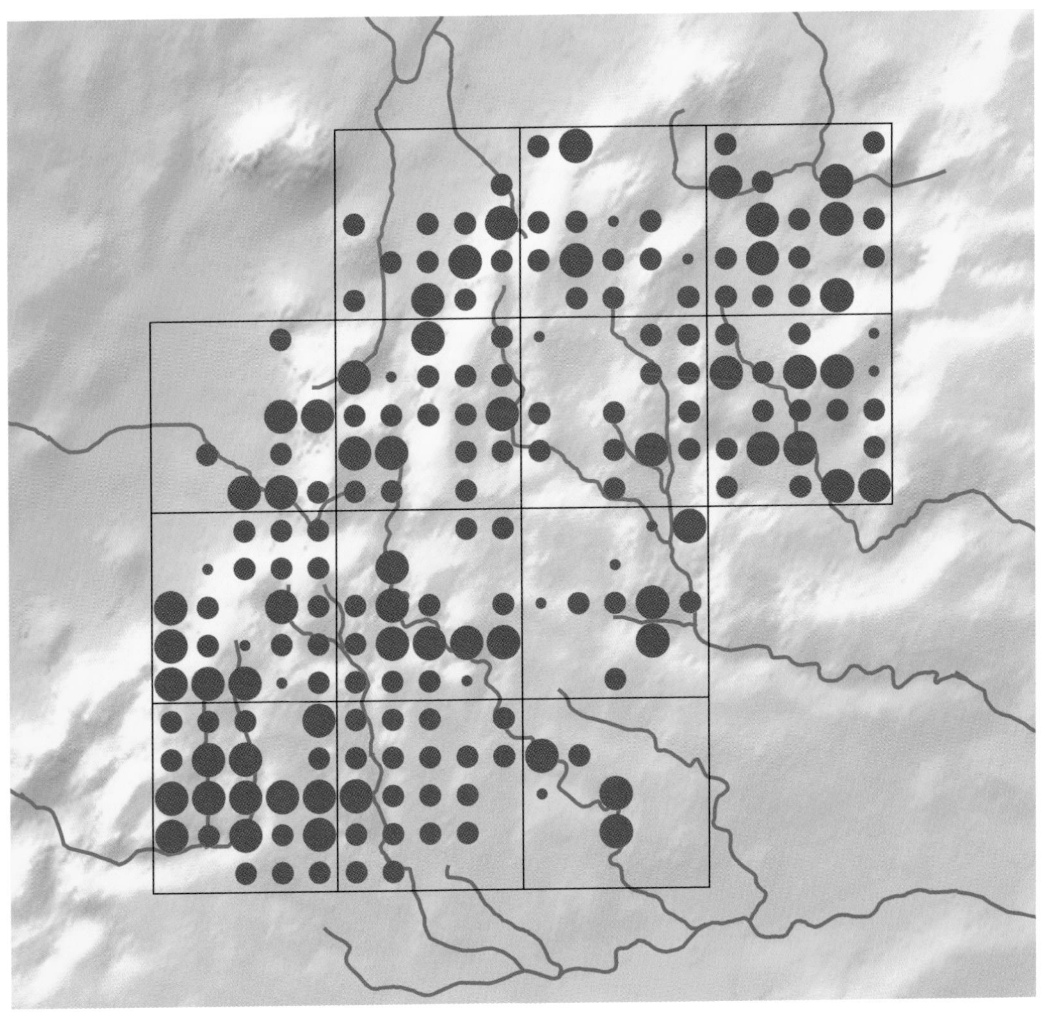

2003–07 survey

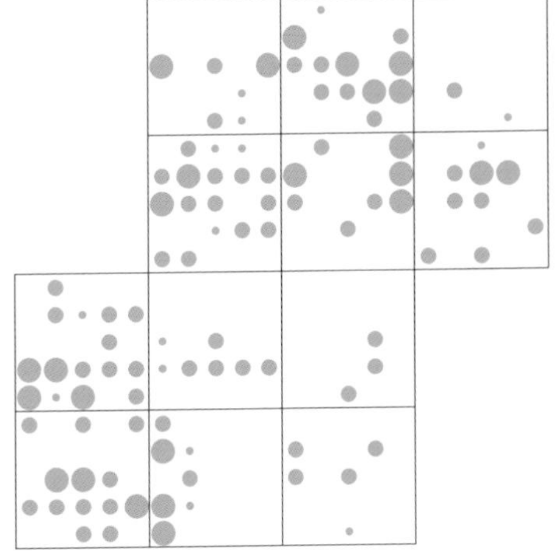

1983–87 survey

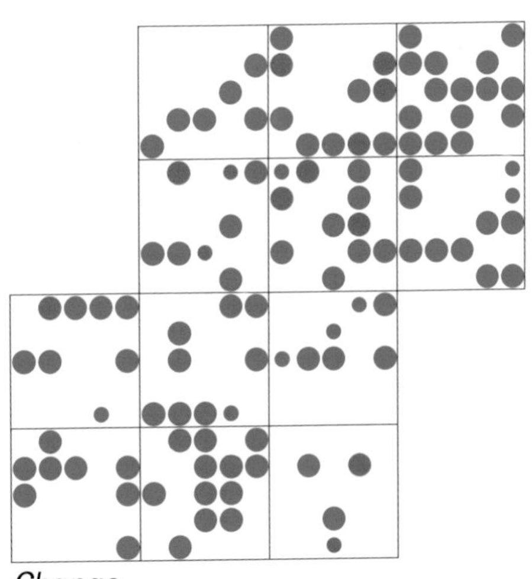

Change

Treecreeper (Eurasian Treecreeper)

Certhia familiaris

GRAHAM WATSON

Resident breeder			
UK conservation status: GREEN			
Long-term UK trend (1970–2006): 0%			
Recent population trend (1994–2007): +14%			
Numbers of tetrads with breeding records:			
	2003–07		1983–87
	13 squares	12 squares	12 squares
Possible	**49**	44	44
Probable	**103**	99	48
Confirmed	**47**	45	19
Totals	**199**	**188**	**111**

Although Treecreepers are rather inconspicuous birds, they sing from early spring through to early May in a range of woodland habitats, and territorial birds were frequently located from their song (though this is quite quiet and can easily be missed). Although occurring in a similar number of tetrads to Nuthatches, the habitat requirements are subtly different, and Treecreepers are often found in copses of smaller trees and bushes, and also overgrown hedgerows. Family parties with distinctly speckled young were frequently seen and heard, and this meant that about a quarter of the records were of confirmed breeding.

BTO survey data indicate that the population of Treecreepers across the UK has been remarkably constant over the years between our two surveys. The general feeling of observers is that this has also been true in the Cotswolds over the same time. The species is therefore a good indicator of the effect of the extra survey effort contributing to this Atlas (see Chapter 9 for details).

The Treecreeper was found in two-thirds more tetrads in 2003–07 than in 1983–87 and there were also more than twice as many confirmed breeding records. In the first Atlas it was suggested that the species was 'considerably under-recorded', especially in the large wooded areas of Cirencester Park and Chedworth, possibly due to the difficulty in finding them in such large woods.

The new survey confirms that Treecreepers are widely distributed, although they are unrecorded in most of the Vale farmland to the north and east of Cheltenham, and records are thinly distributed in the region of the Thames tributaries in the southeast of the recording area.

Species sponsored anonymously

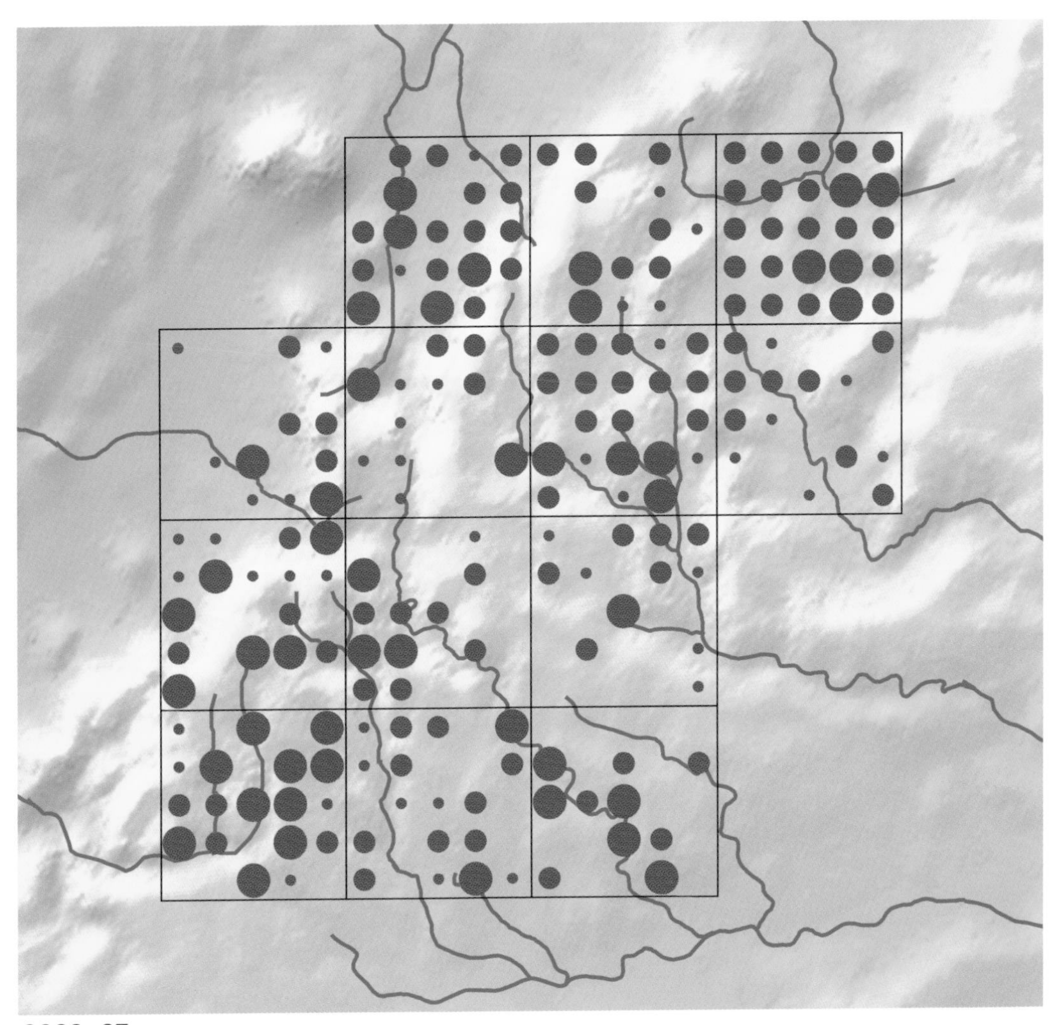

2003–07 survey

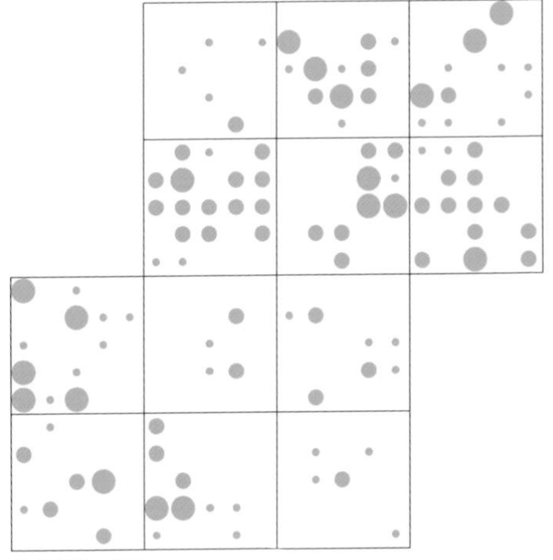

1983–87 survey

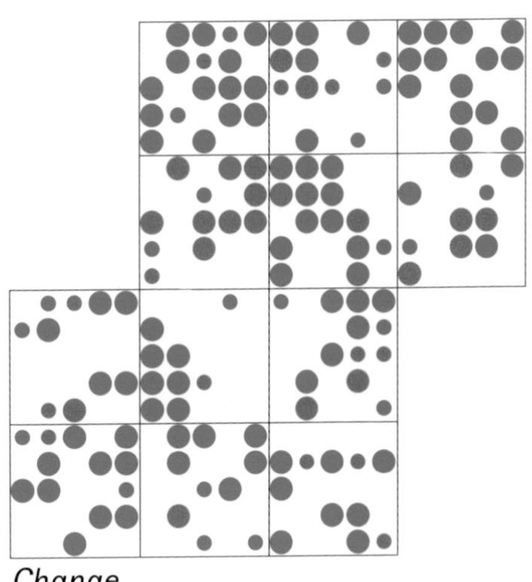

Change

Jay (Eurasian Jay)

Garrulus glandarius

Jays are noisy birds, and their rasping calls in early spring made their location fairly straightforward. However, they become much quieter in the breeding season and may be underrecorded as a result. Since they are predominantly woodland birds and quite secretive, it was difficult to obtain any specific indications of breeding intent, apart from noting pairs of birds in obvious partnership. This explains why 45% of records were only of possible breeding. They are a relatively sedentary species nonetheless and it is likely that the majority of sightings were of genuinely breeding birds. Their secretive breeding behaviour also explains the low number of confirmed breeding records, noisy family groups providing the majority of these.

Jays were found in a variety of woodland habitats, including conifer plantations. The distribution map reflects their preference for woodland, with the majority of confirmed breeding records coming from the scarp. However, they were also regularly seen in less typical habitat, such as amongst lines of trees along streams and field boundaries, and appear to be progressively moving into such areas. They were also found to be quite widespread in suburban gardens in Cheltenham, where the greatest concentration of confirmed breeding records was obtained. As Cheltenham was not included in the 1980s survey, it is not clear whether this is a recent development or whether Jays have been present here for some time, as they have in the suburbs of major cities such as London and Bristol.

The change map shows the overall pattern of distribution to be very similar in the two surveys, the slightly higher number of records in 2003–07 being attributable to greater survey effort.

Resident breeder

UK conservation status: GREEN

Long-term UK trend (1970–2006): -10%

Recent population trend (1994–2007): -7%

Numbers of tetrads with breeding records:

	2003–07		1983–87
	13 squares	12 squares	12 squares
Possible	**91**	85	70
Probable	**84**	77	68
Confirmed	**27**	21	16
Totals	**202**	**183**	**154**

GRAHAM WATSON

Species sponsored by Gill Pearce

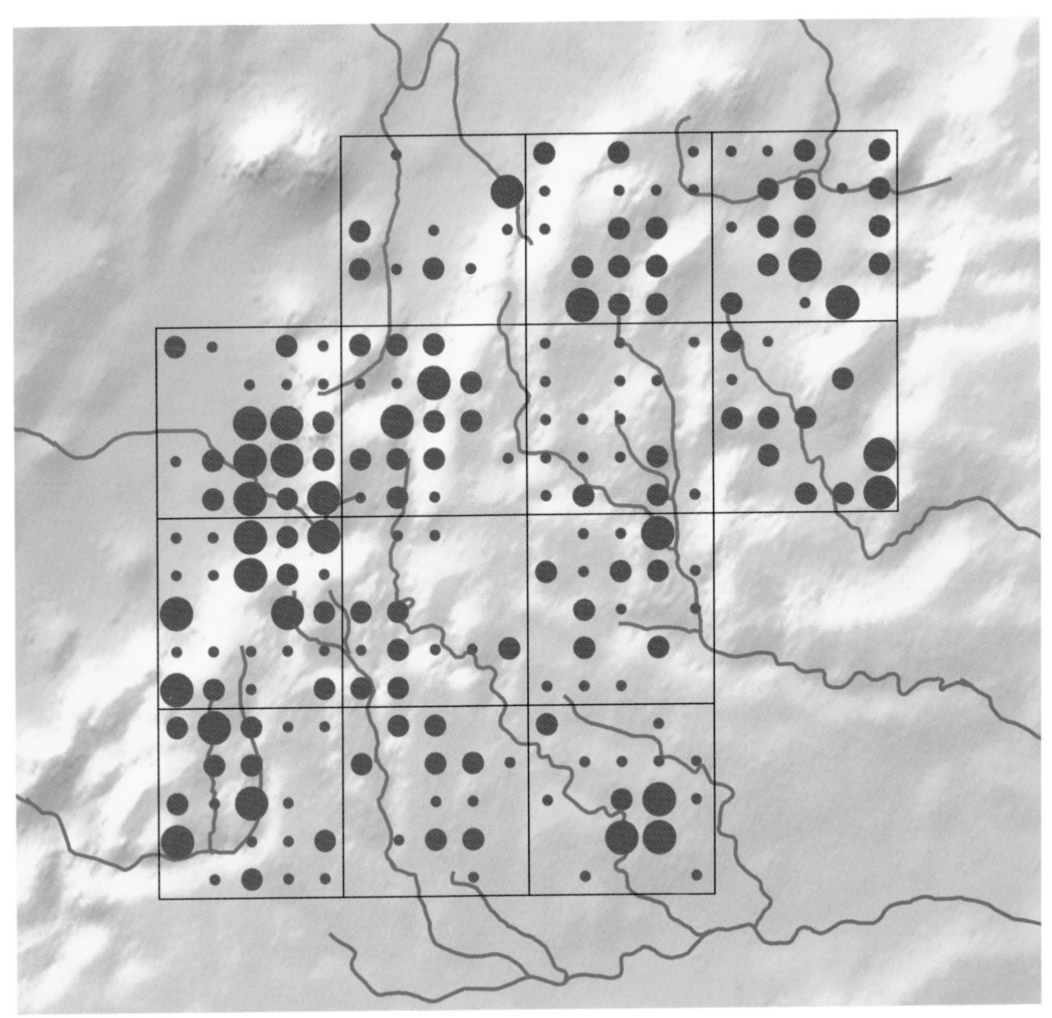
2003–07 survey

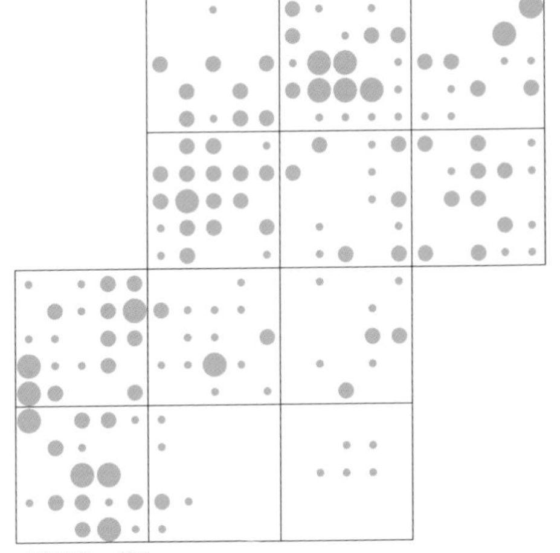
1983–87 survey

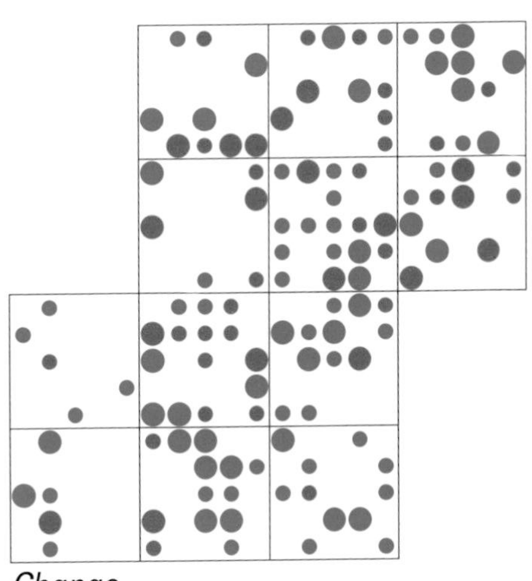
Change

Magpie (Black-billed Magpie)

Pica pica

Although Magpies do not advertise their breeding intentions with any recognizable song, they are very noisy in early spring. The presence of pairs of birds in the vicinity of their very distinctive nests, before leaves were fully developed on trees and bushes, made breeding locations easy to pinpoint. Confirmation of breeding was usually obtained through observation of adults entering a (by then hidden) nest site, or from seeing begging and noisy juveniles in family parties.

Magpies enjoy a wide distribution but are noticeably more common on the edges of towns and villages than in the countryside, where shooting estates may exercise efficient control of their numbers. Over 90% of tetrads were occupied in this survey, which was virtually the same as in 1983–87, and there was no noticeable change in the overall pattern of distribution. However, the number of confirmed breeding records was significantly smaller in the new survey, despite the greater survey effort. This may hint at a thinning out of the population—it was unusual to find more than two or three pairs in a tetrad in rural areas. Another possible explanation for the decline in confirmed breeding records may be the change in categorization of nest building from 'confirmed' in the first survey to 'probable' in this survey (see page 214). The first Atlas stated, 'Confirmation of breeding is not difficult, as the early nest building is easily observed', and it can be assumed that a reasonable proportion of confirmed breeding records in the first survey were derived on the basis of this evidence.

Resident breeder

UK conservation status: **GREEN**

Long-term UK trend (1970–2006): +96%

Recent population trend (1994–2007): +0%

Numbers of tetrads with breeding records:

	2003–07		1983–87
	13 squares	12 squares	12 squares
Possible	**39**	37	55
Probable	**146**	139	85
Confirmed	**116**	100	137
Totals	**301**	**276**	**277**

ROB BROOKES

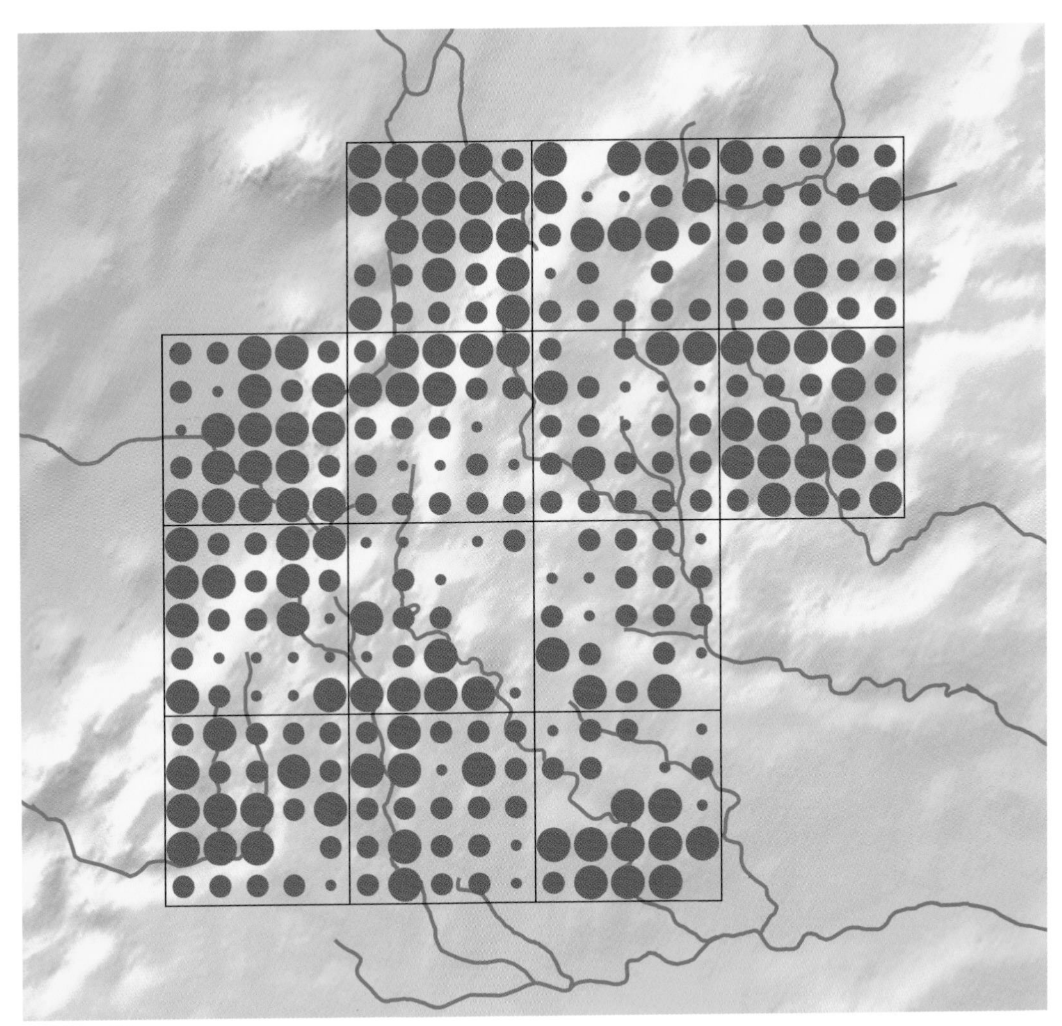

2003–07 survey

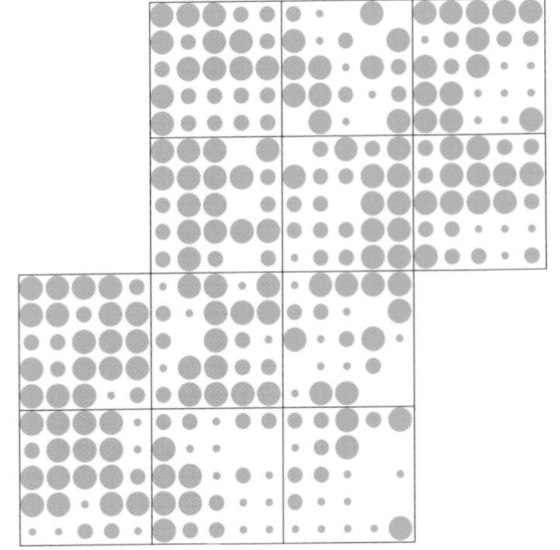

1983–87 survey

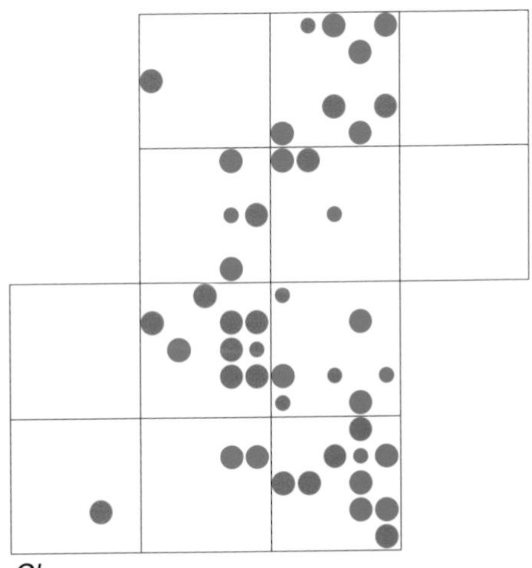

Change

Jackdaw (Eurasian Jackdaw)

Corvus monedula

ROB BROOKES

Resident breeder

UK conservation status: GREEN

Long-term UK trend (1970-2006): +107%

Recent population trend (1994-2007): +40%

Numbers of tetrads with breeding records:

	2003–07		1983–87
	13 squares	12 squares	12 squares
Possible	**13**	10	40
Probable	**70**	58	91
Confirmed	**240**	230	143
Totals	**323**	**298**	**274**

Jackdaws proved to be a very easy species to survey, as their gregarious and noisy habits at breeding sites made them easy to detect. They were found equally commonly in holes in mature trees, and in man-made holes–typically in chimneys, but also in church towers and in both occupied and abandoned farm buildings. They are also known to nest on rock faces. Although Jackdaws do not have a specific breeding call, they could regularly be seen carrying nesting material to sites early in the breeding season (from April onwards) and entering these sites throughout the spring and early summer, making confirmation of breeding easy. Young nestlings could also be heard quite often as the parents returned with food to regurgitate.

Often seen in fairly large groups, Jackdaws were found throughout the region in isolated or scattered trees, farms, villages and towns. Only two tetrads in the survey area had no records of Jackdaws, with breeding confirmed in about 75% of the tetrads. The first Atlas compared the distribution of Jackdaws with that of Rooks and suggested that 'their colonies are usually smaller and more difficult to find'. There was no evidence of this difficulty in the later survey (although it is true that rookeries are *very* easy to locate) and it may be the case that Jackdaws have become more widely distributed (and possibly more abundant) than they were in the 1980s. The change map supports this, suggesting that the Jackdaw has become ubiquitous since 1983–87 by infilling the 25 tetrads which were apparently unoccupied then. However, these may simply have been occupied tetrads that were missed during the first survey.

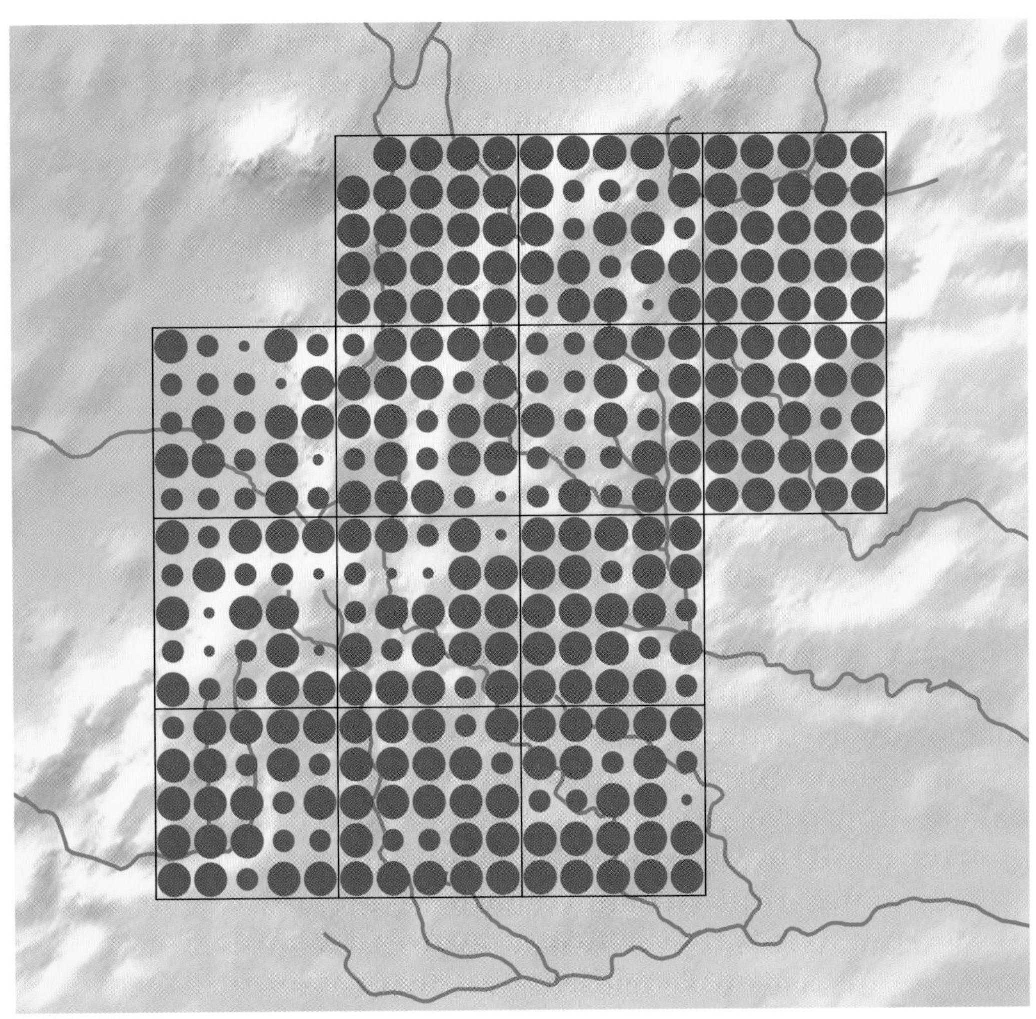

2003–07 survey

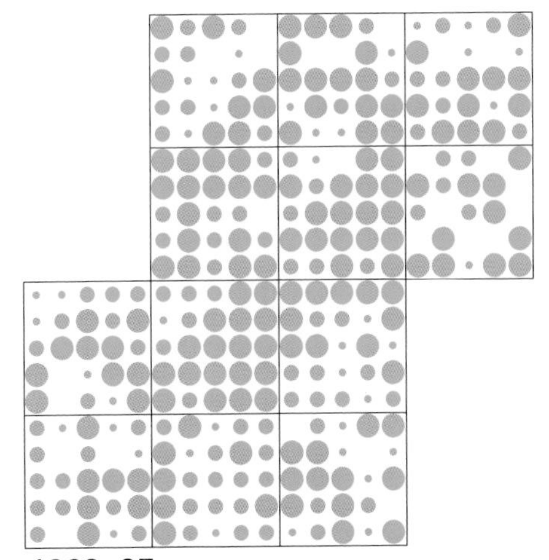

1983–87 survey

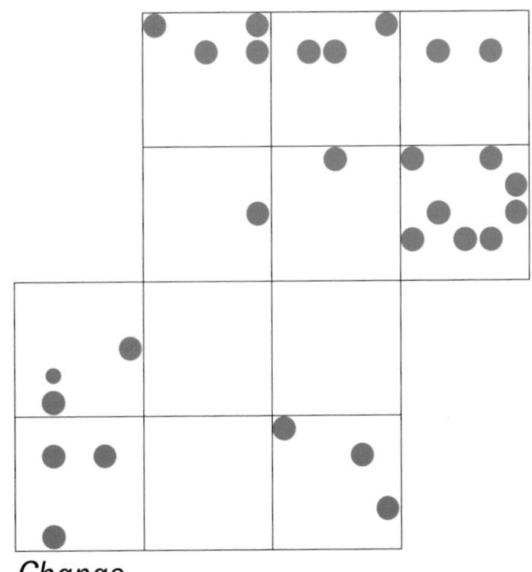

Change

Rook

Corvus frugilegus

Rooks were one of the easiest species to survey. Being both early and generally colonial nesters, and thanks to their very noisy behaviour, it was difficult to overlook their breeding activities. Many rookeries are located in villages and towns, making it a familiar bird to most people. The probable breeding records are the result of finding rookeries late in the season.

Normally built in the upper branches of tall trees, rookeries are visible over considerable distances, making them easy to locate in the winter months. Nests are also usually occupied before leaves begin to appear and so confirmation of breeding was also very easy—about 85% of records were of confirmed breeding, a similar figure to that achieved in the first survey.

The number of occupied tetrads was very similar in the two surveys and it is unlikely that there has been any significant numerical change in the population. However, the change map does suggest that the distribution may not be the same: sites have been lost from the dip slope in the south and west of the region, and gained in the High Wold and Evenlode Valley in the north and east.

Resident breeder

UK conservation status: GREEN

Long-term UK trend: Not available

Recent population trend (1994–2007): -3%

Numbers of tetrads with breeding records:

	2003–07		1983–87
	13 squares	12 squares	12 squares
Possible	**26**	25	21
Probable	**16**	15	9
Confirmed	**230**	215	206
Totals	272	255	236

One further interesting observation was of a number of individual, isolated Rooks' nests, sometimes a little distance from an active rookery and sometimes in apparently total isolation. It would be interesting to follow up some of these sites to see if they become new rookeries and how often an entire population moves from an existing site to a new one (a phenomenon that was noted in one tetrad in the Avon Vale and another near Cheltenham during the survey period).

ROB BROOKES

Species sponsored by Richard and Pat Gunn

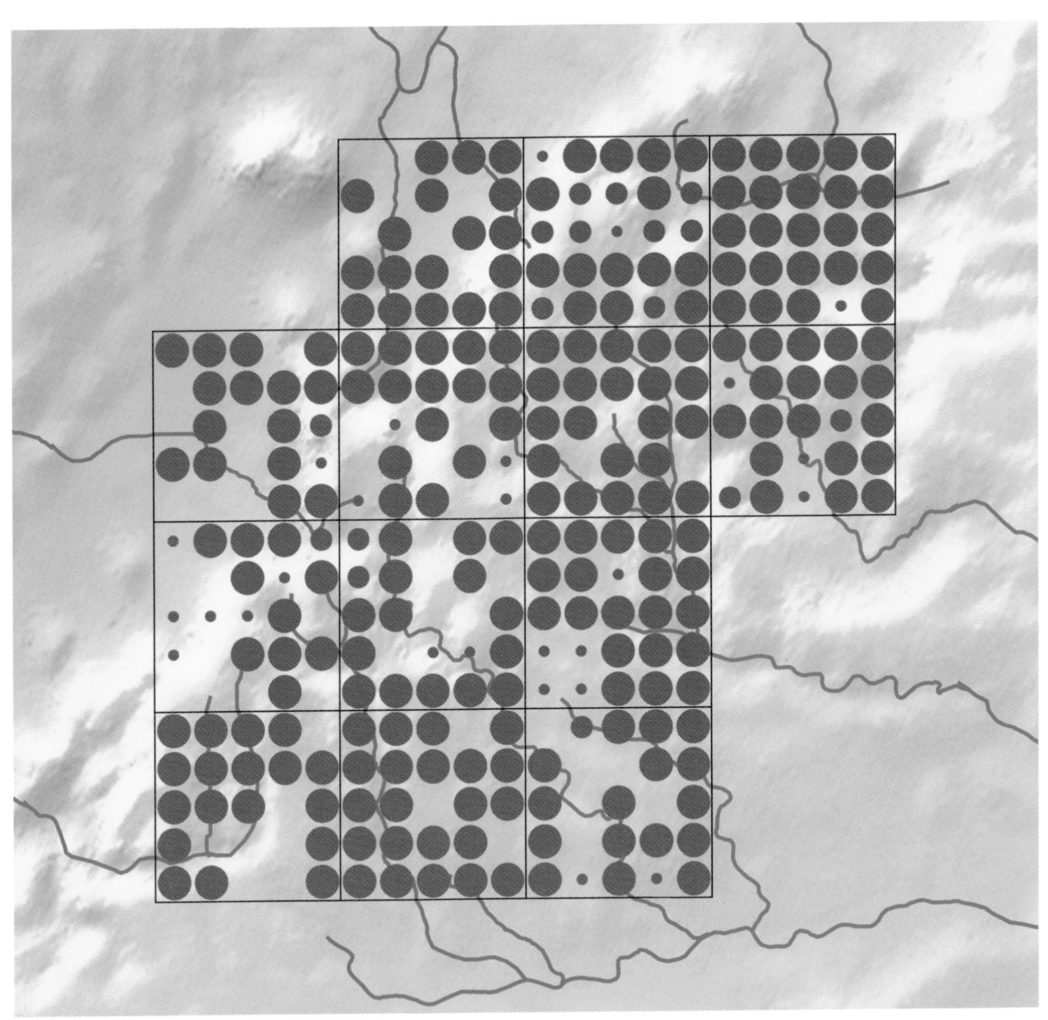

2003–07 survey

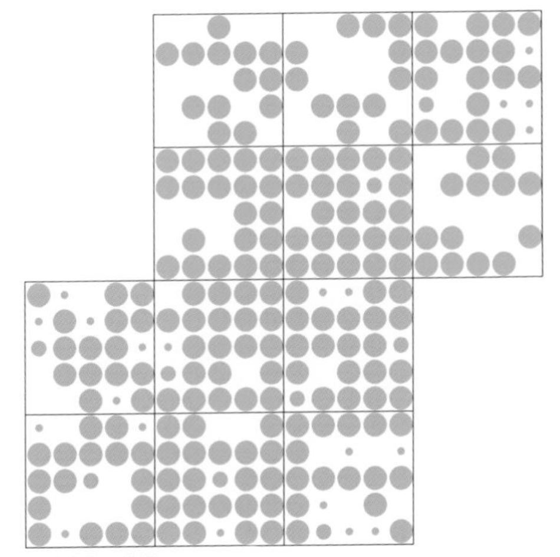

1983–87 survey

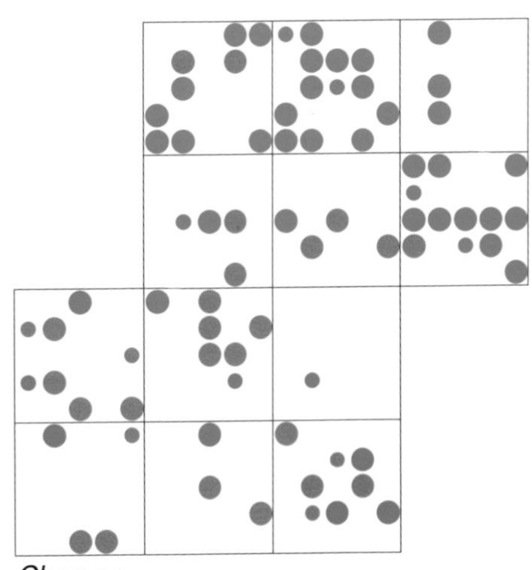

Change

Carrion Crow

Corvus corone

ROB BROOKES

Resident breeder

UK conservation status: GREEN

Long-term UK trend (1970–2006): +83%

Recent population trend (1994–2007): +19%

Numbers of tetrads with breeding records:

	2003–07		1983–87
	13 squares	12 squares	12 squares
Possible	**16**	15	38
Probable	**86**	83	53
Confirmed	**217**	196	200
Totals	**319**	**294**	**291**

Carrion Crows are not colonial breeders, nesting in tall isolated and hedgerow trees rather than in major tracts of woodland. They begin building early in the season before the trees are in leaf, thus making it easy both to locate the nests, and to observe sitting adults. Later the noisy fledglings could be seen or heard begging for food from the canopy so confirmation of breeding was straightforward.

As indicated by the earlier survey, Carrion Crows are probably the most widely and evenly distributed corvid in the region, being found in both rural and urban areas. However, like Magpies, they are well controlled in some keepered estates. The number of occupied tetrads and the number of confirmed breeding records is very similar for the two surveys. As with the other corvids, the fact that nest building was not considered to be an indication of confirmed breeding in the current survey (unlike in the first Atlas) may account for the lower number of confirmed breeding records. However, the increased effort in the later survey will have compensated for this and is probably also the factor behind the apparent filling-in of tetrads in the Thames tributaries region in the southeast of the survey area.

Species sponsored by Pat and Brian Follett

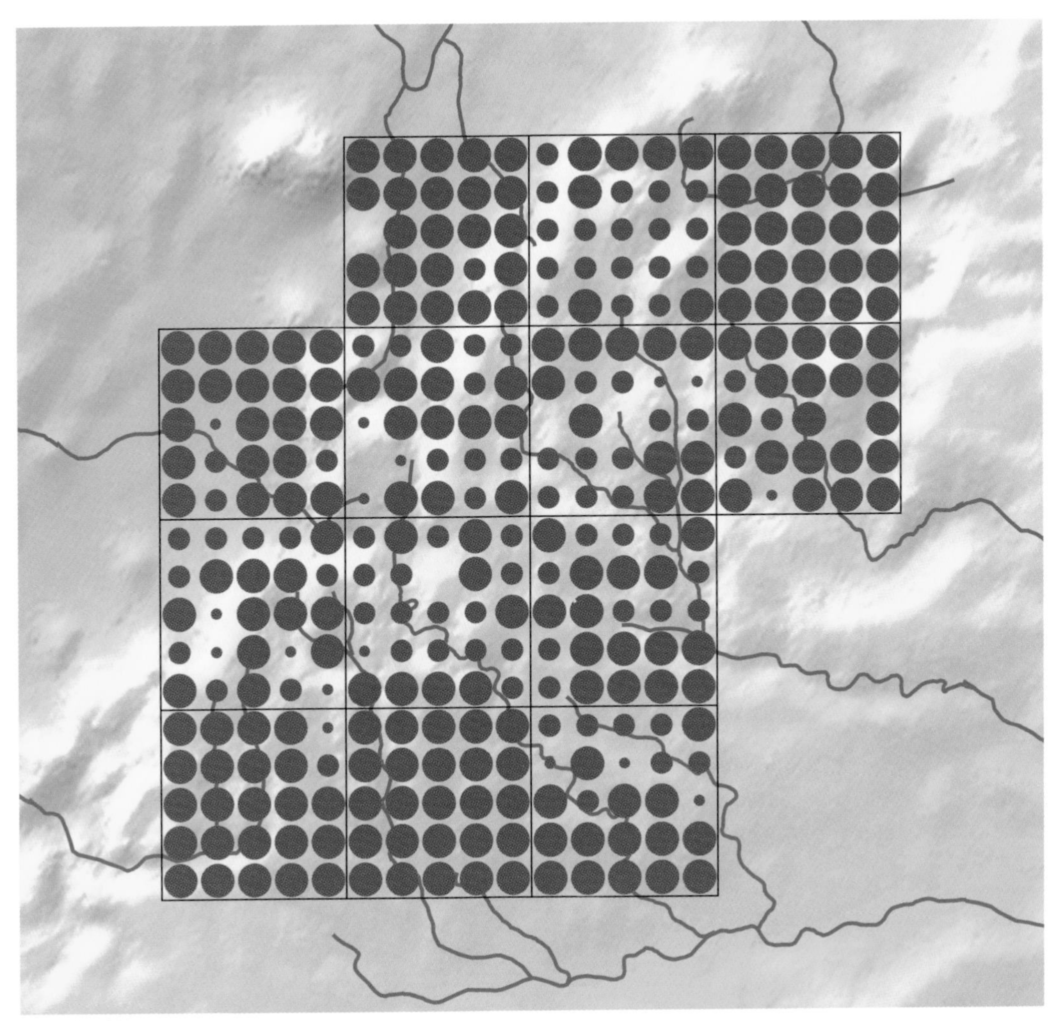

2003–07 survey

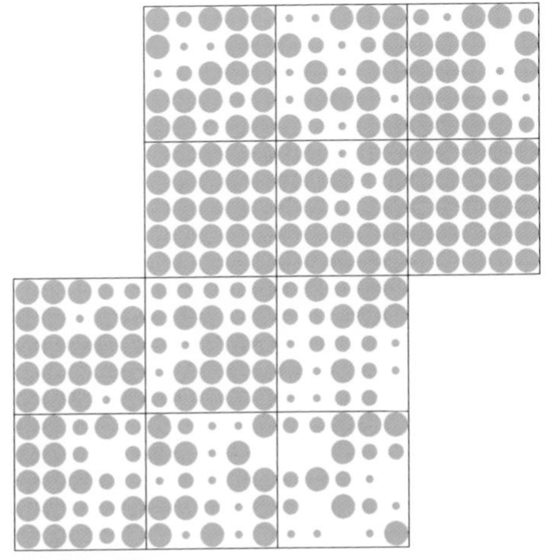

1983–87 survey

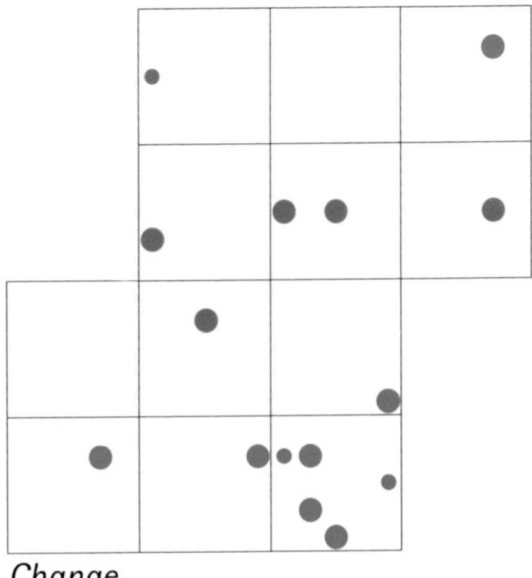

Change

Raven (Common Raven)

Corvus corax

Accurate surveying of Ravens is somewhat difficult. Although they are instantly recognizable from their call, and breeding activity can be gleaned from their aerial antics, such displays are not necessarily near a nest site. Often travelling great distances in pairs, it required a good deal of care to confirm in which tetrad they were actually breeding. In their traditional breeding areas in the west of Britain and Ireland, Ravens nest on mountain crags and seacliffs. There are few of these in the Cotswolds, where Ravens mainly nest in trees (often Scots Pine and Cedar), particularly where clumps of trees occur in elevated positions. Nests were sometimes difficult to pinpoint, sitting birds habitually slipping away unseen. They are increasingly using electricity pylons with a successful nest found in 2004 and several more similar nests located in the last year of the survey (2007).

Regular sightings were at first mostly along the scarp and the dip slope immediately behind it. Although the Raven is now also often seen to the eastern side of the recording area, the majority of the confirmed breeding sites still follow the general line of the scarp. The first recorded nest in the area was in a Scots Pine 3 km east of Cheltenham, and at the time (1998) was the most easterly recorded breeding Raven in Gloucestershire. Now breeding has been confirmed in the east of the Atlas recording area. In most cases there was not more than one pair breeding in any one tetrad, and certainly these tetrads were not all occupied in every year of the survey. Therefore, the actual number of pairs breeding may well be less than suggested by the raw data. However, in line with the picture in Gloucestershire generally, the number of confirmed breeding sites appears to be increasing annually.

The Raven was not recorded during the first Atlas survey during the 1980s, and was an uncommon sight anywhere in the Cotswolds until the late 1990s. This makes the confirmation of breeding in 24 tetrads and the observation of breeding activity in a further 34 all the more remarkable, and it will be interesting to see whether the Raven's distribution continues to expand.

Resident breeder

UK conservation status: **GREEN**

Long-term UK trend: Not available

Recent population trend (1994–2007): +134%

Numbers of tetrads with breeding records:

	2003–07		1983–87
	13 squares	12 squares	12 squares
Possible	**11**	10	
Probable	**34**	32	
Confirmed	**24**	21	
Totals	**69**	**63**	

ROB BROOKES

Species sponsored by Andrew Cleaver

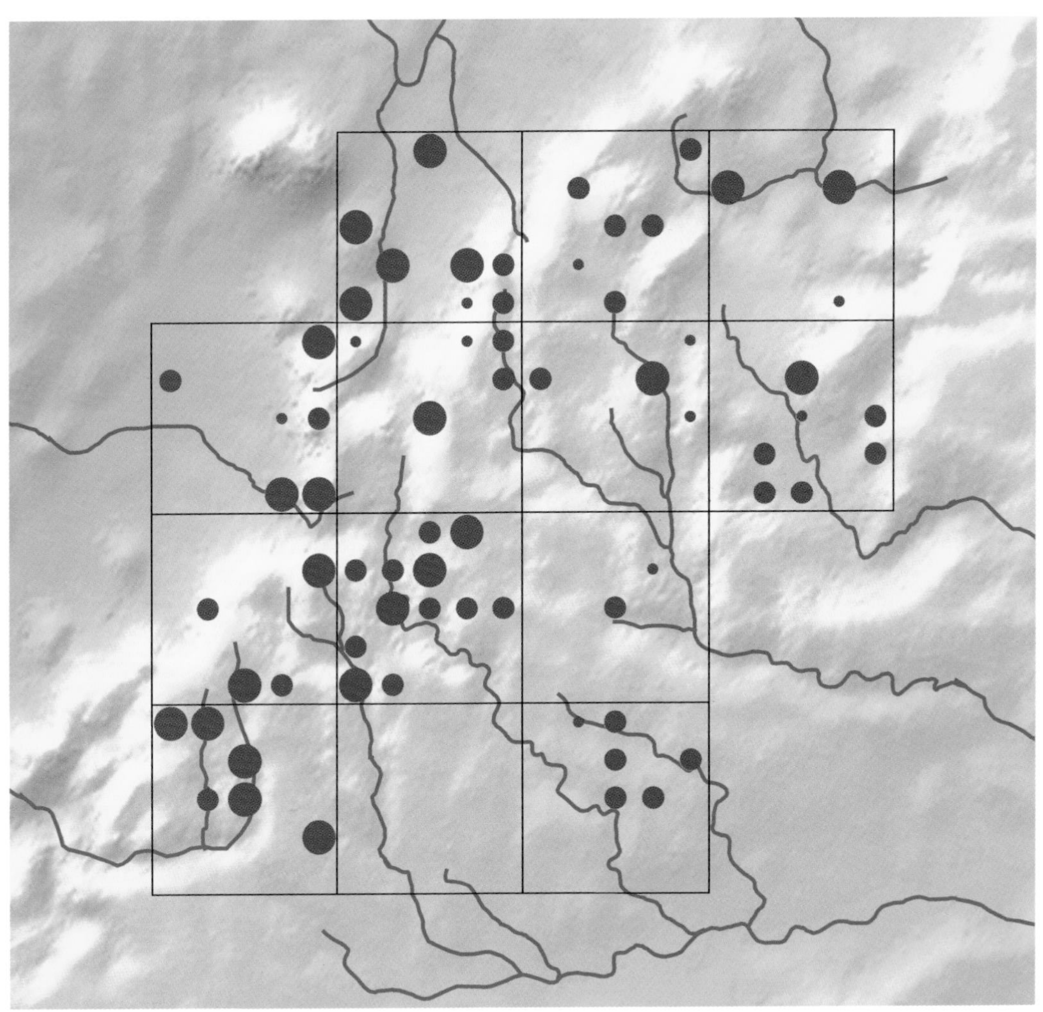

2003–07 survey

Starling (Common Starling)

Sturnus vulgaris

In suburban areas, where they nest in the eaves of houses and similar man-made cavities, Starlings are a very familiar sight, and can be seen singing and wing posturing from early spring. Once a Starling territory has been identified, confirmation of breeding is easy, either by seeing adults openly carrying food from feeding areas to nest sites, by hearing young birds in the nest, or by seeing them on garden lawns soon after fledging. About 80% of occupied tetrads had breeding confirmed.

In the 1980s, breeding Starlings were found throughout most (90%) of the survey area, both in towns and villages, and in woodland and isolated trees, where they use natural holes. Despite the greater survey effort the number of breeding records is down by 11% compared to the first Atlas and the change map shows that they have been largely lost from woodland, rather than urban or suburban areas (see Figures 6 and 7, page 12). They are still numerous in those parts of towns with suitable feeding patches of grassland nearby, but less so in many small villages. The pattern of the new distribution map bears a very striking similarity to that for the House Sparrow (page 185), a species that is traditionally associated with human settlements rather than woodland.

As woodland-breeding Starlings are known to have evicted Great Spotted Woodpeckers from their newly constructed nest holes, the reduced Starling presence in woodland may have had a beneficial effect on the population of Great Spotted Woodpeckers.

Resident breeder

UK conservation status: RED

Long-term UK trend (1970–2006): -73%

Recent population trend (1994–2007): -26%

Numbers of tetrads with breeding records:

	2003–07		1983–87
	13 squares	12 squares	12 squares
Possible	**23**	23	26
Probable	**30**	29	31
Confirmed	**211**	189	213
Totals	**264**	**241**	**270**

GRAHAM WATSON

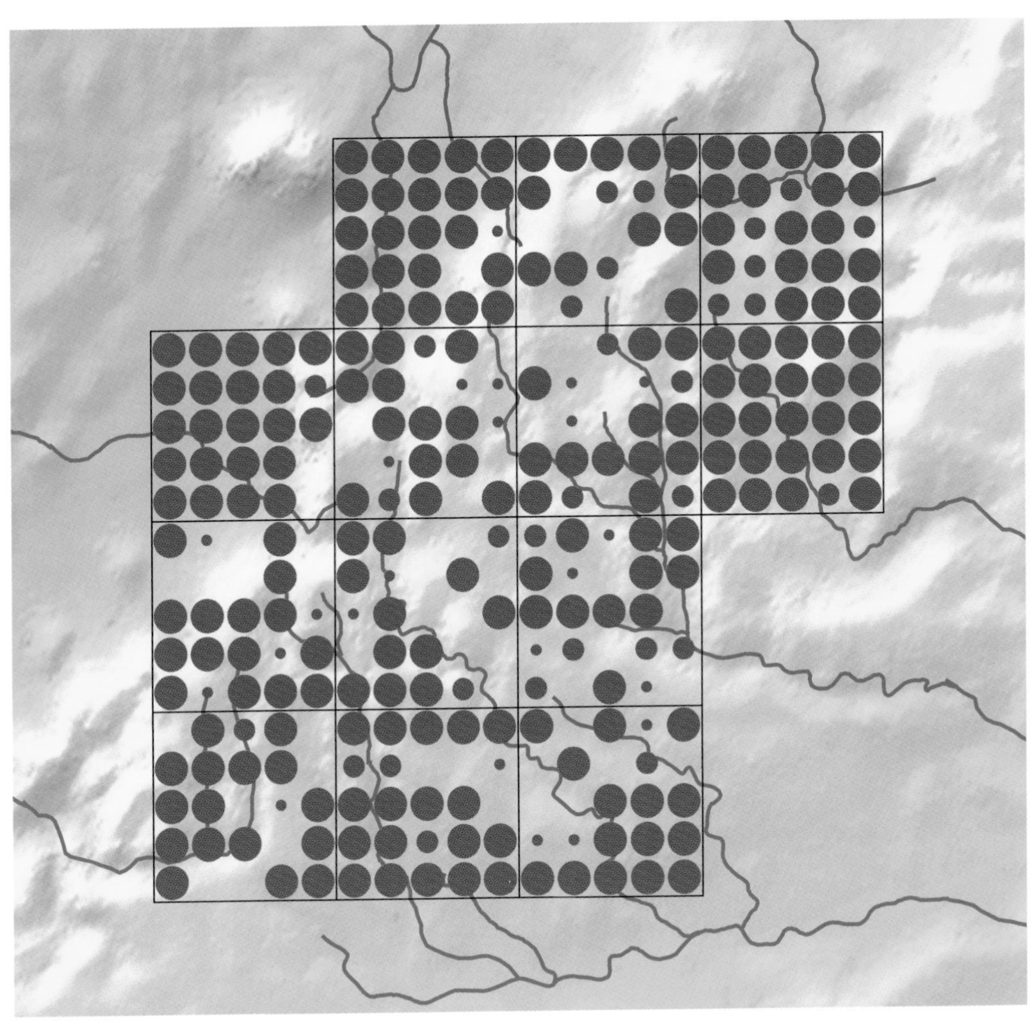
2003–07 survey

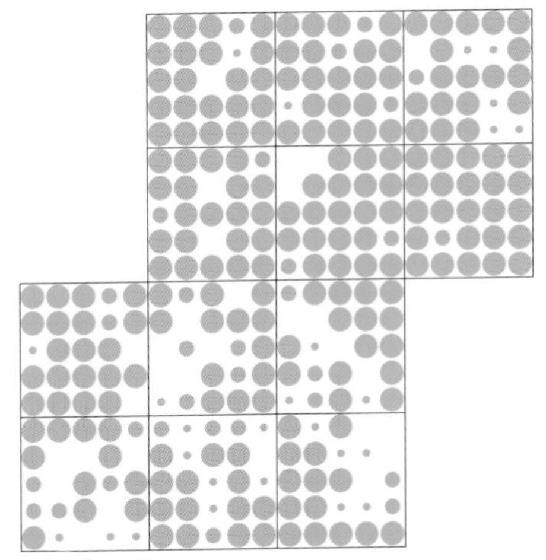
1983–87 survey

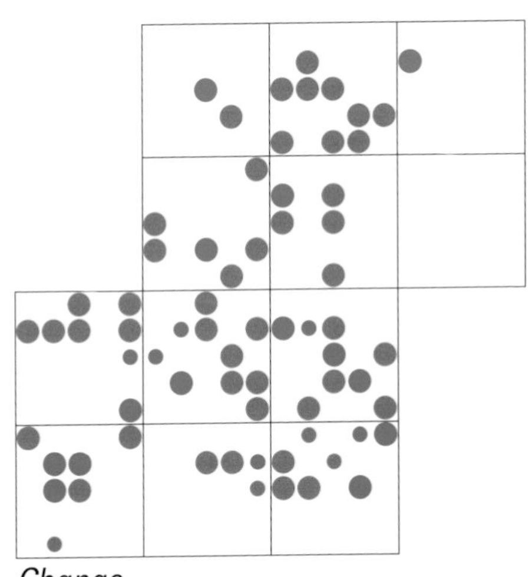
Change

House Sparrow

Passer domesticus

ROB BROOKES

Resident breeder

UK conservation status: RED

Long-term UK trend (1977–2006): -65%

Recent population trend (1994–2007): -10%

Numbers of tetrads with breeding records:

	2003–07		1983–87
	13 squares	12 squares	12 squares
Possible	**5**	5	11
Probable	**64**	58	81
Confirmed	**206**	189	147
Totals	**275**	**252**	**239**

House Sparrows are often found in small rural communities such as farms and isolated groups of houses. Being predominantly colonial in their breeding habits, they were easy to locate, and the data indicate that confirmation of breeding was also easy to obtain, by observation of adults collecting invertebrate food for young, or seeing young fledglings in the colony.

There has been a much-publicized decline in House Sparrow numbers and distribution nationally, but this is not reflected in our survey, where House Sparrow distribution does not seem to be significantly different from that reported in the previous Atlas. However, the feeling was that the species might have been on the increase during the years of the survey, after a decline throughout the 1990s. There are certainly a number of anecdotal reports of House Sparrows having disappeared from some traditional sites, and then having returned after a considerable absence (*eg* 10–20 years). It is also likely that the greater survey effort in this Atlas period masked a slight contraction in distribution (*ie* it may well have been more widely distributed in the 1980s than the original survey suggested).

The tetrads from which House Sparrows were absent are mostly in rural areas with no significant human habitation. Even in a largely rural area with small and thinly scattered human settlements, this species is still widely occurring and fairly numerous.

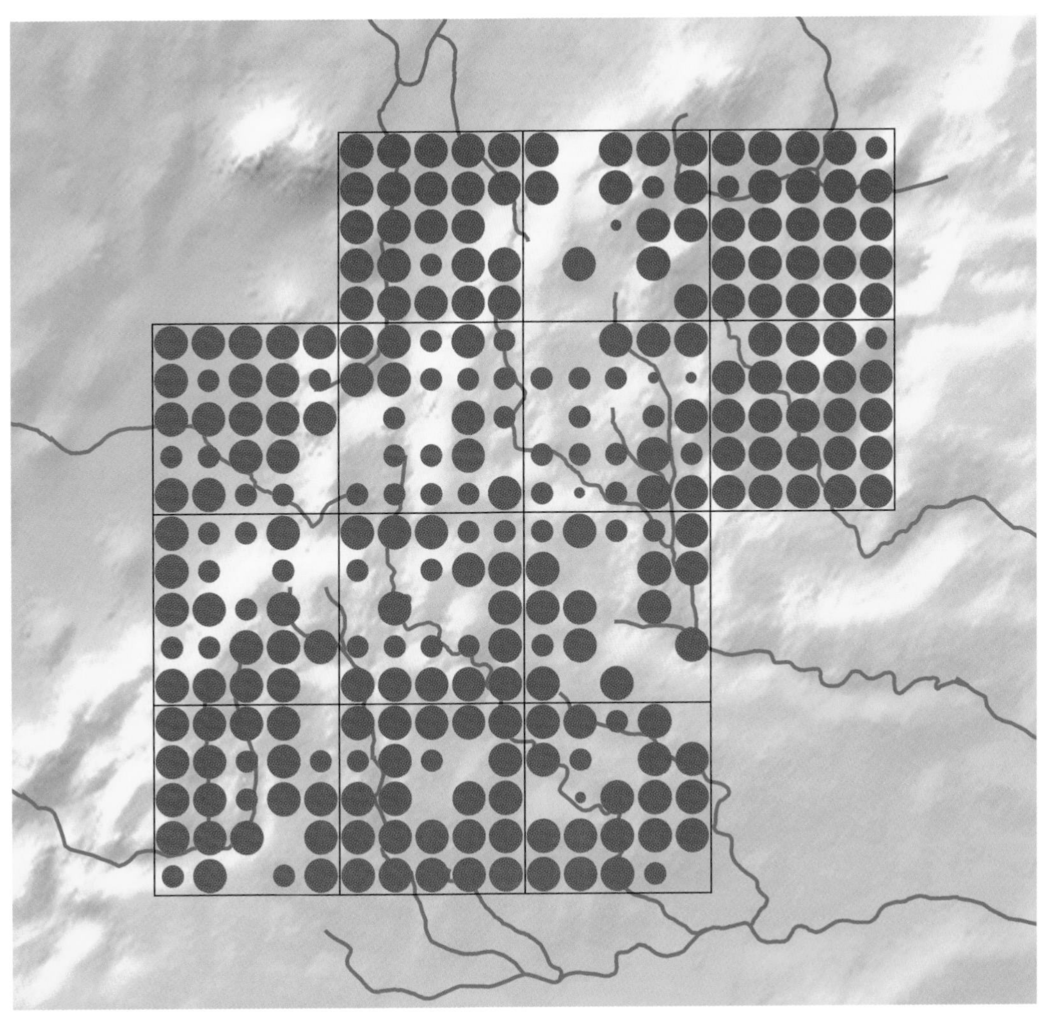
2003–07 survey

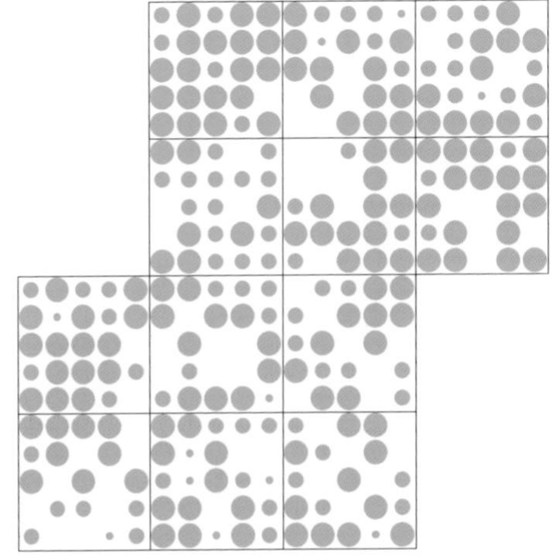
1983–87 survey

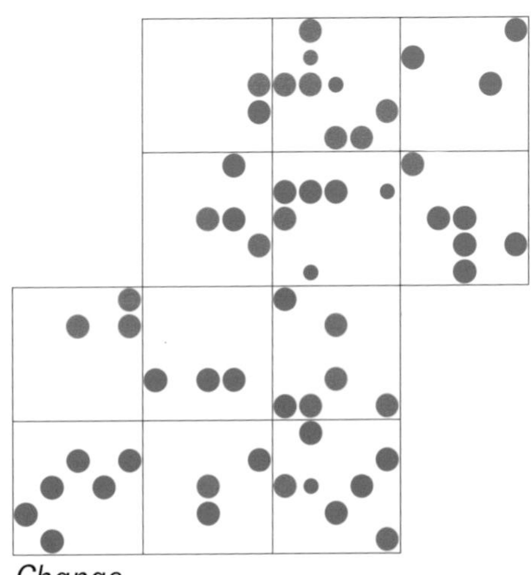
Change

Tree Sparrow (Eurasian Tree Sparrow)

Passer montanus

Tree Sparrows are prone to quite extreme swings in population numbers and distribution—declines of 90% and more, followed by a recovery in numbers, are known. Often associated with supplies of grain, either from farmyards or domestic poultry, they are colonial breeders, nesting in holes in trees and buildings, often in rural locations away from habitation. They rarely compete with House Sparrows for nest sites but will actually take readily to nest boxes. They are usually double or triple brooded, which presumably aids in the rapid recovery in numbers if conditions are favourable.

Resident breeder

Gloucestershire BAP species

UK conservation status: RED

Long-term UK trend (1970–2006): -93%

Recent population trend (1994–2007): +15%

Numbers of tetrads with breeding records:

	2003–07		1983–87
	13 squares	12 squares	12 squares
Possible	**10**	10	13
Probable	**8**	8	26
Confirmed	**25**	25	20
Totals	**43**	**43**	**59**

Records collected by the NCOS in the years between the two surveys have indicated a significant fall in both numbers and range of Tree Sparrows, in line with that seen nationally, though this may be beginning to be reversed. Comparison of the findings from this survey with that of the first Atlas hint at (but do not fully demonstrate) this decline, with a 27% reduction in the number of occupied tetrads. It is worth questioning why in a little more detail. In the first Atlas it was stated that Tree Sparrows 'are unobtrusive and shy, and although their call is distinctive it may not be familiar to all observers. For these reasons, Tree Sparrows are likely to be under-recorded.' It is possible that, in this survey, there was far greater familiarity with the bird and that it has not been underrecorded this time. Overall, there has been a significant contraction in its range. The distribution maps show a distinctly localized distribution, broadly following a line from the northeast of the area to the central south. It has undoubtedly withdrawn from the low-lying Vale areas in the west, as well as much of the Evenlode Valley, but some of its traditional upland strongholds are still occupied, and a new population has become either established or recently discovered in the Stour Valley.

RICHARD TYLER

Species sponsored by Duncan and Rebecca Dine

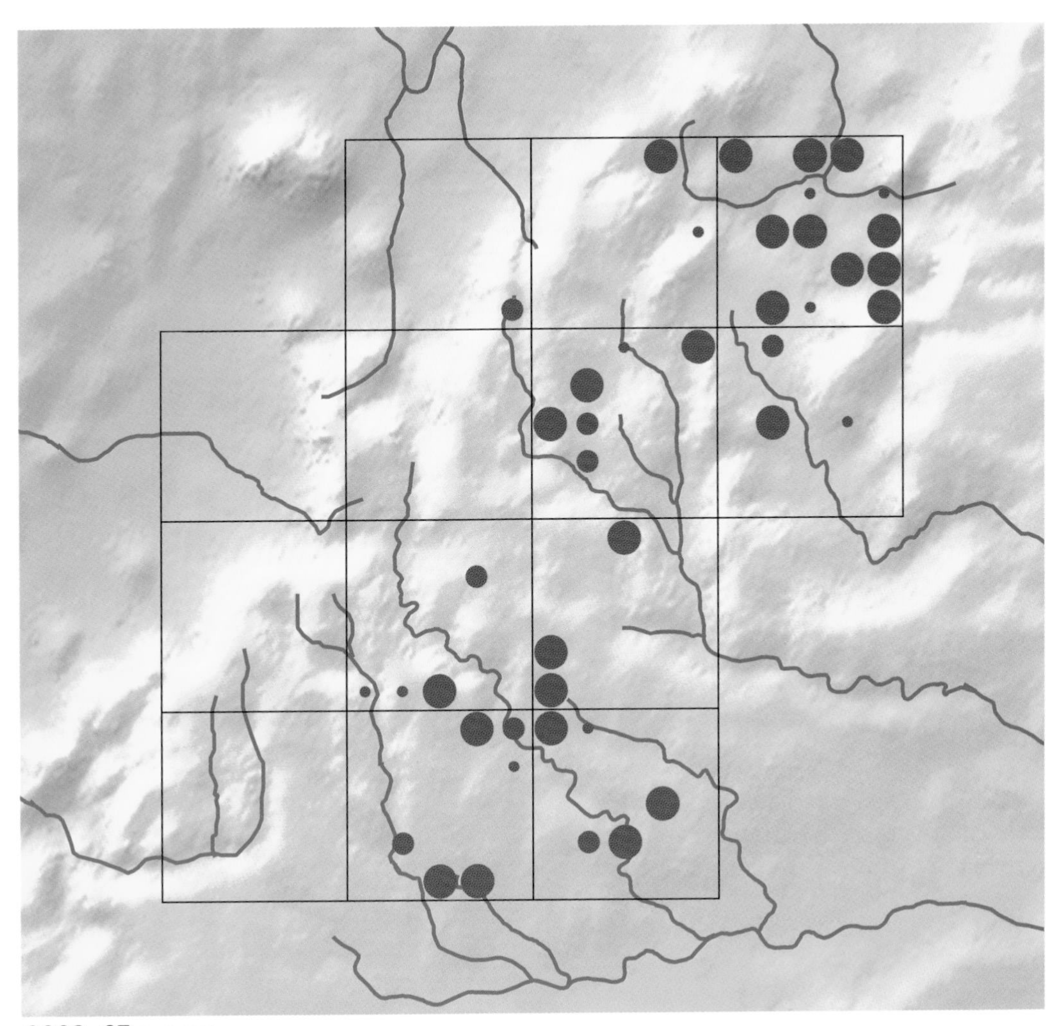

2003–07 survey

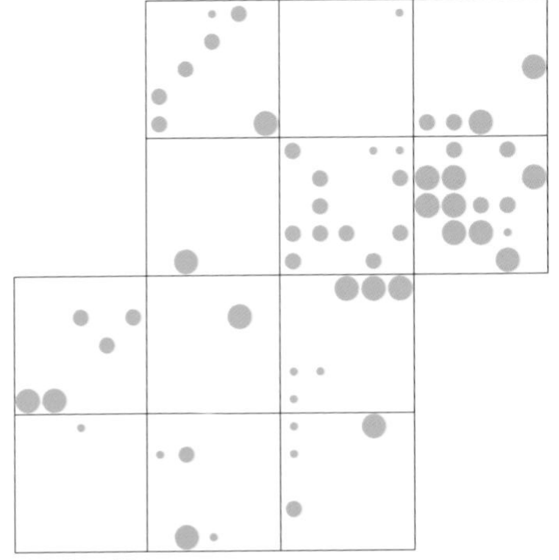

1983–87 survey

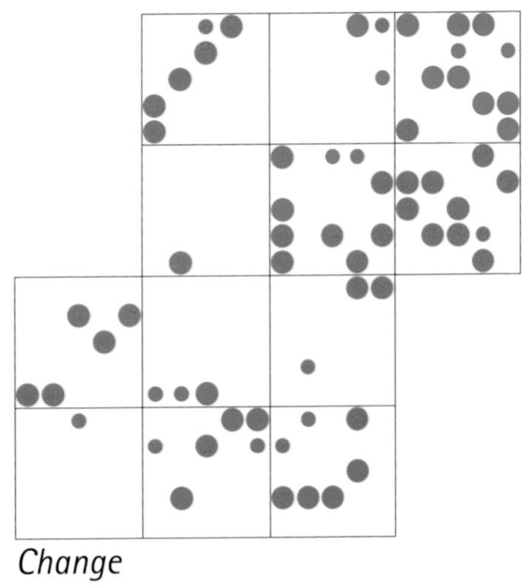

Change

Chaffinch

Fringilla coelebs

ROB BROOKES

Resident breeder

UK conservation status: GREEN

Long-term UK trend (1970–2006): +39%

Recent population trend (1994–2007): +14%

Numbers of tetrads with breeding records:

	2003–07		1983–87
	13 squares	12 squares	12 squares
Possible	**1**	1	2
Probable	**122**	114	185
Confirmed	**202**	185	110
Totals	**325**	**300**	**297**

Chaffinches are generally single-brooded, but they have a fairly prolonged breeding season, with singing starting in February, and egg-laying occurring from late April to mid-June. The familiarity of their song and the ubiquitous nature of the bird made it one of the easiest species to survey. It is also the only finch of the region that visibly carries food to its young, making confirmation of breeding straightforward. In contrast, the nests are usually quite well concealed, often in a thick hedgerow, and confirming breeding by finding nests was less easy.

Among the most widespread and numerous birds in the Cotswolds, all Chaffinches require are a few trees as song posts and a hedge or bushes in which to nest. These requirements can be met in farmland, both broad-leaved and coniferous woodlands, villages and towns. Consequently they were found in virtually every tetrad in the 1983–87 survey and were ubiquitous in this survey. They almost certainly bred successfully in every tetrad in the survey area each year.

The higher proportion of confirmed breeding reports during this survey compared with that in the 1980s (over 60%, compared with 37%) probably reflects the greater survey effort rather than any increase in population—more emphasis having been put on confirming breeding this time. In the first Atlas, it was suggested that broad-leaved woodlands held the highest densities of Chaffinches in the survey area and there is no evidence of any change in this situation.

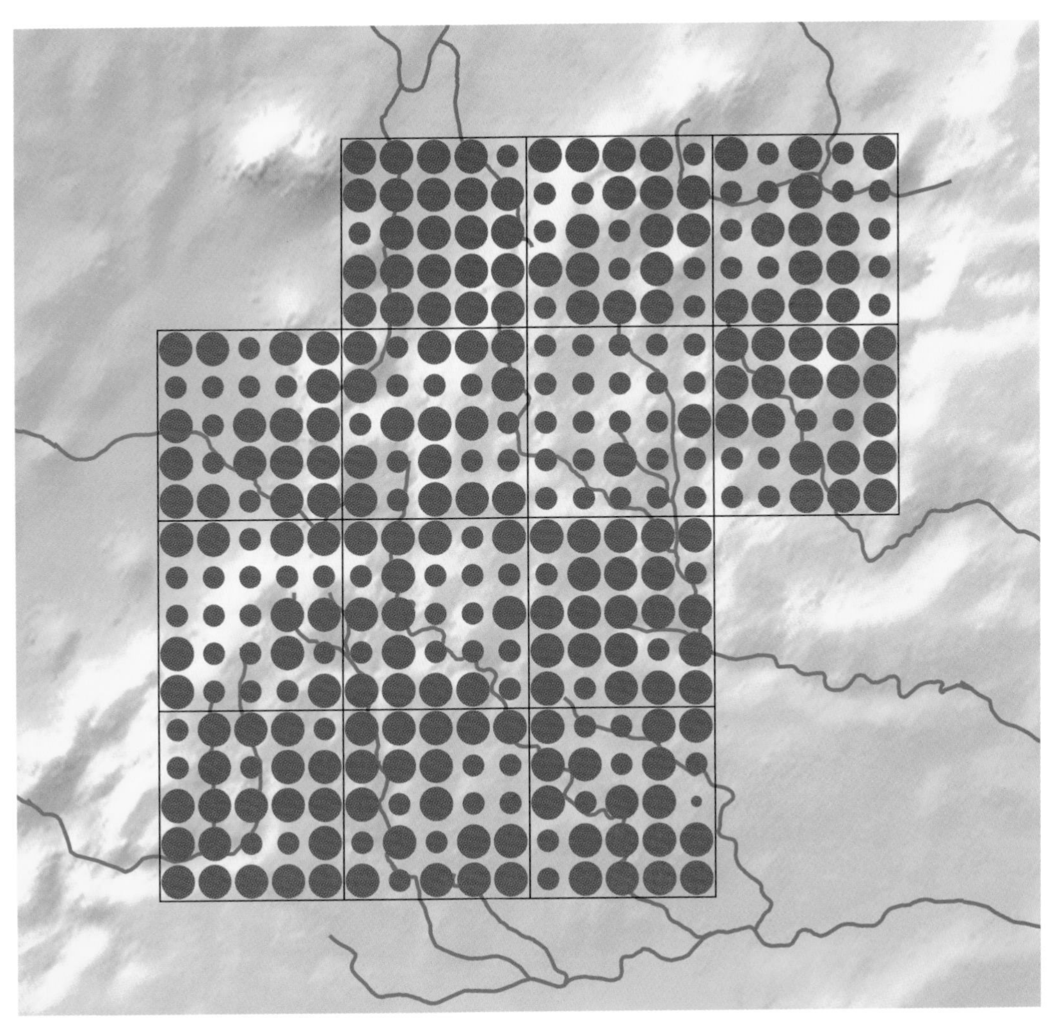

2003–07 survey

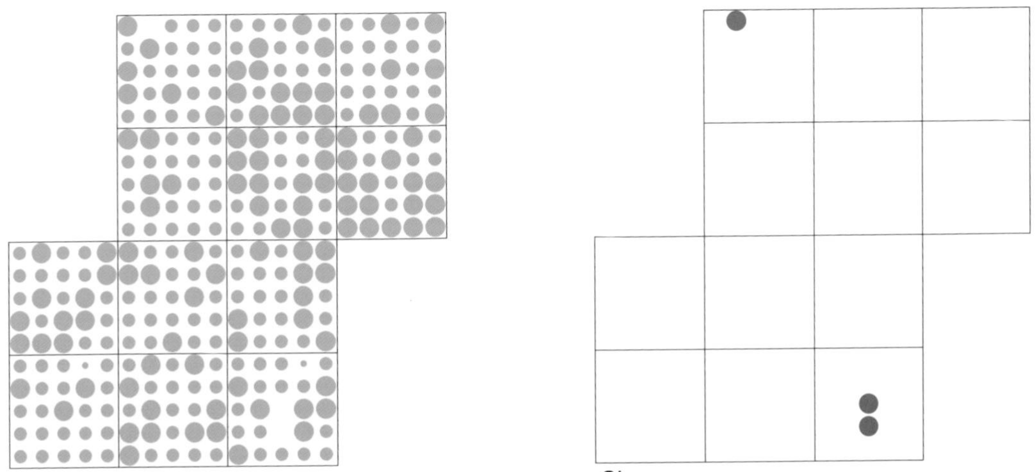

1983–87 survey

Change

Greenfinch (European Greenfinch)

Carduelis chloris

ROB BROOKES

Resident breeder

UK conservation status: GREEN

Long-term UK trend (1970–2006): +30%

Recent population trend (1994–2007): +27%

Numbers of tetrads with breeding records:

	2003–07		1983–87
	13 squares	12 squares	12 squares
Possible	**2**	2	28
Probable	**193**	182	191
Confirmed	**125**	112	37
Totals	**320**	**296**	**256**

Like Chaffinches, Greenfinches have an extended breeding season, from late April to mid-August in the survey area, and more than one brood is usual. The display flight and song of the Greenfinch are very recognizable and start early in the season; this, combined with the fact that the bird is loosely colonial, makes detection of breeding intent very easy. However, in common with most British finches, it feeds its young by regurgitation. Therefore, confirmation of breeding is slightly more of a problem—observation of noisy food-begging fledglings is the usual method.

Greenfinches, being found in all but five tetrads, are nearly as widely distributed as Chaffinches. This gives a slightly misleading picture, however, as Greenfinches are much more associated with human habitation, from towns and villages to farms and isolated houses. In some of the tetrads in which they were recorded it is possible that this refers to one specific site only.

In the 1980s survey, Greenfinches were unrecorded in 44 tetrads; in this survey, they were only absent from four tetrads in the original twelve 10 km squares, and the number of confirmed breeding records was trebled. Although greater survey effort might explain much of this apparent increase, a moderate expansion in distribution seems possible. There is no doubt that it continues to be a very successful colonizer of its preferred habitats.

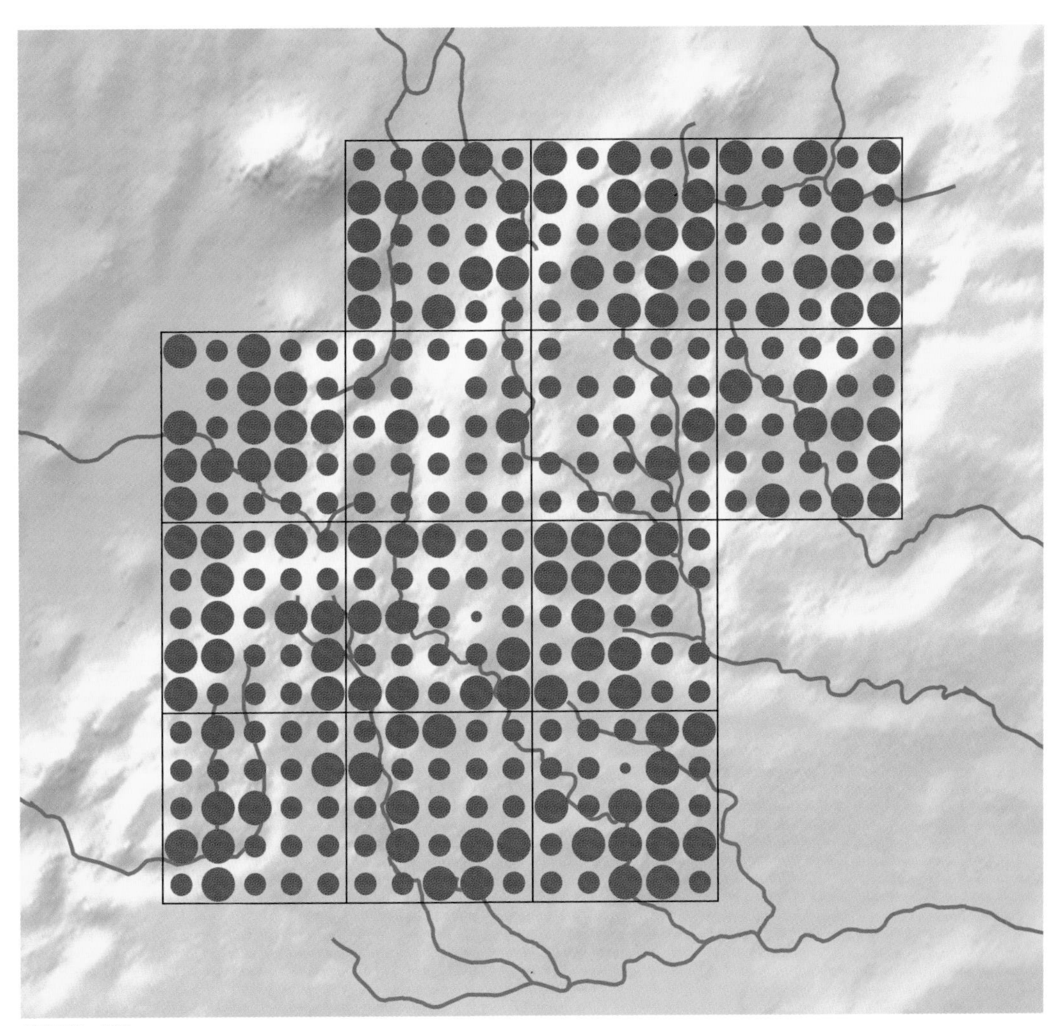
2003–07 survey

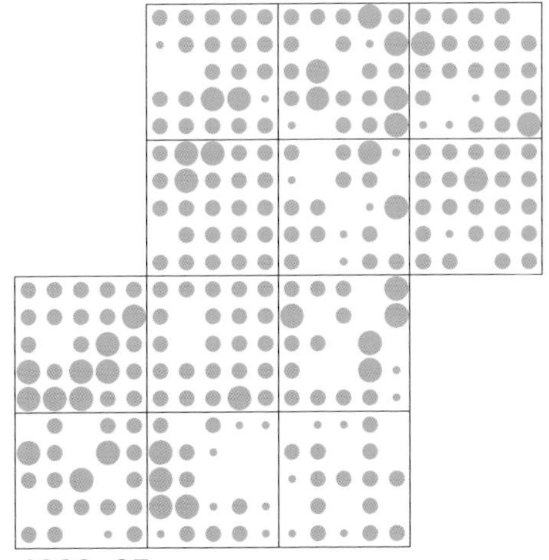
1983–87 survey

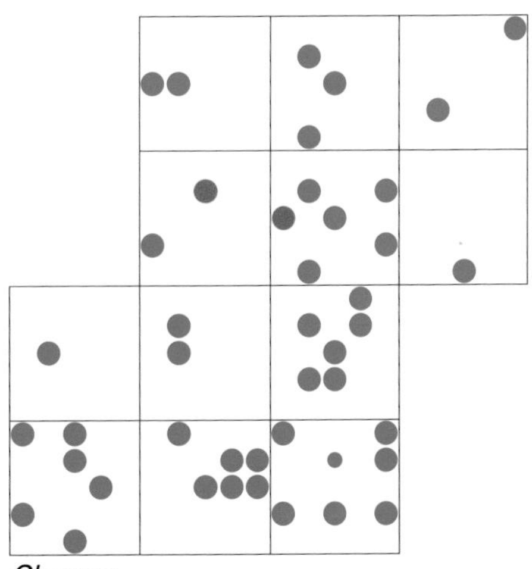
Change

Goldfinch (European Goldfinch)

Carduelis carduelis

The Goldfinch has become a very familiar bird, especially in villages, and any areas with low-density human habitation. Its metallic twittering song is very distinctive and can be heard from April onwards, making location of territorial birds very easy although, being a relatively late breeder, it tends to be heard later in the season than many other song birds. Goldfinch nests are usually constructed within the foliage of trees, and are not occupied until that is fully developed, cover seeming to be the most important requirement. Confirmation of breeding, therefore, usually depended on finding family parties, with easily identifiable young birds begging for food. Young are fed by regurgitation and so food carrying cannot be directly observed. Although the peak fledging time tends to be around late June and early July, young birds were regularly seen well before the end of May in some areas.

Goldfinches have become so widespread in the Cotswolds that it may surprise some that, in the mid-1980s survey, it was recorded in only 61% of tetrads. Indeed the first Atlas commented that its distribution was 'patchy', being somewhat tied, like the Greenfinch, to human habitation. It still shows a marked preference for villages, but was also regularly encountered in farmland habitats. As illustrated in the change map, the expansion in distribution to nearly all the tetrads in the later survey is significant, and possibly due to changing farming practices such as set-aside and leaving headlands uncultivated. This is in agreement both with the general feeling of observers that Goldfinch numbers have risen in the area and suggest a 60% increase in the population since the 1980s.

Resident breeder

UK conservation status: GREEN

Long-term UK trend (1970–2006): +55%

Recent population trend (1994–2007): +39%

Numbers of tetrads with breeding records:

	2003–07		1983–87
	13 squares	12 squares	12 squares
Possible	**10**	7	34
Probable	**187**	179	124
Confirmed	**122**	110	26
Totals	**319**	**296**	**184**

ROB BROOKES

Species sponsored by Keith White

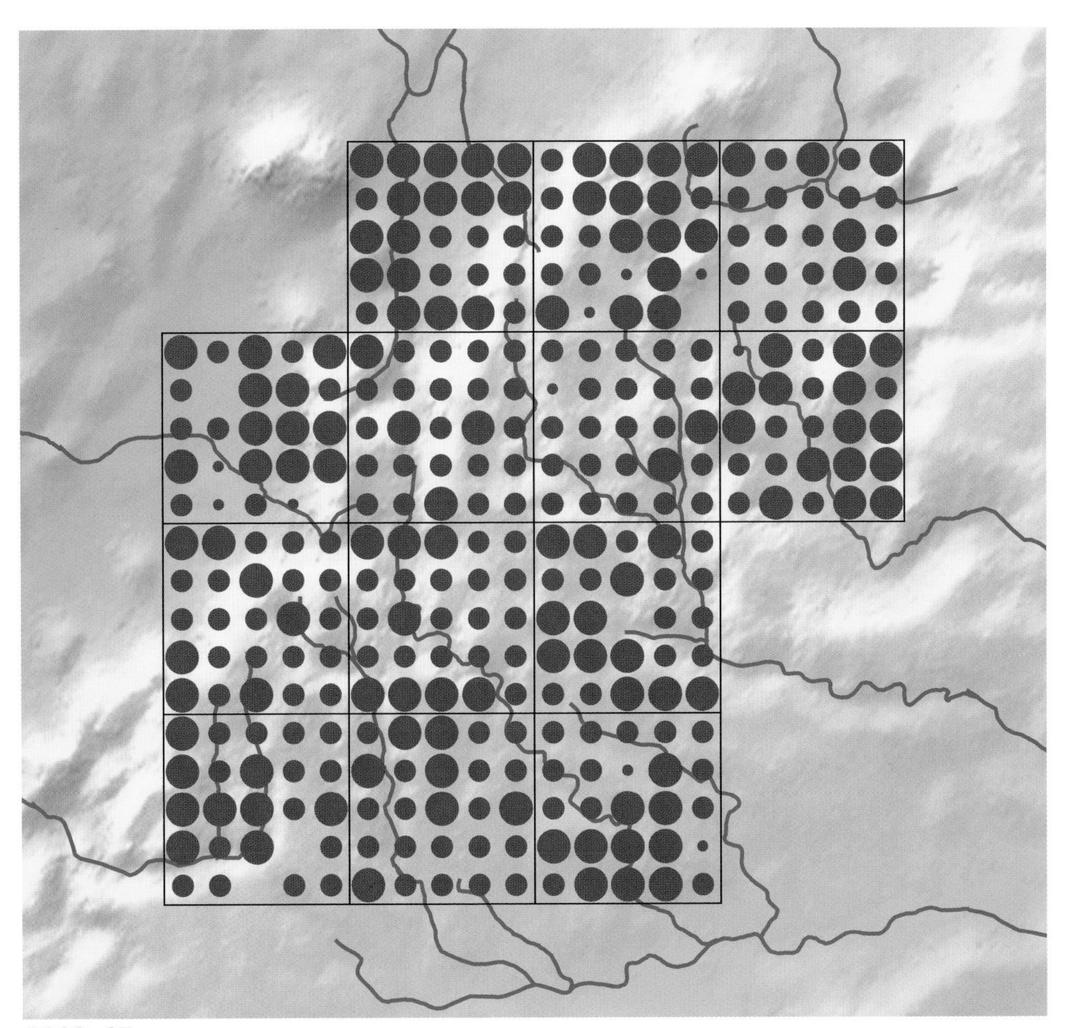
2003–07 survey

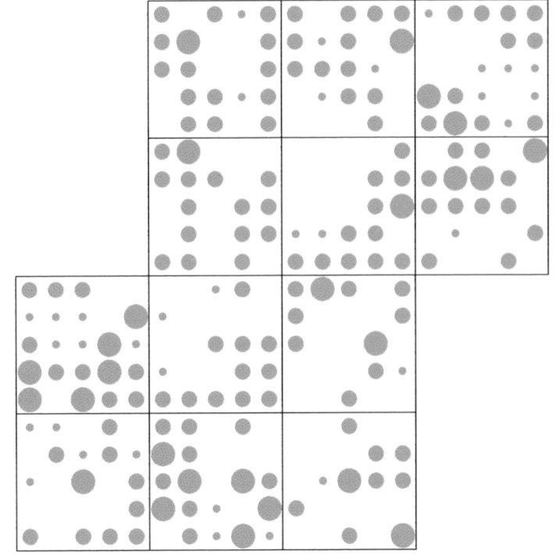
1983–87 survey

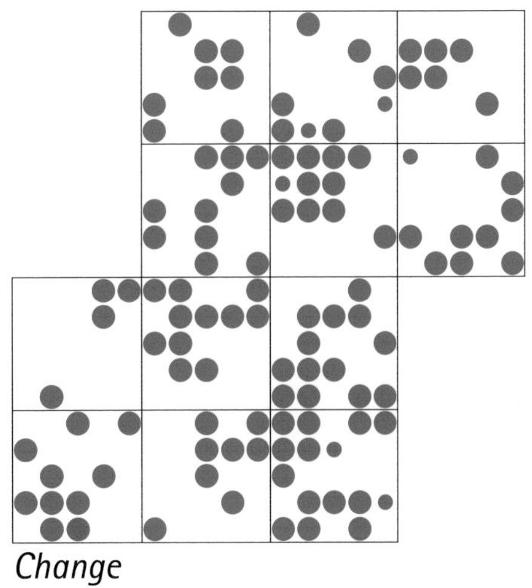
Change

Linnet (Common Linnet)

Carduelis cannabina

A species of open country, the Linnet favours areas where it can feed on weed seeds and which contain scattered bushes or hedgerows in which to nest. It is particularly suited to the arable farmland that is found throughout much of the survey area, and other open habitats such as abandoned airfields and old quarries are also utilized. Its distinctive song and its non-secretive nature help to make location of breeding birds fairly straightforward. They typically sing from April through to midsummer, but recorders had to be cautious in the early spring, as Linnets tend to sing whilst still in winter flocks. Confirmation of breeding was less easy as Linnets do not openly carry food to young, and the young themselves are not strikingly different in plumage from adult females.

Breeding Linnets are widespread, being found in 90% of tetrads, with the only major gap in its distribution being in urban Cheltenham. As in the first survey, the general impression was that it was particularly numerous on Cleeve Common, where there is an extensive area of scattered clumps of gorse.

Resident breeder

Gloucestershire BAP species

UK conservation status: RED

Long-term UK trend (1970–2006): -57%

Recent population trend (1994–2007): -27%

Numbers of tetrads with breeding records:

	2003–07		1983–87
	13 squares	12 squares	12 squares
Possible	**15**	14	25
Probable	**212**	202	98
Confirmed	**70**	64	30
Totals	**297**	**280**	**153**

In the 1980s survey Linnets were recorded in only half of the tetrads, although the ease of locating singing males led the authors of that Atlas to comment that this was probably a fair reflection of the species' distribution. The change map illustrates a significant expansion, which is in marked contrast to the national picture of a species in decline (at least in terms of abundance), particularly in the farmland habitat which constitutes most of the survey area. It may be that the Linnet is one of the few species that have benefited from the farming practice of set-aside in recent years.

RICHARD TYLER

Species sponsored by Geoff Bailey

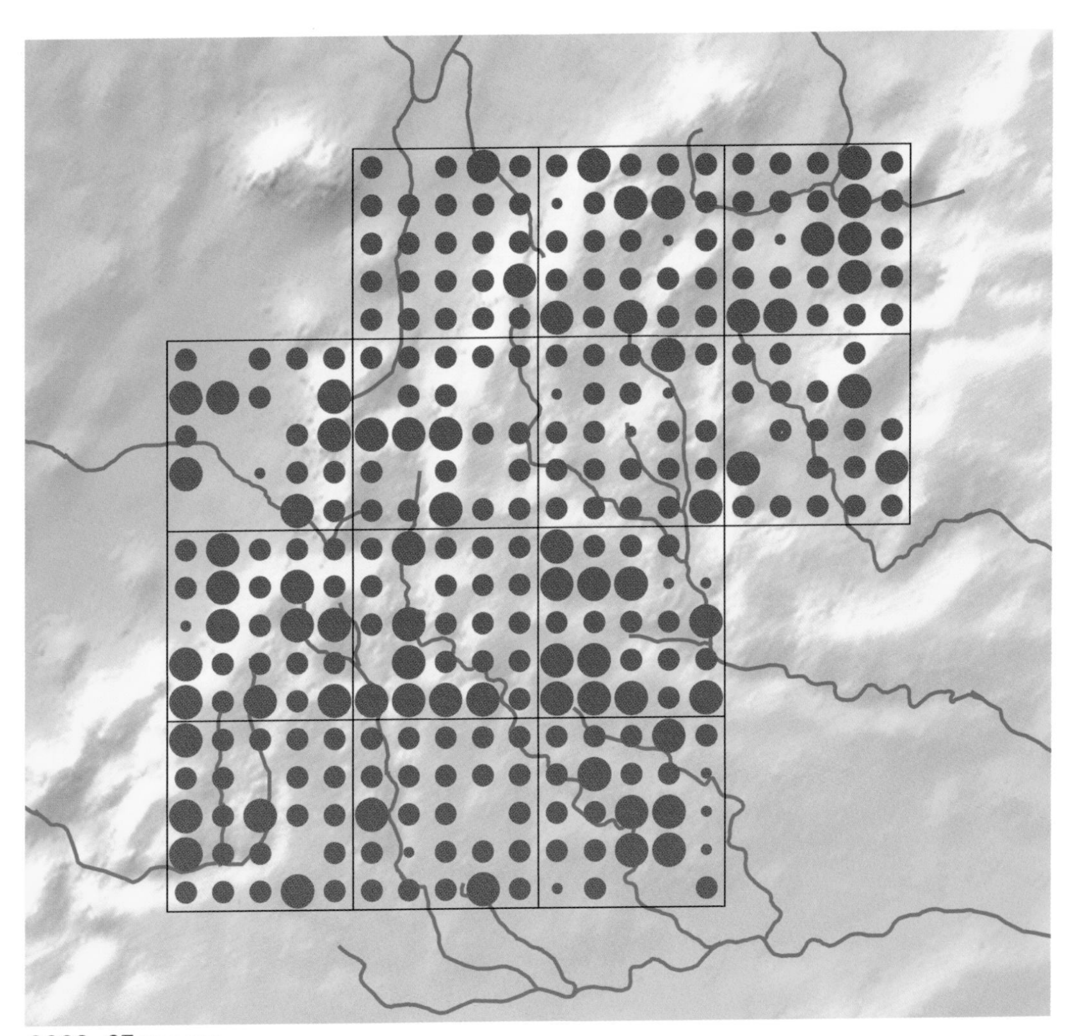

2003–07 survey

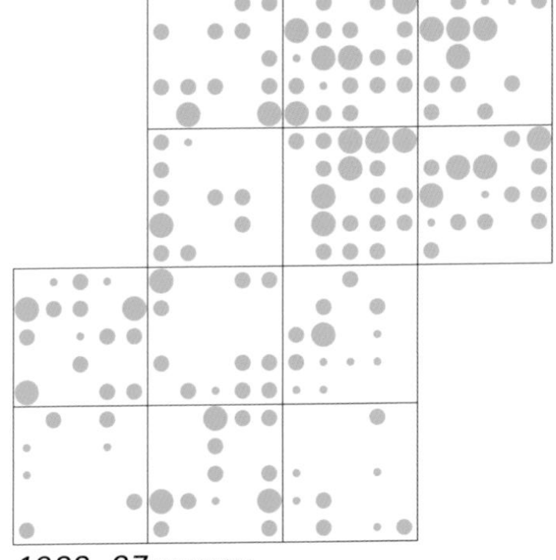

1983–87 survey

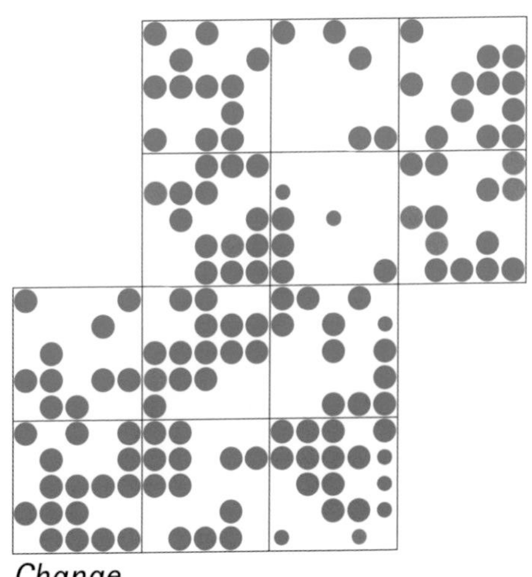

Change

Crossbill (Common Crossbill)

Loxia curvirostra

Crossbills are uncommon and irregular breeders in the Cotswolds, sporadic invasions occurring during irruptions into this country. They require groups of large mature conifers, which are relatively scarce in the Cotswolds. The birds are best located from their loud and persistent 'chip-chip' flight call. However, their breeding patterns are strongly determined by food availability, and can occur at almost any time of year, and so determining whether breeding has occurred in an area is not easy. In the 1980s survey, this was achieved by observation of family parties with young.

During this survey, there were no significant Crossbill irruptions and no regular colonies were located. The only areas in which breeding activity was recorded were in the southwest, in the Stroud Valleys, but there were no confirmed breeding records. This contrasts with the situation in the 1980s survey, when breeding was confirmed in the heavily wooded area between Moreton-in-Marsh and Broadway. During that survey, about six pairs were regularly noted in this woodland during most winters. In 1985–86 numbers rose to about 50 individuals, with an estimate of 10 breeding pairs, and young birds were observed being fed at the tops of larches in May 1986.

Resident breeder

UK conservation status: **GREEN**

Long-term UK trend: Not available

Recent population trend (1994–2007): -37%

Numbers of tetrads with breeding records:

	2003–07		1983–87
	13 squares	12 squares	12 squares
Possible	**3**	3	2
Probable	**3**	3	4
Confirmed	**0**	0	2
Totals	**6**	**6**	**8**

ANDREW CAREY

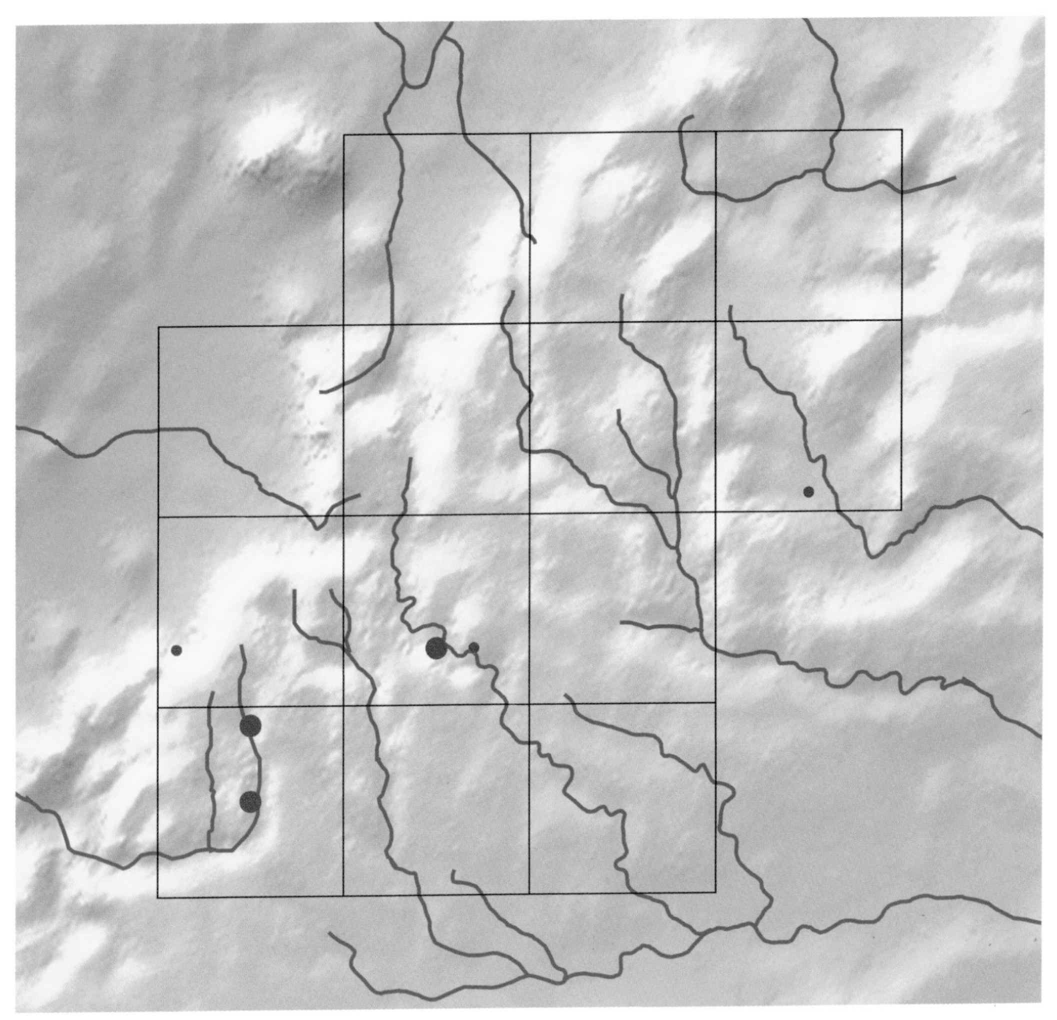

2003–07 survey

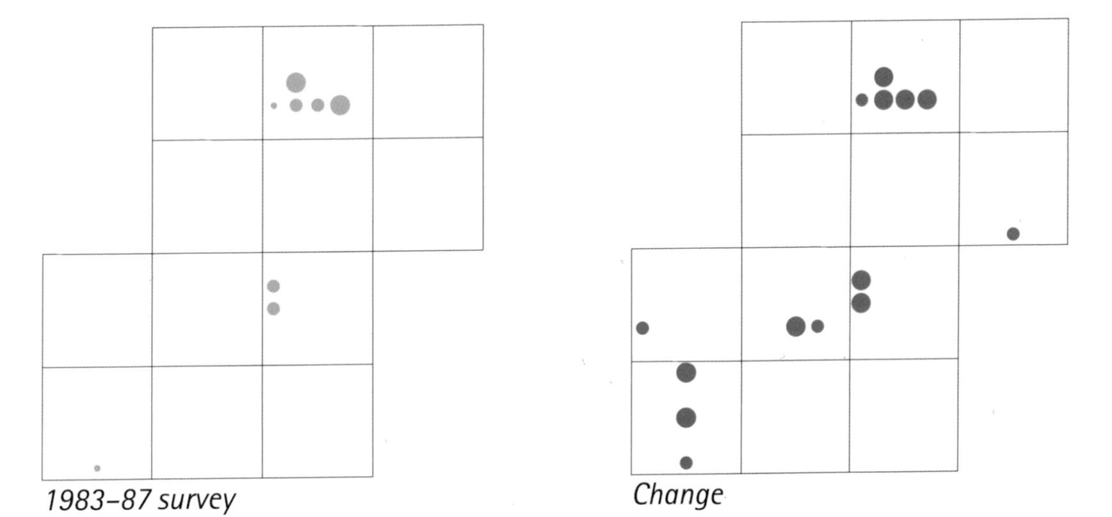

1983–87 survey

Change

Bullfinch (Common Bullfinch)

Pyrrhula pyrrhula

GRAHAM WATSON

Resident breeder

Gloucestershire BAP species

UK conservation status: RED

Long-term UK trend (1970–2006): -51%

Recent population trend (1994–2007): -18%

Numbers of tetrads with breeding records:

	2003–07		1983–87
	13 squares	12 squares	12 squares
Possible	**41**	36	43
Probable	**169**	163	125
Confirmed	**53**	45	22
Totals	**263**	**244**	**190**

Despite the male's very colourful plumage, the Bullfinch is an easy bird to overlook in a breeding survey. Its soft call, subtle song and secretive behaviour can make detection difficult in the thick scrub or woodland in which it breeds. In common with most finches, it feeds its young by regurgitation, and the most usual means of confirming breeding was by observation of family groups with newly fledged young.

Bullfinches were located in about 80% of tetrads, although breeding was only confirmed in one-fifth of these. In general Bullfinches were thinly scattered throughout the survey area, and nowhere were they particularly common. There were a few discrete areas in which they were not found, most notably in the open countryside of the High Wold, the dip slope around Withington and Chedworth, and to the south of Northleach.

Compared with the 1980s survey, the 28% increase in number of occupied tetrads is almost certainly due to the greater survey effort this time, and reflects the difficulty in accurately recording a species with a relatively low density. The change map shows a redistribution but no significant contraction or expansion in any area. The first Atlas commented that it had probably been underrecorded and, as it is likely that the distribution and density have not changed markedly, this may still be the case.

Species sponsored by Michael Williams

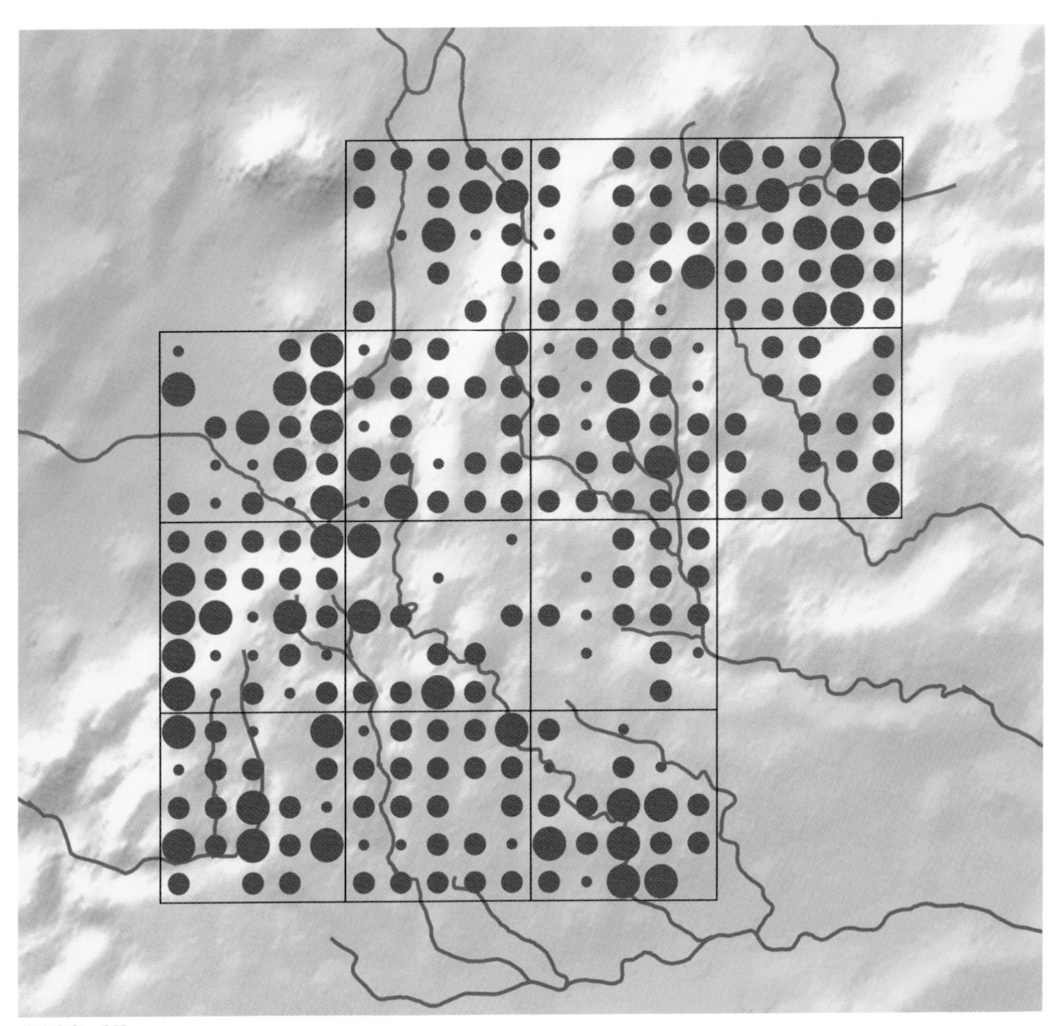

2003–07 survey

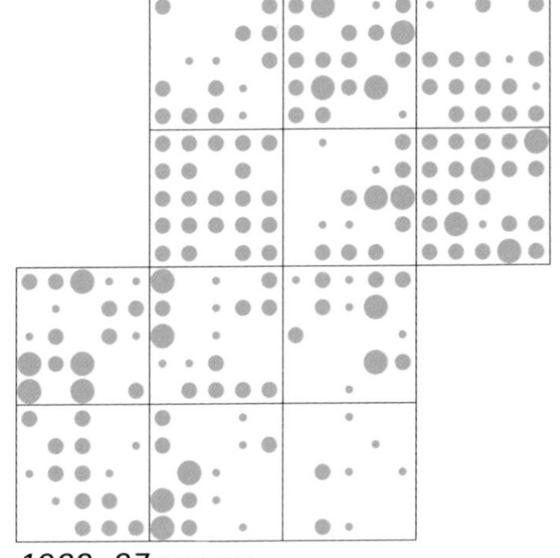

1983–87 survey

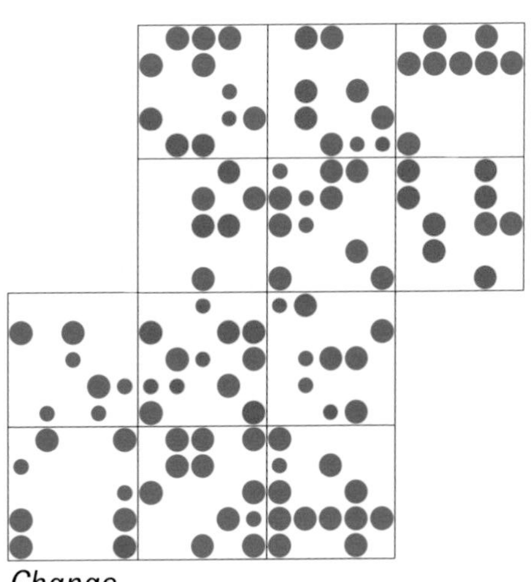

Change

Yellowhammer

Emberiza citrinella

The Yellowhammer's breeding season is very protracted–singing was regularly heard from the beginning of April and continued throughout the summer until late August. Therefore, location of birds holding territory was very easy. Birds carrying food could be seen from the beginning of June, but the peak of feeding activity seems to be around mid-July and into August. Although surveying activity tended to be winding down during the Yellowhammer's main fledging period, there was a reasonably high number of confirmed breeding records.

It remains one of the most familiar and widespread species in the study area, occurring wherever there is open countryside for feeding, hedges or small trees for song posts, and bushes or hedgerow for nesting. As elsewhere in the UK, it is rarely found associated with urban areas or woodland, and it is the tetrads in which this type of habitat predominates that yielded no breeding records. The highest densities are to be found in the upland arable areas of the Cotswolds, with deep valleys being largely avoided.

Nationally, a decline in Yellowhammer numbers and a contraction in range have been suggested, but there does not appear to be any evidence of such a trend in our survey area. While these surveys do not measure population density, there have been no anecdotal reports of a local decline, and the Yellowhammer, along with the Skylark, retains its place among the most characteristic birds of the upland Cotswolds.

Resident breeder

UK conservation status: RED

Long-term UK trend (1970–2006): -54%

Recent population trend (1994–2007): -19%

Numbers of tetrads with breeding records:

	2003–07		1983–87
	13 squares	12 squares	12 squares
Possible	**5**	2	9
Probable	**183**	175	217
Confirmed	**124**	121	63
Totals	**312**	**298**	**289**

ROB BROOKES

Species sponsored by Tony and Pam Perry

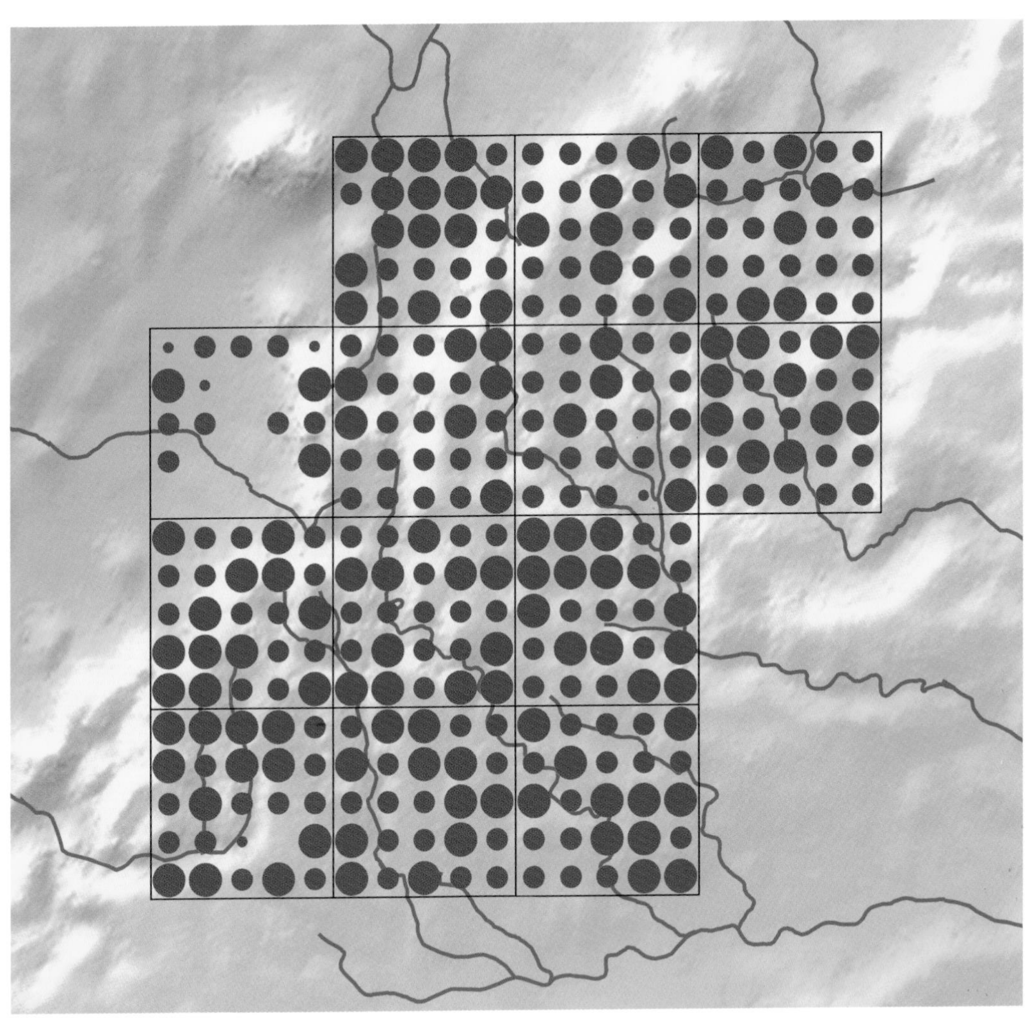

2003–07 survey

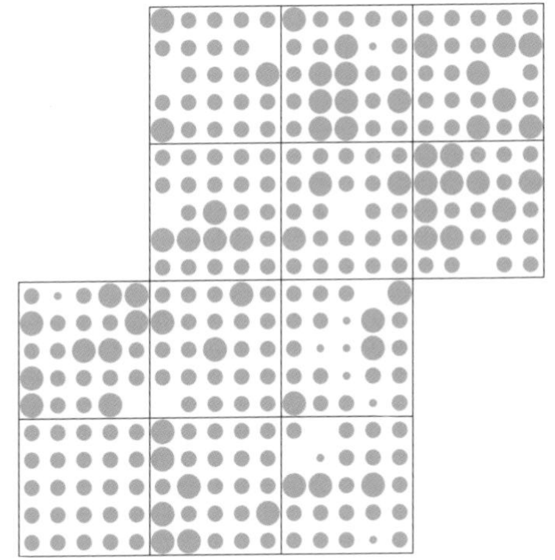

1983–87 survey

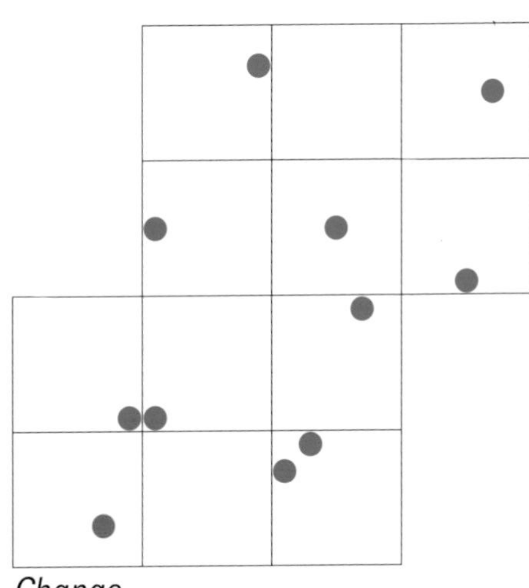

Change

Reed Bunting

Emberiza schoeniclus

RICHARD TYLER

Resident breeder

Gloucestershire BAP species

UK conservation status: RED

Long-term UK trend (1970–2006): -33%

Recent population trend (1994–2007): +31%

Numbers of tetrads with breeding records:

	2003–07		1983–87
	13 squares	12 squares	12 squares
Possible	5	5	8
Probable	62	58	20
Confirmed	4	4	9
Totals	71	67	37

Reed Bunting distribution in the survey area is somewhat unusual, in that it appears to occupy three distinctly different habitats. It is typically a bird of vegetation around wetlands: the low-lying meadows of the wide river valleys of the Windrush, Evenlode and Stour, and the areas around Bourton and Fairford Pits provide significant expanses of this particular habitat, and are home to a reasonably sized population.

The second habitat type is found in the arable farmland of the Avon Vale, where the regular planting of large areas of oilseed rape proved to be to the Reed Bunting's liking (although it was also found associated with cereal crops). This probably represents a remnant of a national increase in Reed Bunting numbers seen in the 1970s, when they moved into farmland. The third type of habitat is the gorse on Cleeve Common, where one or two pairs regularly breed.

The total number of occupied tetrads increased by 80% on that found in the 1980s survey. The change map shows this as an expansion in distribution, with the Thames Vale (Fairford), Stour Valley and Avon Vale populations being new. However, it is known that Reed Buntings have been breeding around the Fairford area and in the Avon Vale at least since the beginning of the 1990s, so they may have been overlooked there previously, especially as coverage of these two particular areas was limited for the first Atlas. What this later survey has undoubtedly shown is the distinct habitats in which breeding Reed Buntings can be found.

Species sponsored by Mick and Jo Jones

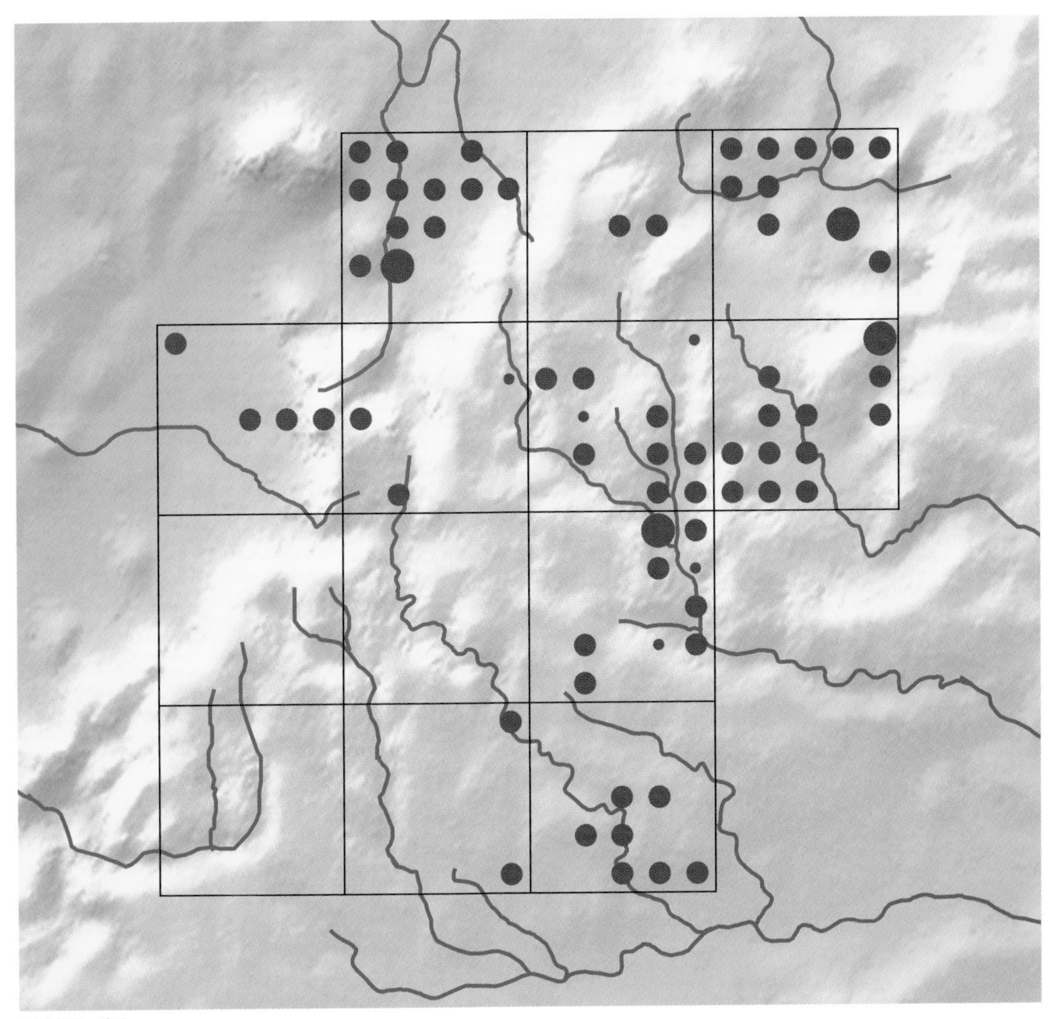

2003–07 survey

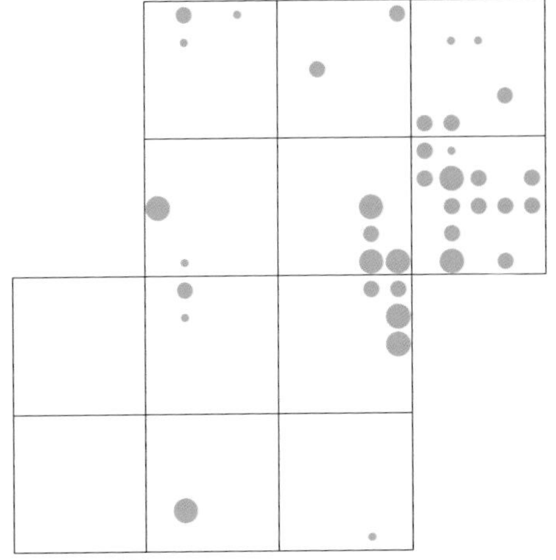

1983–87 survey

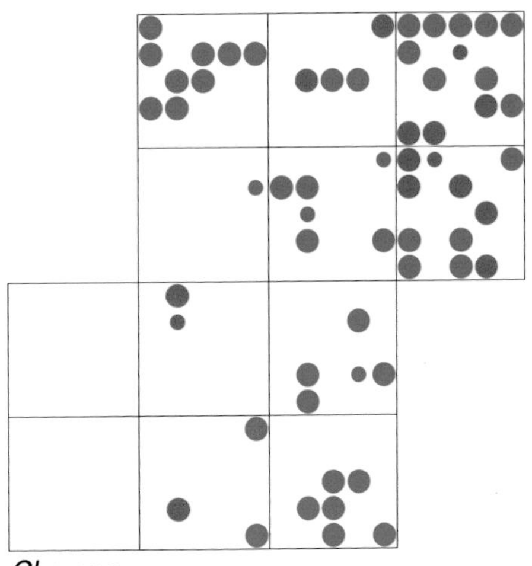

Change

Corn Bunting

Emberiza calandra

As the song of the Corn Bunting is distinctive and usually audible for a considerable distance, it is unlikely that many occupied areas went unnoticed. However, the bird is relatively uncommon and often builds its nest in the middle of a cereal field, which makes confirmation of breeding difficult. The easiest way to achieve this is to see an adult carrying food to a nest site.

The Corn Bunting requires rather specialized habitat for breeding–open areas with uncluttered horizons and the presence of scattered tall bushes and small trees or overhead wires from which to sing. Typical habitat for them in the Cotswolds is provided by cereal-growing farmland with little in the way of other features. As in the survey in the 1980s, the cereal fields of the High Wold and dip slope around Condicote and Hawling–Turkdean provided a significant percentage of the records, and it is in these areas that the highest density of breeding activity was again noted. A stronghold that was not identified in the first survey is the area between Aldsworth and Sherborne near the Thames Valley, though it was discovered in a subsequent special survey shortly afterwards.

Resident breeder

Gloucestershire BAP species

UK conservation status: RED

Long-term UK trend (1970–2006): -89%

Recent population trend (1994–2007): -36%

Numbers of tetrads with breeding records:

	2003–07		1983–87
	13 squares	12 squares	12 squares
Possible	**3**	3	10
Probable	**47**	47	60
Confirmed	**4**	4	10
Totals	**54**	**54**	**80**

Corn Buntings are quite mobile, not moving into some of their breeding sites until June and therefore some of the records may refer to birds moving around or through the area. Even without allowing for this, it is clear that the Corn Bunting's breeding range has contracted, especially from the High Wold in the north of the region around Chipping Campden, and from the countryside due south of Cheltenham, around Leckhampton Hill and Birdlip. It now appears to be absent from both of these areas. Whilst the numerical data indicate that the number of occupied tetrads has decreased by a third over the past 20 years, the general feeling of observers is that the greater survey effort may mask an even larger contraction in range.

ROB BROOKES

Species sponsored by Iain and Jill Main

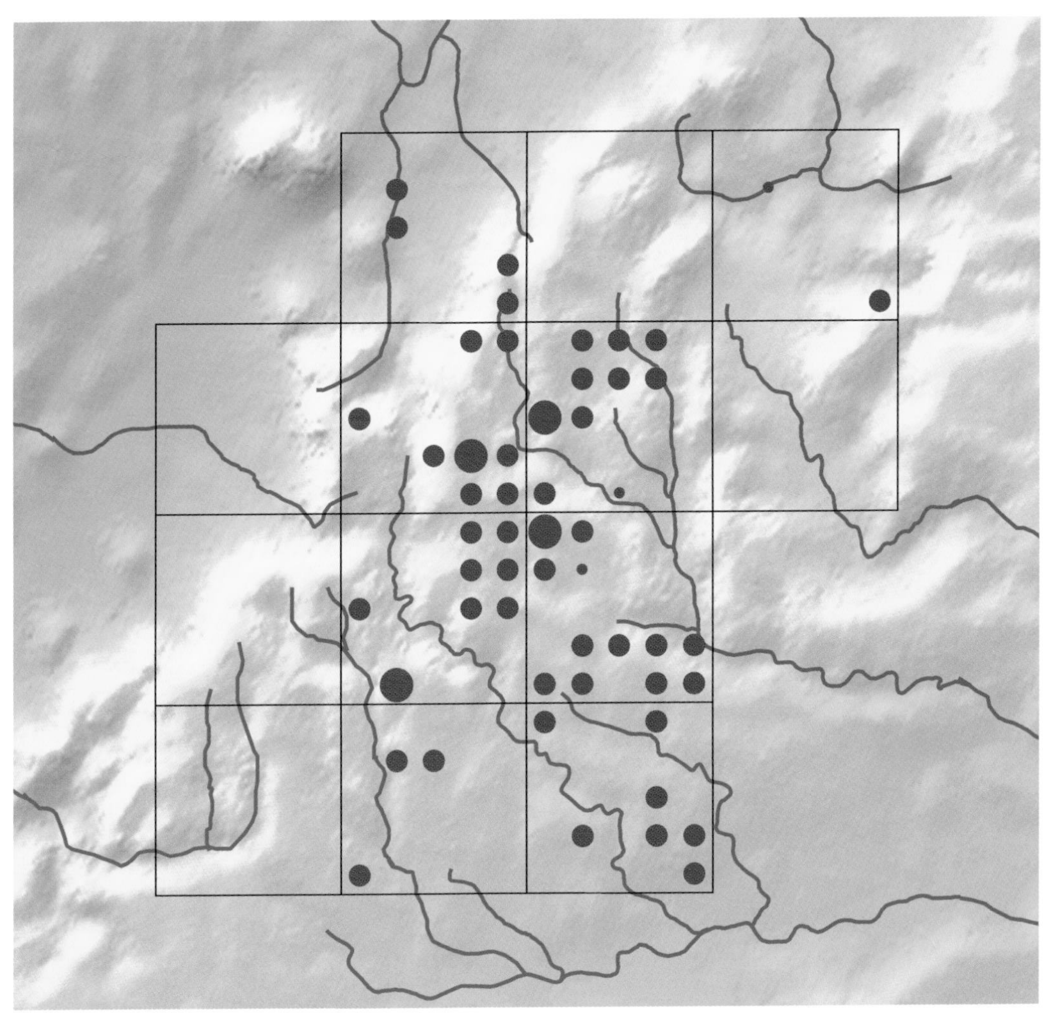

2003–07 survey

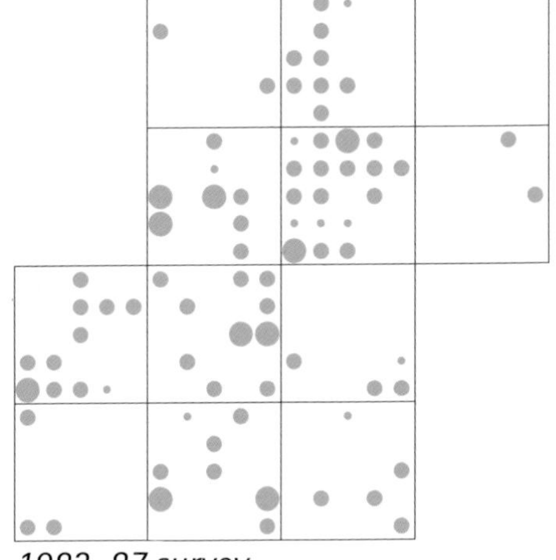

1983–87 survey

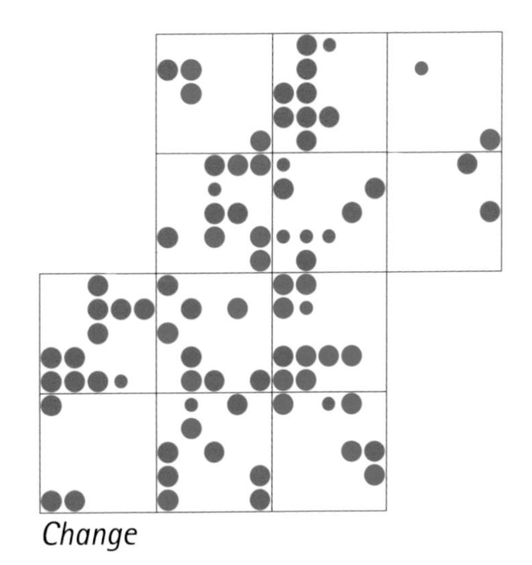

Change

7. Brief species accounts

Shelduck (Common Shelduck)
Tadorna tadorna

Shelducks are uncommon in the survey area, and are limited to an occasional winter or passage sighting. However, a pair were noted in the Stoke Orchard landfill site northwest of Cheltenham, and breeding was subsequently confirmed by the finding of a nest with young. This particular area lies in 10 km square SO92, which was not included in the 1980s survey. Breeding has never been suspected in any other part of the region.

Gadwall
Anas strepera

Gadwall ducklings were found by the River Coln between Fairford and Quenington. This species is regularly seen in winter in suitable habitat throughout the survey region and there have been regular spring and summer records, particularly in the Fairford area, so it is perhaps surprising that no other breeding activity has been observed.

Red-crested Pochard
Netta rufina

This species has been a regular visitor to the Cotswold Water Park since the publishing of the first Atlas. It was observed during the new survey on those few lakes within the recording area, and breeding was confirmed at Fairford Pits and also at Bourton Pits, where several ducklings were seen in two successive years. These are believed to be the first confirmed breeding records for the survey area.

Ruddy Duck
Oxyura jamaicensis

The North American Ruddy Duck is one of the more controversial species in the UK avifauna. The UK population is descended from birds which escaped from captive wildfowl collections in the 1950s and early 1960s. Because it is closely related to and hybridizes with the endangered White-headed Duck (*O. leucocephala*) which breeds in southern Spain, there have been efforts to control its spread. Although it was not reported in the 1983–87 survey it was regularly seen in subsequent years, particularly at Bourton and Fairford Pits. Sightings have been significantly less frequent since the turn of the century. However, breeding was confirmed during the survey in the Little Wolford area, east of Moreton-in-Marsh, and at Bourton Pits where three small young were seen.

Cormorant (Great Cormorant)
Phalacrocorax carbo

There were many records of Cormorant over the survey period, and the species is now a regular non-breeding summer visitor, as well as a more numerous winter visitor. Birds are often seen at Bourton Pits, Sherborne National Trust nature reserve and Fairford Pits, and it is quite plausible that breeding may occur there in the near future. However, there was no evidence of breeding activity during this survey period. The species was not recorded during the survey in the 1980s.

Little Egret
Egretta garzetta

At the time of the first survey in 1980s, the Little Egret was very much a rarity in Britain and Ireland, and was unknown in the Cotswolds. In the new survey, Little Egrets were occasionally reported as summering along the River Churn north of Cirencester and are present there more frequently in winter. In the year after the survey had been completed, several birds were regularly seen in the Fairford area, and it seems likely that this species will colonize the area further in the near future.

Red Kite
Milvus milvus

Sightings of Red Kite have become increasingly frequent in the survey area since the mid-1990s. In the 1980s it would have been considered to be a notable rarity (indeed it remained on the Gloucestershire Rarities Committee's list well into the 1990s). Most sightings are between March and May, and there were several records of birds in suitable breeding habitat; these were mainly in the east of the region (to the west of Chipping Norton), but no breeding attempts were observed. However, a pair was seen copulating and nest building in woodland southeast of Cheltenham, although the breeding attempt was unsuccessful. The Cotswolds would seem to offer ideal habitat for Red Kites and it is perhaps surprising that they have not yet colonized the area. It seems only a matter of time before this species continues its spread, either from its reintroduction areas or from native populations, to become established in the survey area.

ROB BROOKES

Red Kite.

Goshawk (Northern Goshawk)
Accipiter gentilis

The Goshawk has long been a sought-after species in the Cotswolds and so there was naturally great excitement when two breeding pairs were found during the final year of the survey in woodland southeast of Cheltenham. One of the pairs was successful in raising at least two fledglings, but the fate of the other pair is unknown. The year after the survey ended, another two pairs were observed in well-separated areas, and it seems likely that the Goshawk will become a regular breeder in the Cotswolds, particularly since there is a healthy breeding population about 40 km away, in the Forest of Dean.

Peregrine (Peregrine Falcon)
Falco peregrinus

The Peregrine population has been spreading significantly in southern Britain, and sightings in general have become much more regular in the survey area in recent years. In this survey, breeding was confirmed at two rock faces (where young were observed): one in the southwest of the region on the dip slope near the Stroud Valleys, and the other in the north on the High Wold. Peregrines are regularly seen over the Cotswolds, and immediately after the survey had ended a pair was seen regularly on the two tallest buildings in Cheltenham.

Water Rail
Rallus aquaticus

In the 1980s Water Rails were regular, if uncommon, winter visitors to the survey area. Although no evidence of breeding was found during the 1983–87 survey, breeding was known to have occurred a few years before, and it was suspected that it might have continued during the survey period. The regularity of sightings has subsequently declined. During the 2003–07 survey there were sightings of Water Rails in the north of the region near Chipping Campden, where they were also heard during the summer months, and further investigation is warranted to see if breeding is occurring or being attempted.

Oystercatcher (Eurasian Oystercatcher)
Haematopus ostralegus

There was a very surprising confirmed record towards the end of the survey of Oystercatchers breeding, when several chicks were seen at Fairford Pits. Adult Oystercatchers subsequently returned to the site and displayed in the following year. This area provided excellent habitat for passage waders in the 1980s, when Little Ringed Plovers bred regularly, but there were no reports of Oystercatchers during that time.

Little Ringed Plover (Little Plover)
Charadrius dubius

The Little Ringed Plover cannot be considered to be a regular breeding bird of the survey area. Its typical breeding habitat is on freshwater gravel/shingle areas; this type of habitat can result from gravel excavations and, in the 1980s, gravel workings in the Fairford area led to regular breeding at this site. However, the habitat there has subsequently become less suitable. Although gravel extraction continues in the region, it is just outside the recording area. However, suitable habitat was available in the Stoke Orchard area northwest of Cheltenham, where breeding took place during the 2003–07 survey.

Snipe (Common Snipe)
Gallinago gallinago

In the survey in the 1980s there were a handful of Snipe sightings, and one drumming male was found. It was believed that, although all of the

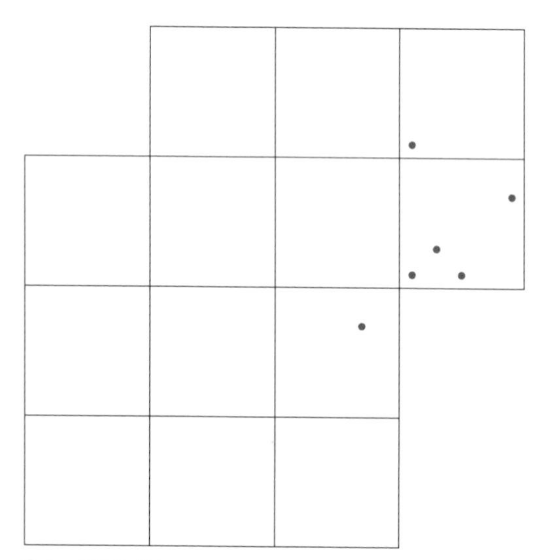

Snipe 2003–07 survey

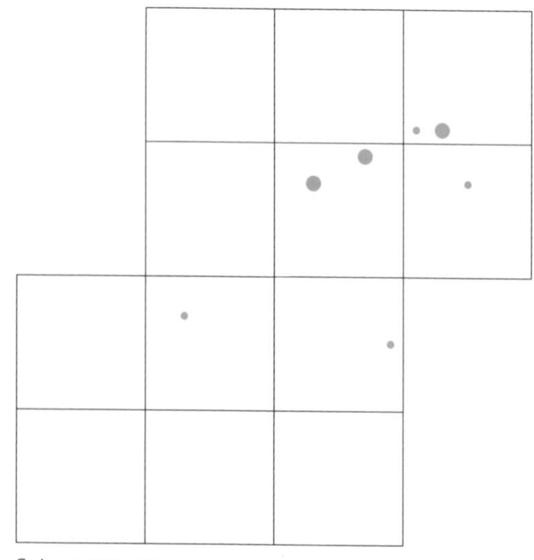

Snipe 1983–87 survey

records obtained were from suitable breeding habitat, they most probably related to late or early winter visitors rather than to genuine breeding attempts. The new survey again yielded a handful of records of birds in suitable breeding habitat—to the east of Bourton-on-the-Water—but none of definite breeding activity. It seems unlikely that Snipe breed in the survey area, although historically they have probably done so.

Woodcock (Eurasian Woodcock)

Scolopax rusticola

In the 1980s survey, roding Woodcock were found in several locations, including Bourton Woods near Blockley, Wolford Wood near Moreton-in-Marsh, Foxholes Nature Reserve area south of Kingham, and Cirencester Park. Although the bird's secretive habits did not permit breeding to be confirmed and it was considered that the species was underrecorded, the fact that roding birds were in the same places over many years suggests that breeding occurred regularly. In the new survey, despite the greater observer effort, only one breeding season record was forthcoming, of an 'agitated' bird between Fairford and Coln St Aldwyns in June 2005. It is clear that the species has declined seriously in the survey area, and may be close to extinction as a breeding bird.

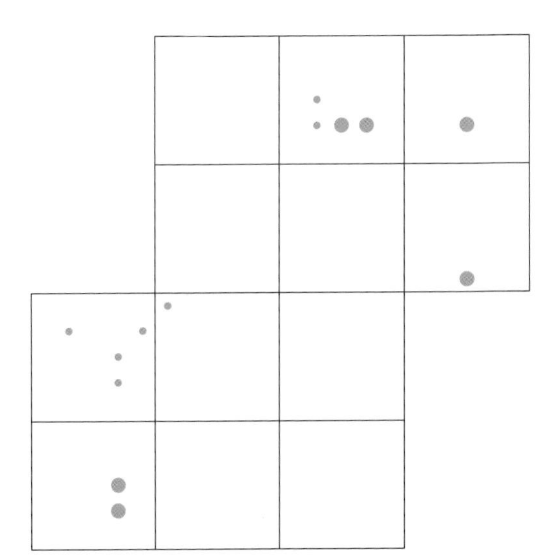

Woodcock 1983–87 survey

Redshank (Common Redshank)

Tringa totanus

In the first survey, Redshank breeding activity was noted at Fairford Pits and in the National Trust nature reserve at Sherborne, where breeding was confirmed. However, in the later survey, no Redshanks were found anywhere, despite the restoration of wet meadows at Sherborne, where the habitat is still suitable.

Lesser Black-backed Gull

Larus fuscus

With a large breeding population in nearby Gloucester, Lesser Black-backed Gulls are a common sight in many parts of the survey area throughout the year. However, they breed only in the Cheltenham area, usually on the flat roofs of the town's large factory units, where they are in fact quite numerous; they also nest between chimney pots on older buildings and against ventilators on the sloping roofs of industrial buildings. Breeding was confirmed in only two tetrads in Cheltenham and may have occurred in a third. However, as nests are hard to see from ground level, it is probable that breeding occurred in more than those two tetrads. Cheltenham Borough Council stated that there were over 150 breeding pairs of Herring and Lesser Black-backed Gulls in the town in 2003, and a four-year programme of egg-oiling was initiated. Since this is in an area that was not included in the 1980s survey, the species was not mentioned in the previous Atlas. However, it is unlikely that any gulls bred in Cheltenham in the 1980s.

Herring Gull
Larus argentatus

Like Lesser Black-backed Gulls, Herring Gulls breed only in Cheltenham and nesting was confirmed in two tetrads there. Owing to the inaccessibility of many nest sites, the relative breeding status of the two species is difficult to determine. However, Herring Gulls are much less common than Lesser Black-backs, although their numbers are increasing. Data from the NCOS Winter Random Square Surveys that have taken place since the mid-1990s suggest that Herring Gulls are 14 times less numerous in winter than Lesser Black-backs. They are frequent visitors to Dowdeswell and Witcombe Reservoirs—the two main waterbodies close to Cheltenham—but are less regularly encountered outside these areas.

Common Tern
Sterna hirundo

Common Terns do not breed in the survey area. However, they do breed nearby, in part of the Cotswold Water Park lying outside our recording area, and make regular feeding visits to the recording area during the summer months. It seems quite likely that, if they were provided with suitable nesting rafts, breeding might be attempted.

Long-eared Owl
Asio otus

Long-eared Owls have had winter roosts in the survey area in some years, but they have become an unusual sight since the late 1990s. Such roosts were often on private estates, and it was not established whether birds stayed on to breed.

However, during the survey there was an exceptional report of confirmed breeding in the Badgeworth–Shurdington area just southwest of Cheltenham, where an underweight young bird with a small amount of down on its head was picked up but died shortly afterwards (Gloucestershire Bird Report 2003). To our knowledge this is the first confirmed breeding of this species in the survey area.

Woodlark (Wood Lark)
Lullula arborea

Although the species did breed in the survey area until 1961, a record in June 2004 of a singing Woodlark just outside Cleeve Common was the first since 1981. This individual held territory for 10 days from 6 to 16 June. Given the spread of the species in southern Britain and the existence of suitable habitat, there is a reasonable chance that it will recolonize the Cotswolds in the future.

Sand Martin
Riparia riparia

Sand Martins easily excavate sandy banks in which to create nest holes. These are fairly uncommon in the survey area, except around the Stoke Orchard landfill site northwest of Cheltenham, where breeding was strongly suspected, and at Huntsman's Quarry north of Naunton. The latter site builds up large piles of sand each year and since the mid-1990s Sand Martins have bred annually under the benevolent eye of the quarry owners with a total of 65 active nest holes in 2006. There were no records of Sand Martin in the 1983–87 survey.

RICHARD TYLER

Sand Martin.

Cetti's Warbler
Cettia cetti

Towards the end of the survey period, a male Cetti's Warbler was heard and seen singing at Fairford Pits. Although initially observed in the autumn, the bird was subsequently found singing the following spring and summer, and at the time of writing three were believed to be there. This is only the second record of Cetti's Warbler in the survey area, but there are several sites in the region that would seem to be suitable for it.

Wood Warbler
Phylloscopus sibilatrix

The Wood Warbler is another species that has been lost to the survey area as a breeding bird. In the 1980s, there were several records of singing birds in the beech woods on or near the scarp, with the largest number of records coming from the Spring Hill area on the High Wold, and Witcombe and Cranham woods south of Cheltenham. Breeding was confirmed at the southern end of the escarpment, and was believed to have occurred at Spring Hill and Cranham. In the new survey there were just three or four records, all probably of passage migrants.

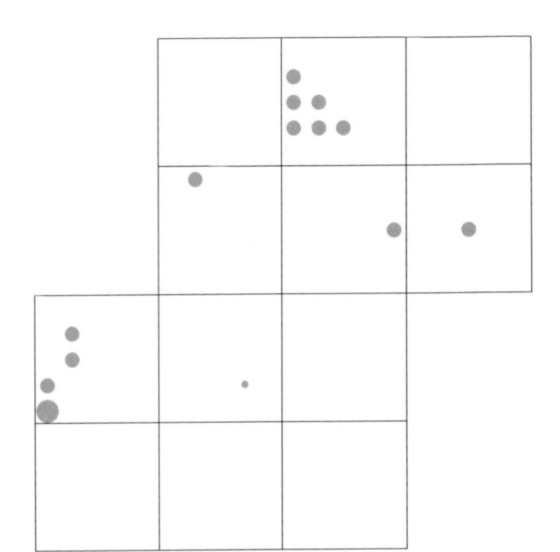

Wood Warbler 1983–87 survey

Siskin (Eurasian Siskin)
Carduelis spinus

The Siskin has never been considered to be a breeding bird of the survey area, although it is a regular winter visitor in variable numbers. However, in 2004 courtship behaviour over an extended period was noted until 12 May in an Atlas observer's garden in the Pittville area of Cheltenham, although there was no sign of a successful outcome. Siskins were noted as late-staying winter visitors in the 1980s survey, although breeding was not considered likely, and so it is debatable whether the 2004 observation can be regarded as a genuine breeding record.

Hawfinch
Coccothraustes coccothraustes

In the first survey, Hawfinches were confirmed as breeding in parkland at Adlestrop in the Evenlode Valley east of Stow-on-the-Wold, and at Moreton-in-Marsh. However, this colony had disappeared by the early 1990s, and there were no Hawfinch reports during the survey. During the two winters following the end of the survey there was a small but welcome influx of Hawfinches into the area and it is possible they will reappear as a breeding bird in the future.

Escaped species

Of the escaped birds encountered during the survey (which included Helmeted Guineafowl *Numida meleagris* and Hooded Merganser *Lophodytes cucullatus*), the only one believed to have bred outside captivity was the Australian Black Swan *Cygnus atratus*. A pair regularly raised young in the Bibury area, and there were a few other sightings of individual birds, including a very long-lived one at Bourton Pits.

8. The survey in detail

The recording area

For surveying a geographical area lacking either natural limits (such as an island) or a well-defined boundary (such as a county or other administrative district), the use of a coordinate system such as a national grid has several advantages. Not being subject to changes over time is important when, as here, the results of surveys decades apart are to be compared. Map reading and, increasingly, the use of GPS (Global Positioning by Satellite) handsets is easier, and the occurrence of grid squares overlapping the boundary is avoided. The recording area for the 1983–87 survey was defined by twelve 10 km squares of the British National Grid, all lying within the area covered by the OS one-inch-to-a mile Touring Map and Guide 8, 'The Cotswolds' at revision A, which is now obsolete—most field-workers now use the OS 1:25,000 scale maps. The same recording area has been used for all subsequent NCOS general bird recording and special surveys, which have been continuous for over a quarter of a century. In 1991 the area was extended by the addition of square SO92. Although introducing a contrasting urban habitat by including the town of Cheltenham, this had the merit of completing coverage of the escarpment at the edge of Cleeve Common in the northwest. This thirteen 10 km square area was adopted for the new survey. It lies almost entirely within the present county of Gloucestershire, with small

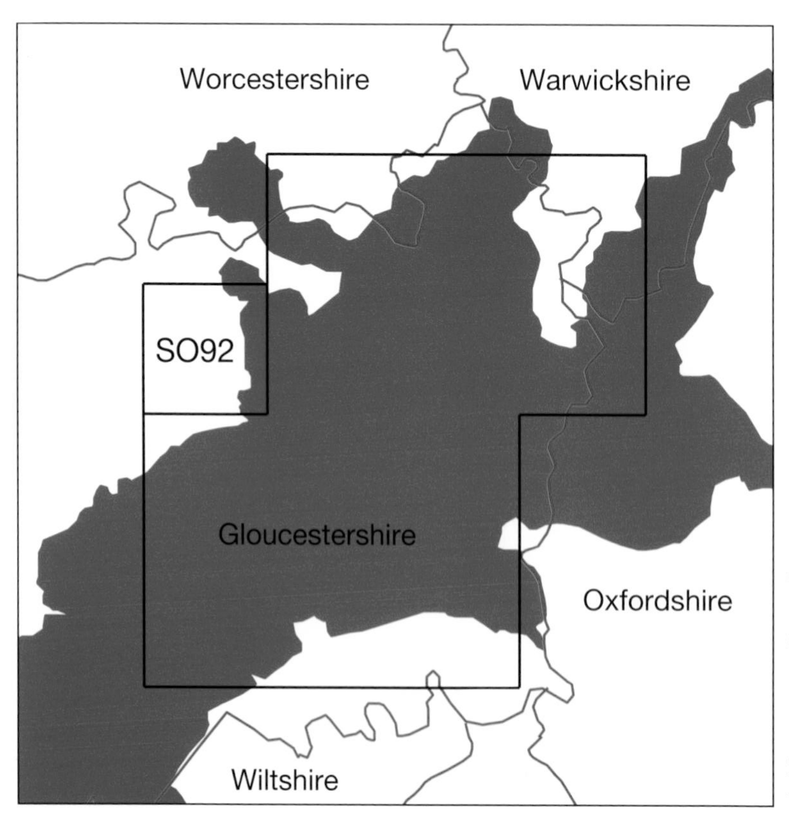

Figure 14. Cotswolds Area of Outstanding Natural Beauty (AONB). Green area denotes the AONB. Red: County boundaries. Black: Atlas recording areas (the isolated 10 km square SO92 was not covered in the 1983-87 survey).

incursions in the north and east into Worcestershire, Warwickshire and Oxfordshire. It covers the area bounded by Cirencester Park and Fairford to the south, and Chipping Campden and Shipston-on-Stour to the north. This includes the high ground of the Cotswolds, the escarpment north and east of Cheltenham, parts of the Severn Vale and the Vale of Evesham, and the upper reaches of the Thames Valley. Approximately 70% of the 13-square recording area lies within the Cotswolds Area of Outstanding Natural Beauty (Figure 14).

The mapping unit

The size of mapping unit to be adopted for a natural history atlas will depend on the area to be surveyed and the number of fieldworkers available. Atlases covering a whole country, based on fieldwork by large numbers of mostly volunteer observers, have used 10 km by 10 km squares of the appropriate national grids (Sharrock 1976, Lack 1986, Robertson *et al* 2007, Gibbons *et al* 1993). An atlas covering a relatively small area, based on fieldwork by full-time professional observers, can afford to use a 1 km by 1 km grid square (Sharpe 2007). At an intermediate level are the large and increasing numbers of local and regional atlases relying entirely or almost entirely on volunteer fieldworkers, where the use of a 2 km by 2 km grid square (a 'tetrad') is usual. These include the previous Atlas (Wright *et al* 1990) and this present Cotswolds Atlas. One person can conveniently survey a tetrad in half a day (usually a morning). The thirteen 10 km grid squares in our recording area comprise 325 tetrads (25 per 10 km square) and these could be covered in the five years of the survey, with two or more visits each at different times in the breeding season, by 50 or more volunteers. The 25 tetrads were labelled conventionally with the 26 letters of the alphabet, omitting the letter O.

Organization

A square steward was appointed for each of the thirteen 10 km squares. Inevitably, the stewards tended to shoulder the greater part of the fieldwork, but they were supported by other observers who could concentrate on particular tetrads (in some cases those in which they lived) but might also help by visiting those that were underrecorded. The steward was responsible for ensuring that the records for his square were forwarded to the data coordinator at the end of the breeding season. He was also responsible for ensuring that the coverage in his 10 km square was as complete as possible, checking the accuracy of the records submitted by other observers and confirming that the records collected by the data coordinator were accurate.

Survey methods

The methods used in both surveys were in general accordance with those outlined by the European Ornithological Atlas Committee (EOAC), the body established in 1971 to coordinate all bird atlas projects in Europe. The EOAC scheme (Hagenmeijer and Blair 1997) defines a hierarchy of 16 criteria, assigning records to three categories labelled 'possible breeding', 'probable breeding' and 'confirmed breeding'. The present survey, like its predecessor (Wright *et al* 1990), identifies the criteria by 16 different alphabetical labels, and the breeding categories by three numerical labels.

The two criteria S (singing) and B (building a nest), at the weak and the strong limits of the probable breeding category respectively, have in practice been accorded different weights by the organizers of different surveys. The former is described in the EOAC scheme as 'singing male(s) present (or breeding calls heard) in the breeding season', and taken as an indicator of only 'possible breeding'. For the first (1968–72) atlas for Britain and Ireland (Sharrock 1976), singing males present in the same place *on more than one date* were accepted as 'probably breeding'. In the 1988–91 survey (Gibbons *et al* 1993), singing was merely accepted as evidence for the category 'seen', since that survey allowed for single visits to tetrads, making it not always possible to record singing males on two separate occasions. Other evidence of territorial behaviour was required for acceptance as 'breeding' (the probable and confirmed categories were not separated in that survey).

There is a case for restricting criterion S (male singing in suitable breeding habitat) to the possible category and using the stronger criterion T (territorial behaviour) when song was recorded on separate days for probable breeding. In practice, the 1983–87 Cotswolds survey admitted song (in suitable habitat and with appropriate care to exclude singing migrants) as evidence of probable breeding. In reviewing the data gathered for the present Atlas, we made the same decision, since it was not apparent that all recorders had observed the distinction between S and T consistently. Observers had, however, been asked to beware of birds singing on migration, something which could be determined on the basis of the observation date and knowledge of the species concerned, as well as by visiting the site on different days. A critical review was carried out after the survey had finished, in an attempt to ensure that migrants were excluded.

With the aim of making the interpretation of the criteria by all observers as uniform as possible, a document was drawn up and circulated before the start of the second breeding season. This described what should be regarded as, for example, a call, a song, or territorial behaviour for each species. The ability to identify song and calls was important, since for many observers this was the first way they detected a new bird species. Although this might have been better done at the outset, we do not believe that any lack of uniformity among observers has significantly affected the results overall. Possible exceptions for certain species are discussed in the species accounts.

In the previous Cotswolds survey, criterion B (nest building or excavating a nest hole) was promoted from probable to confirmed breeding. This was not done for the present Atlas, a decision which is not expected to affect significantly the comparison between previous and present species distributions, since relatively few 'probable' records were derived from criterion B.

Observers were asked to visit each tetrad several times, and at different dates during the breeding season, in order to locate both early and late breeders. An early visit was required for those breeders such as some corvids and for species such as owls and woodpeckers, which set up their territories in the early months of the year. Certain songbirds such as thrushes are also very active at this time. Certainly a visit was required after the summer migrants had arrived in late April and early May. Some species had more than one brood so a late visit in the season was also useful, in particular to upgrade to 'confirmed breeding'. A late survey visit was also required to pick up open-country species, and confirm breeding of birds such as Whitethroat, Yellowhammer and Corn Bunting. In fact most tetrads were surveyed many times and over the full range of the breeding season. It is difficult to be precise but the average time spent in a tetrad over the five years of the survey is estimated to be in the range 8–15 recorder-hours. Thus a total of approximately 5,000 recorder-hours of fieldwork contributed to the results of this survey.

It is important at this point to emphasize that the purpose of the survey was to identify the potential breeding species and their breeding categories. Only one record was required for each species—the aim was to upgrade records to higher breeding categories. Although no attempt was made to estimate population densities, which would have required timed visits, in certain cases information can be deduced as to whether a species is common, local or scarce, based on the number of tetrads in which it occurs, and in such circumstances distribution changes between surveys may indicate population changes.

Different observers carried out their tetrad surveys in different ways. Some observers spent each survey year dividing their time approximately equally in all 25 tetrads in a 10 km square. Others carried out intensive visits to only a small number of tetrads each year, and rarely revisited them. In a few cases, observers who lived in the tetrad recorded only in that tetrad and provided records throughout the survey period. These observers generally returned a high proportion of 'confirmed breeding' species.

Some 60 observers recorded data for the survey, submitting their records to the data coordinator electronically or, in a few cases, on paper forms. The data coordinator merged all the records for each tetrad and at the conclusion of the survey produced maps for each species covering the thirteen 10 km squares. At intervals throughout the survey, and at the end of each breeding season, the data coordinator circulated a table listing all the species recorded to date, with the numbers in each breeding category, and a set of schematic maps. The latter showed the numbers of breeding records obtained in each tetrad, in the three categories separately and combined. These (particularly the totals for the probable and confirmed breeding combined, and confirmed breeding on its own) gave an indication of where any additional effort could be most usefully directed in the following season. The figures at the end of the survey are shown in Appendix C (page 224).

Coverage

Figure 15 reveals that satisfactory overall coverage was achieved in the five breeding seasons allocated to the survey. It may be seen that the relatively small number of 'possible' records (bottom curve) increased more slowly (and even decreased between 2006 and 2007) as new records were balanced (or outweighed) by those being upgraded to 'probable' or 'confirmed'. Similarly with upgrades from 'probable' (top curve) to 'confirmed'. It is also apparent that extending the survey to another year would not have provided any net increase in 'probable' records. Although further fieldwork might have promoted a few of these to 'confirmed', the overall gain would not have justified the effort.

A nominal target of about 40 species was adopted to try to ensure that a given tetrad was covered with adequate rigour. In fact the final results generally ranged from lows of 40 species in a tetrad to highs of 60 although there were some exceptions (see Appendix C). There will of course be a variation in the number of species to be found breeding in a tetrad, depending on the range of habitats it contains, so that complete uniformity is not to be sought. Tetrads which have little woodland and mainly large fields, and those containing mainly woodland and no habitations, would be expected to have the fewest breeding species; those with diverse habitats of water, houses and woods, along with a range of field sizes and crops, should have the greatest variety.

In the earlier Atlas it was suggested that a time limit should have been placed on the total amount of fieldwork in each tetrad, to avoid too much time being spent in less productive ones; this might have biased the data in favour of these tetrads, making the distribution of some species with restricted habitat requirements appear more uniform than was actually the case. However, no such limit was adopted for this new Atlas. Although in principle the observer could keep a count of the number of hours spent in each tetrad, in practice it is difficult to do this if the observer lives in the tetrad. Moreover, all observers could supply breeding records for all tetrads, and not only those for which they had specific responsibility. It is difficult to assign a time to a casual observation of a bird carrying food, seen while an observer drives through a tetrad. Consequently, data reported in the present Atlas may contain a similar bias.

Living in a tetrad will certainly influence the number of species found in that tetrad, and

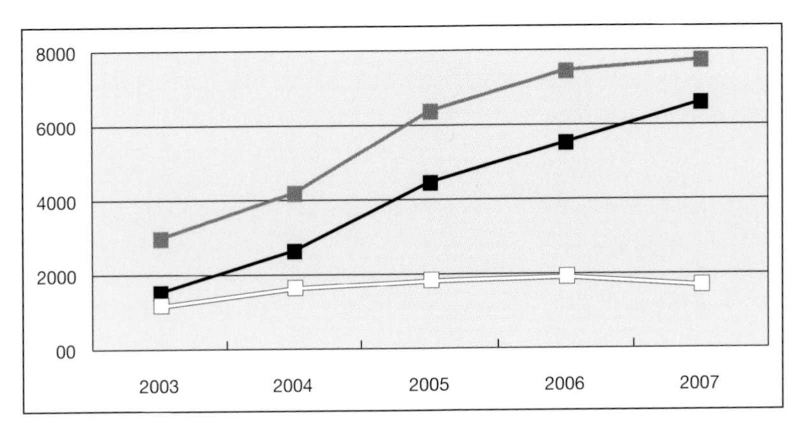

Figure 15. The cumulative numbers of records in each breeding category at the end of each breeding season. White—possible breeding; grey—probable breeding; black—confirmed breeding.

in particular the number of confirmed breeding species. Variability of access to different tetrads may also distort the apparent distribution of species since permission to roam over a farmer's land, for example, will increase the number of species above those in surrounding tetrads, even if they are of similar habitat. The table in Appendix C shows the number of species found in each tetrad, in the various breeding categories.

Data handling and analysis

Observers submitted their records by email to the data coordinator, initially at intervals during the breeding season to assess the workings of the system, and thereafter at the end of each season. The recording form consisted of a Microsoft® Excel worksheet (page 18). When completing the form, the observer entered the appropriate alphabetical symbols for the breeding criteria in the second column, *not* the numerical breeding categories. The latter appeared automatically in the third column, whose cells were locked to prevent hand entry. This had two advantages: first, the nature of the breeding evidence was retained with the data, which could be valuable if there was any reassessment of the data at a later stage. Secondly, it allowed the conversion of breeding criteria to categories to be changed at any later time, by making one simple alteration in a lookup table. A printout of the worksheet was provided for those few field recorders who preferred to submit their data on paper. Such data, however, had to be manually transferred to electronic forms by the data coordinator, which was not only time-consuming but also introduced an extra source of possible error into the data handling chain.

All incoming recording forms were saved as read-only files without modification in any way (except occasionally after discussion of possible keying errors with the observer). Paper recording forms were filed conventionally, but this was inevitably less secure than electronic storage. Data manipulation was performed on 325 *tetrad files*. These were worksheets identical to the recording form, but were updated with both systematic and casual records throughout the survey to provide a cumulative data set. Records from several observers reporting from the same tetrad could be distinguished by colour coding.

The tetrad files served as sources for two *master files* which cross-tabulated the breeding categories for all species and tetrads, and a number of *summary files*. One master file stored the cumulative information in a workbook containing 13 worksheets (one for each 10 km square). A given worksheet showed the breeding categories recorded for all species (rows) in all tetrads (columns) of the relevant square, and provided the most convenient way of providing the data to the square stewards for checking annually and at the end of the survey. The other master file cross-tabulated the same information in a single large worksheet, which was more convenient for generating the species maps. This file also tabulated the number of tetrads in which each species was recorded, and the number of species recorded in each tetrad, broken down by breeding category in each case. The production of both master files from the tetrad files was completely automatic, except for the relatively few extra species written into the recording forms. Because their positions in the corresponding tetrad files were not consistent, their numerical breeding categories had to be entered into one of the master files by hand (they were then automatically transferred to the other).

The two most important summary files showed the numbers of breeding records of each category for each species, and the number of breeding records of each category in each of the 325 tetrads. These provided a convenient indication of year-by-year progress. In addition to these cumulative summaries, similar files could be produced for records obtained in a given breeding season. The production of all these summary files from the master files was completely automatic. The species and tetrad summary data for the complete survey are given in Table 1 and Appendix C respectively.

The basic software was developed and tested during the winter prior to the beginning of the fieldwork. A workbook similar to the one used for the real data (but with only 12 squares) was constructed by reading off breeding categories from the published maps in the 1983–87 Atlas. This provided a data set for testing all subsequent stages in the analysis, the ultimate test being a check that these procedures led correctly back to the source maps. Stringent backup and other security procedures were adopted to guard against data loss due to computer failure, fire or theft at the data coordinator's premises, accidental saving of erroneous edits, and incapacity of the data coordinator. A detailed description of all data handling and analysis procedures was written, and revised regularly. All updated versions were lodged with another NCOS member.

Although the software proved to be error-free and robust in practice, it was apparent by the end of the project that in any later surveys the method of collecting the data could be improved. Specifically, the form did not allow the dates of observations to be recorded. This had the disadvantage that assessment of the data during or at the end of the survey was made more difficult than it might have been. As a specific example, when revisiting the question of whether a singing bird might have been a migrant it was necessary to reconcile the record with the society's database of general records, which could be laborious and in fact not always possible, so that an occasional record might have to be rejected on less than firm grounds. A separate consequence, though not affecting the Atlas directly, was that records submitted to the Atlas data coordinator were not always sent also to the society's bird recorder, although observers were asked to do this. Free-standing Atlas records could not be added to the general database since although they contained space information (to the nearest tetrad) they lacked a date.

All distribution maps were produced by the program DMAP. Sheets of provisional thumbnail maps were provided for discussion and checking, and final maps prepared for the printer as Encapsulated PostScript files.

9. Compatibility with the 1983–87 survey

The influence of survey effort

The first priority for this Atlas is to establish the distribution of the breeding species which were present during the period 2003–07. A second aim, however, is to elucidate the changes that have occurred over the previous two decades by comparing the distributions with those in the previous Atlas. However, the members who carried out fieldwork for both atlases are sure that the present survey was carried out more comprehensively than that for the previous Atlas, since more observers had been available. Clearly, by searching for longer there is a greater chance of recording the scarcer species, even if their population has declined or their distribution has contracted. It is therefore important to try to consider the extent to which differing survey effort (observer-hours in the field) might distort any apparent differences in the results.

As a crude initial comparison we may simply compare the total numbers of the records in each category for the 12 squares surveyed both times, counting only species which were reported in the previous Atlas. That comparison shows that, like for like, the new survey produced 18% more records (see Table 4).

In view of the known serious national decline in many bird species, and farmland birds in particular, it seems certain that greater survey effort has masked a real overall decrease. It is also likely that the recorded increase in the number of occupied tetrads for some species (more green than red in the change map) may be at least partially due to the greater survey effort. While it is not possible to calculate the difference in the survey effort directly since timed logs were not kept, a rough estimate may be obtained from the data with the help of other evidence.

The effect of extra survey effort will vary from species to species. At one extreme, a widespread and easily observed species (such as the Blackbird) will already have been recorded in most tetrads, and any increase in the true number of tetrads occupied will be undetectable, whatever the survey effort. Scarcer and more thinly spread species (such as the Lesser Spotted Woodpecker) will be more susceptible to increased fieldwork, and here it will be necessary to separate, if possible, the influence of survey effort from any real distribution expansion. At the other extreme would be species which are ubiquitous, or nearly so, but hard to find (such as the Bullfinch). For such a species extra fieldwork will simply pull in further records which cannot be taken to indicate a real distribution expansion. However, the data for such a species might provide an estimate of the increase in survey effort.

An example of the latter type of species is the Treecreeper. Not every birdwatcher can hear or recognize its call with confidence and claim its presence. It is difficult to see because of its

Table 4. Numbers of tetrads with breeding records in each category. The figures in the second column are for the new survey, counting only species reported in the previous atlas and omitting the 10 km square SO92 not covered in the previous survey.

	1983–1987 (300 tetrads)*	2003–2007 (300 tetrads)	New/old ratio
Possible	1924 (15%)	1500 (10%)	0.78
Probable	6649 (53%)	7180 (49%)	1.08
Confirmed	3929 (31%)	6026 (41%)	1.53
Totals	12502	14706	1.18

*Because of rounding errors, the percentages in this column do not add up to 100%.

cryptic marking, although it does have the merit of always being on the move seeking out food. Its breeding habitats of tall hedgerows, copses and woods are present in most tetrads in the recording area (see Chapter 3). Moreover, while the relationship between population and distribution is tenuous, it may be noted that this is one of the few species whose population has remained stable during the period between the two surveys: the CBC/BBS data for both England and Wales show no significant change between 1980 and 2005 (Baillie *et al* 2007). Despite our conclusion that Treecreepers probably breed in virtually every tetrad, they were recorded in only 111 tetrads in the 1983–87 survey, and in only 188 tetrads (out of 300) in the 2003–07 survey. This appreciable increase (almost 70%) may be attributed largely to the extra survey effort.

Assuming a 70% increase in survey effort, we can confidently conclude that a recorded increase significantly greater than this in the number of tetrads occupied by a given species reflects, at least in part, a real expansion of that species' distribution. In practice, for the reasons suggested above, a somewhat smaller increase should be acceptable for most species (those lacking the atypical ubiquitous-but-hard-to-find characteristic of the Treecreeper). While the precise number is of little significance, we shall adopt, as a conservative estimate, the figure of 50%.

Species meeting this threshold were listed in Table 2 (page 22). It will be legitimate to claim a real, if unquantifiable, range expansion since the previous survey for those species (apart from Treecreeper). For the species not listed in the table, with increases below 50%, such a claim will not usually be justified.

Of course any apparent *contraction* in a species' distribution (more red than green in the change map) can confidently be assumed to be real, and possibly more extreme than that suggested by the data.

Uniformity of coverage

As well as the greater survey effort, the later survey achieved a more uniform coverage. This may be deduced from a plot of the raw totals recorded in each 10 km square (Figure 16). For the 12 squares covered in both surveys, the earlier survey shows greater variability (the standard variation from the mean number of records per square is 99.4 for the earlier survey and 62.3 for the later survey). There is of course no logical reason to attribute the difference entirely to survey technique, though it is not thought that habitats or species richness within the different squares differed significantly between the two surveys.

The 2003–07 survey produced a markedly greater number of 'confirmed' records, a further indication that this survey was more comprehensive since, in general, being present to observe the brief event which provides a breeding record (*eg* an adult carrying food) does require more

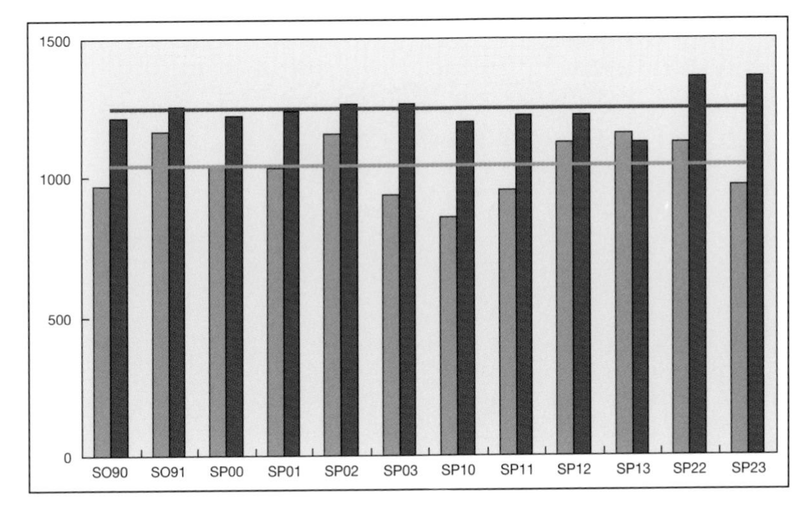

Figure 16. Records in each 10 km square, from the 1983–87 survey (pale) and the 2003–07 survey (dark). Horizontal lines indicate arithmetic means (1042 and 1247 respectively).

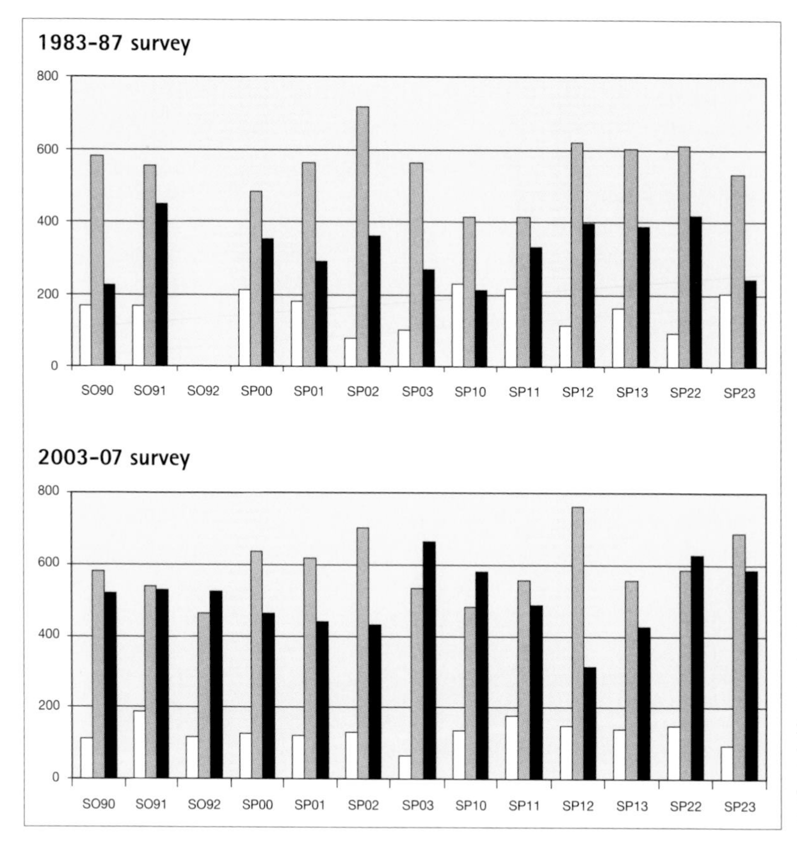

Figure 17. Numbers of tetrads providing records in the breeding categories shown, in the two surveys. White–possible; grey–probable; black–confirmed.

time in the field. In both surveys the proportion of 'confirmed' records varies across the recording area (Figure 17). In the 2003–07 survey it ranged from 26% in square SP12 to 53% in SP03. While this could reflect a genuine difference in the nature of the habitats in the various squares, it is also likely that, as argued above, the time spent by recorders in the tetrads of the square and their different recording methods had a major influence.

Appendix A. Species league table

The top 70 most widespread breeding species in the recording area in 2003–07, ranked by the number of tetrads containing breeding records in any category.

Rank	Species	Number of tetrads
1=	Woodpigeon	325
1=	Wren	325
1=	Dunnock	325
1=	Robin	325
1=	Blackbird	325
1=	Blue Tit	325
1=	Chaffinch	325
8=	Great Tit	323
8=	Jackdaw	323
10	Greenfinch	320
11=	Carrion Crow	319
11=	Goldfinch	319
13	Song Thrush	318
14	Pheasant	317
15	Blackcap	315
16	Yellowhammer	312
17	Chiffchaff	311
18	Buzzard	310
19	Skylark	309
20	Swallow	306
21	Magpie	301
22	Great Spotted Woodpecker	298
23=	Whitethroat	297
23=	Linnet	297
25=	Stock Dove	296
25=	Mistle Thrush	296
27	Goldcrest	290
28	Kestrel	289
29	Long-tailed Tit	284
30	Pied Wagtail	281
31	House Sparrow	275
32=	Green Woodpecker	272
32=	Rook	272
34	Collared Dove	268
35	Starling	264
36	Bullfinch	263
37	Willow Warbler	255
38	House Martin	254
39	Coal Tit	251
40	Red-legged Partridge	242
41	Mallard	224
42	Moorhen	206
43	Jay	202
44	Sparrowhawk	200
45	Treecreeper	199
46	Marsh Tit	196
47	Nuthatch	195
48	Swift	187
49	Lesser Whitethroat	155
50	Spotted Flycatcher	148
51	Garden Warbler	135
52	Tawny Owl	131
53	Little Owl	126
54	Cuckoo	119
55	Lapwing	112
56	Grey Wagtail	111
57	Coot	104
58	Feral Pigeon	99
59	Canada Goose	86
60	Grey Heron	76
61	Tufted Duck	73
62	Reed Bunting	71
63	Raven	69
64=	Grey Partridge	65
64=	Little Grebe	65
64=	Barn Owl	65
67	Redstart	58
68=	Mute Swan	54
68=	Corn Bunting	54
70	Kingfisher	49

Appendix B. Gazetteer

Place	Grid ref
Ablington	SP1007
Adlestrop	SP2427
Aldsworth	SP1510
Ampney Crucis	SP0702
Andoversford	SP0219
Aston Magna	SP1935
Aston Somerville	SP0438
Badgeworth	SO9019
Barnsley	SP0705
Bibury	SP1106
Birdlip	SO9214
Bishop's Cleeve	SO9427
Bledington	SP2422
Blockley	SP1634
Bourton Pits	SP1719
Bourton Woods	SP1633
Bourton-on-the-Water	SP1721
Brailes Hill	SP2938
Bredon	SO9334
Brimpsfield	SO9312
Broadway	SP1037
Buckland	SP0836
Burford	SP2412
Chalford	SO9003
Chedworth	SP0610
Chedworth Woods	SP0513
Cheltenham	SO9423
Childswickham	SP0736
Chipping Campden	SP1539
Chipping Norton	SP3127
Churchill	SP2824
Cirencester	SP0201
Cirencester Park	SO9703
Cleeve Common	SO9925
Coberley	SO9616
Coln St Aldwyns	SP1405
Coln St Dennis	SP0810
Condicote	SP1528
Cotswold Water Park	SP1600
Cranham Wood	SO9012

Place	Grid ref
Cricklade	SU0993
Crickley Hill	SO9216
Daglingworth	SO9905
Donnington	SP1928
Donnington Brewery	SP1727
Dowdeswell Reservoir	SO9919
Duntisbourne	SO9707
Evesham	SP0443
Fairford	SP1500
Fairford Pits	SP1600
Ford	SP0829
Fossebridge	SP0711
Foxholes	SP2520
Gloucester	SO8318
Gotherington	SO9629
Guiting Wood	SP0726
Hawling	SP0622
Huntsman's Quarry	SP1225
Kingham	SP2522
Lechlade	SP2100
Leckhampton	SO9418
Lineover Wood	SO9818
Little Wolford	SP2635
Longborough	SP1729
Lower Harford	SP1322
Lower Slaughter	SP1622
Mickleton	SP1143
Misarden Park	SO9308
Miserden	SO9308
Moreton-in-Marsh	SP2032
Naunton	SP1123
North Cerney	SP0108
Northleach	SP1114
Nosehill Farm	SP1226
Paxford	SP1837
Pittville Park	SO9523
Prestbury	SO9624
Quenington	SP1404
Rendcomb Park	SP0110
Salperton	SP0720

Place	Grid ref
Sapperton	SO9403
Sherborne	SP1417
Sherborne Brook	SP1615
Shipston-on-Stour	SP2640
Shurdington	SO9118
Snowshill	SP0933
Spring Hill	SP1233
Stanway	SP0632
Stoke Orchard	SO9128
Stow-on-the-Wold	SP1925
Stroud	SO8405
Taddington	SP0831
Tewkesbury	SO8933

Place	Grid ref
Toddington	SP0332
Todenham	SP2436
Turkdean	SP1017
Upper Slaughter	SP1523
Whittington	SP0120
Winchcombe	SP0228
Witcombe Reservoir	SO9014
Witcombe Wood	SO9113
Withington	SP0315
Withington Woods	SP0313
Wolford Wood	SP2333
Wormington	SP0336

Appendix C. Distribution of records by tetrad

Total breeding records (categories 1, 2 and 3)

					45	52	53	52	47	42	46	46	51	61	61	56	55	54	53
					48	49	54	52	57	39	31	49	48	48	54	60	55	59	61
					49	49	59	53	53	39	39	48	58	51	51	60	50	63	50
					48	47	51	53	52	39	45	45	47	42	50	49	53	50	56
					54	43	49	48	47	39	40	46	44	40	58	47	51	53	54
46	36	39	51	53	53	57	55	43	50	42	38	51	52	63	51	57	51	45	56
41	37	45	39	49	53	60	45	45	49	47	48	48	63	43	50	61	59	57	57
37	42	46	57	61	51	62	51	44	49	45	64	41	48	46	50	60	62	53	54
44	39	45	52	45	54	53	46	46	47	45	49	53	61	40	56	46	52	51	60
34	35	44	46	41	57	66	41	47	39	43	42	44	66	43	57	52	59	53	49
44	45	49	48	59	46	48	42	43	49	44	43	50	67	57					
52	40	46	53	52	43	49	43	49	43	47	44	44	59	49					
62	50	53	58	47	45	59	50	26	51	51	47	53	60	62					
56	44	49	55	52	50	44	47	51	46	40	57	43	47	53					
53	47	48	40	42	73	69	52	55	50	43	45	42	38	38					
48	53	56	43	46	59	56	56	41	62	49	38	42	47	40					
48	52	48	45	51	56	50	45	48	48	51	42	40	40	46					
49	56	61	47	49	44	49	48	44	44	48	46	60	55	46					
54	49	51	33	50	50	53	42	47	39	45	48	70	46	48					
42	43	40	52	50	50	45	39	56	51	46	44	49	69	47					

Square	Records
SO90	1216
SO91	1244
SO92	1104
SP00	1222
SP01	1223
SP02	1263
SP03	1264
SP10	1202
SP11	1223
SP12	1225
SP13	1123
SP22	1358
SP23	1363
Total	**16030**

Probable breeding records (category 2)

					20	17	23	10	18	17	20	21	16	32	12	37	22	26	23
					21	20	25	22	18	18	23	18	24	15	36	35	34	26	28
					21	22	30	23	27	26	25	24	18	23	34	39	25	19	30
					25	22	27	23	25	23	20	25	26	24	30	30	24	17	36
					22	14	13	24	20	25	25	21	22	25	33	22	18	20	32
21	19	22	32	26	27	30	28	24	26	29	24	34	23	33	24	22	23	18	18
21	24	21	12	20	20	28	27	30	30	32	33	34	33	29	21	34	24	21	26
21	18	13	17	17	33	37	38	20	25	29	36	32	34	22	19	30	34	22	24
19	15	8	16	26	25	29	31	30	25	35	34	31	22	26	28	21	22	24	23
16	15	16	18	8	27	39	22	30	21	31	34	36	28	26	25	25	23	24	19
18	21	26	25	16	19	14	20	26	30	19	21	23	26	30					
20	14	24	26	28	32	29	24	25	27	22	15	16	28	25					
17	23	31	12	23	29	28	37	21	30	20	22	23	34	30					
26	19	20	24	27	26	29	26	35	23	18	23	22	27	24					
9	26	27	16	21	32	20	23	19	22	20	20	17	17	16					
26	19	22	16	27	32	24	24	22	28	25	24	29	11	16					
26	25	25	22	30	24	31	25	30	29	17	23	25	21	30					
19	26	15	32	21	20	27	30	24	23	26	33	7	11	11					
17	25	27	24	24	23	22	28	25	20	21	21	5	11	15					
18	15	26	26	28	26	21	16	29	32	25	23	11	15	28					

Square	Records
SO90	581
SO91	539
SO92	461
SP00	635
SP01	646
SP02	702
SP03	532
SP10	484
SP11	558
SP12	760
SP13	556
SP22	594
SP23	688
Total	**7736**

Probable or confirmed breeding records (categories 2 and 3)

					44	46	50	48	45	36	44	42	47	55	59	53	51	51	48
					48	46	52	48	57	35	27	39	40	38	53	52	51	53	55
					46	45	57	50	50	35	35	40	53	40	47	57	48	61	46
					48	44	50	50	47	37	40	39	42	36	48	48	51	45	54
					51	38	48	47	44	31	38	42	35	37	52	45	47	46	51
39	33	36	45	48	46	55	53	40	50	36	37	44	45	56	44	51	49	40	51
38	33	43	33	40	51	51	39	42	44	40	40	43	51	35	44	55	54	53	52
32	35	43	55	61	47	54	48	38	43	38	49	36	40	40	44	53	52	47	50
36	32	39	49	38	45	42	39	41	39	41	42	51	60	35	51	43	48	46	51
30	30	39	39	39	48	61	38	43	35	39	41	38	58	39	53	47	51	50	43
36	39	45	45	54	39	43	38	37	42	40	38	39	62	50					
41	33	40	44	43	41	38	34	45	39	43	34	36	49	41					
58	40	46	53	41	40	59	46	25	47	45	42	45	54	53					
49	32	37	46	43	40	43	40	46	41	33	48	38	41	43					
49	38	43	32	38	69	61	50	50	46	36	38	35	32	31					
46	47	49	37	42	52	51	52	37	56	41	34	35	44	35					
42	48	46	41	45	52	45	41	43	42	40	40	27	34	39					
47	53	60	43	41	35	42	44	38	39	41	41	57	55	28					
50	45	46	31	46	44	46	38	44	36	39	44	67	46	41					
37	36	36	44	44	45	39	37	52	48	40	40	49	67	42					

Square	Records
SO90	1102
SO91	1065
SO92	985
SP00	1098
SP01	1099
SP02	1132
SP03	1199
SP10	1066
SP11	1046
SP12	1074
SP13	983
SP22	1222
SP23	1272
Total	**14343**

Confirmed breeding records (category 3)

					24	29	27	38	27	19	24	21	31	23	47	16	29	25	25
					27	26	27	26	39	17	4	21	16	23	17	17	17	27	27
					25	23	27	27	23	9	10	16	35	17	13	18	23	42	16
					23	22	23	27	22	14	20	14	16	12	18	18	27	28	18
					29	24	35	23	24	6	13	21	13	12	19	23	29	26	19
18	14	14	13	22	19	25	25	16	24	7	13	10	22	23	20	29	26	22	33
17	9	22	21	20	31	23	12	12	14	8	7	9	18	6	23	21	30	32	26
11	17	30	38	44	14	17	10	18	18	9	13	4	6	18	25	23	18	25	26
17	17	31	33	12	20	13	8	11	14	6	8	20	38	9	23	22	26	22	28
14	15	23	21	31	21	22	16	13	14	8	7	2	30	13	28	22	28	26	24
18	18	19	20	38	20	29	18	11	12	21	17	16	36	20					
21	19	16	18	15	9	9	10	20	12	21	19	20	21	16					
41	17	15	41	18	11	31	9	4	17	25	20	22	20	23					
23	13	17	22	16	14	14	14	11	18	15	25	16	14	19					
40	12	16	16	17	37	41	27	31	24	16	18	18	15	15					
20	28	27	21	15	20	27	28	15	28	16	10	6	33	19					
16	23	21	19	15	28	14	16	13	13	23	17	2	13	9					
28	27	45	11	20	15	15	14	14	16	15	8	50	44	17					
33	20	19	7	22	21	24	10	19	16	18	23	62	35	26					
19	21	10	18	16	19	18	21	23	16	15	17	38	52	14					

Square	Records
SO90	521
SO91	526
SO92	524
SP00	463
SP01	453
SP02	430
SP03	667
SP10	582
SP11	488
SP12	314
SP13	427
SP22	628
SP23	584
Total	**6607**

Acknowledgements

The editors wish to emphasize that this book is a cooperative effort involving many NCOS members and others, and particularly those who, with great dedication, spent five seasons in the field as volunteers collecting the data. Their names are listed on page x. In preparing the Atlas we have also called upon the help of other NCOS members. We are particularly grateful to Andy Lewis and Tony Perry, who read the manuscript, identified some mistakes and made a number of other helpful suggestions. Jill Main checked dot maps, quotations, tables and map references. If any errors remain they are ours alone.

We are grateful to Andrew Carey for generously donating his photograph of a Redstart for the front cover, and the picture of a Crossbill on page 196. Dave Morgan donated the Lesser Whitethroat picture on page 142. All other photographs are by NCOS members. The photo credits are shown individually beside the images.

The coverage of the survey recording area would have been less complete than it was without the generous assistance from various landowners and farmers in allowing access to their land.

The editors were greatly encouraged by the support of Anthony Cond, our commissioning editor at Liverpool University Press, who showed enthusiasm for the project from the beginning. We benefited greatly from expert advice and painstaking attention to detail, not only during production but while the manuscript was being prepared, from Chris Reed and Amanda Thompson of BBR. Fruitful discussions were had with David Norman, author of *Birds in Cheshire and Wirral*, which was in preparation at about the same time as this book. BTO staff members Chris Wernham and Jacquie Clark helped by providing lists of the linguistic and typographical conventions which were defined for *The Migration Atlas*, and which have guided us when writing this Atlas.

Alan Morton, the author of DMAP and DMAP Digitizer, provided prompt and efficient software support at all stages.

Sponsors

Sincere thanks are due to the following for their financial support which enabled the Atlas to be published in this form:

Geoff Bailey
Lady Bamford
Siobhan Barker and Christopher Main
The Batsford Foundation
Dr W. Bechtolsheimer
Nigel Birch
Rozza Birch
Ian Boyd
Wendy Bridgman
The Brookes family
Patrick Buxton
Andrew Cleaver
Debbie Colbourne and Julian Miles
Cotswold District Council
Ian Cox
Duncan and Rebecca Dine
Dursley Birdwatching and Preservation Society
Mark Farmer
Pat and Brian Follett
Rita Gerry
Richard and Pat Gunn
Sarah and Dennis Gornall
Mrs Heber-Percy
Tim Holland-Martin Charitable Trust
Huntsmans Quarries Limited
Tim Hutton

Mr D. Jenks
Charles and Jackie Johnson
Mick and Jo Jones
Mr S. Keswick
Gordon and Jenny Kirk
Andy Lewis
Jill Lewis
The John Spedan Lewis Foundation
Iain and Jill Main
Matthew Main
Charlotte and Jessica Nash
Alexander Nice
Chris Oldershaw
Peter Ormerod
Dave Pearce
Gill Pearce
Constance Perry
Tony and Pam Perry
John and Viv Phillips
Sarah Rouche
Severn Trent Water Limited
Mrs K.E. Sloane
Beryl Smith
Talland School of Equitation
Ian and Val Tucker
Lord and Lady Vestey
Keith White
Michael Williams
Mr C.N. Wright
and a number of anonymous donors

The John Spedan Lewis Foundation

The John Spedan Lewis Foundation was set up in 1964 in memory of the founder of the John Lewis Partnership. The focus of support is on educational projects relating to entomology, horticulture and ornithology.

The Cotswold District covers approximately 450 square miles, and includes some of the most valuable natural and historic assets in southern England, with 70% of the District within the Cotswolds AONB and another 7% within the Cotswold Water Park.

There are over 120 Sites of Special Scientific Interest, more than 260 Key Wildlife Sites, and many hundreds of acres of land under agri-environment schemes in the Cotswolds—providing great habitats for a wide variety of bird species.

With such a rich and diverse natural environment, Cotswold District Council takes seriously its role in promoting the protection and enhancement of biodiversity: for example, by supporting and funding the Gloucestershire Biodiversity Action Partnership and the Cotswold Water Park Biodiversity Action Plan; through the work of the Council's Biodiversity Officer in assessing development proposals and ensuring that opportunities are taken to provide new habitats for wildlife, as well as protecting existing ones; via the Community Strategy and in many other ways.

Severn Trent Water

Severn Trent is unique in the water industry in two respects: it covers parts of two countries (England and Wales) and has no coastline. With an area of 8,000 square miles and a land ownership of 20,000 hectares—of which half comprises the Lake Vyrnwy estate in Powys, an important RSPB reserve—the Severn Trent region is diverse. In the north lie the Upper Derwent Valley reservoirs, famous as the training ground for the Dambusters, whilst in the south there are sewage treatment works along the Severn and the Wye. In the west Llyn Clywedog nestles in the Cambrian Mountains and to the east Stanford Reservoir sits in rolling Northamptonshire farmland.

With 6,000 operational sites, the bird-life is equally diverse, and larger reservoirs such as Carsington and Ogston in the Derbyshire Dales, and Draycote Water near Rugby, have recorded over 200 species since they were built. A major success for Severn Trent has been a unique project to reintroduce Black Grouse to the Peak District. Extinct in 2000, Black Grouse are now thriving on the moorlands and the project is being replicated in other European countries.

In the North Cotswolds Severn Trent has a number of rural sewage works, where small treatment reedbeds provide winter roost sites for Reed Buntings and other small passerines. One site, just outside the NCOS recording area, has breeding Reed Warblers, as well as occasional wintering Snipe and Water Rails.

In supporting this major piece of ornithological work, Severn Trent Water acknowledge the tremendous voluntary effort put in by members of NCOS over the years, and wish them every success in the future.

As a Cotswolds company with more than 80 years of history and a long commitment to wildlife, we are delighted that we are a partner in this publication.

For those who may not know of us, we are a local business in an industry dominated by national and multinational operators. Our base is at Naunton and we supply most of our natural and recycled aggregates, building stone, ready-mixed concrete and roofing tiles in the area around Bourton-on-the-Water, Moreton-in-Marsh and Stow-on-the-Wold. Although relatively small we are an important employer in the Cotswolds.

We have family origins and take very seriously our commitments to our neighbours and friends in the local community. That means not just minimizing our day-to-day environmental impacts but restoring our sites with pride. We regularly open our gates for open days and often welcome smaller groups and school parties. Many people come not just to see what we do but to enjoy the geological Site of Special Scientific Interest on the edge of our site.

We have been working with Plantlife and English Nature for some years now on a site-specific biodiversity action plan to preserve and enhance plant, bird and mammal species. We already look after BAP or endangered plant species such as Cotswold Pennycress, Basil, Thyme, Corn Gromwell and Prickly Poppy. In addition, we provide a home for **RED**-listed bird species such as Corn Bunting, Linnet, Skylark, Yellowhammer, and the **AMBER**-listed Sand Martin.

There is greater potential still to be realized and we are looking forward to further improving the habitats in and around our quarry. We hope you will come to see what we are achieving—and that you enjoy this Atlas in the meantime.

Further reading and references

Simple descriptions of the bird species mentioned in this Atlas may be found in any good Field Guide, such as the *Collins Bird Guide* (Mullarney *et al* 1999). For more details, the standard reference work is the nine-volume handbook by Cramp *et al* (1977–94). Flegg (2004) provides a readable account of the migrational and other seasonal movements of the birds of Britain and Ireland derived from ringing data. For all the details see *The Migration Atlas* (Wernham *et al* 2002). A convenient account of the fortunes of the birds of Britain and Ireland at the start of the 2003–07 survey is given by Mead (2000). Swaine (1982) gives a corresponding account of the local situation just before the 1983–87 survey.

Baillie, S.R., Marchant, J.H., Crick, H.Q.P., Noble, D.G., Balmer, D.E., Barimore, C., Coombes, R.H., Downie, I.S., Freeman, S.N., Joys, A.C., Leech, D.I., Raven, M.J., Robinson, R.A. & Thewlis, R.M. (2007), *Breeding Birds in the Wider Countryside: their conservation status 2007*. BTO Research Report 487 (British Trust for Ornithology, Thetford); http://www.bto.org/birdtrends2007.

Cramp, S. & Perrins, C.M. (eds.) (1977–94), *The Birds of the Western Palearctic*, vols. I–IX (Oxford University Press, Oxford).

Dudley, S.P., Gee, M., Kehoe, C. & Melling, T.M. (2006), The British Ornithologists' Union Records Committee (BOURC), The British List: A Checklist of Birds of Britain (7th edition), *Ibis* 148 (3), 526–63; http://www.bou.org.uk/recbrlst.html.

Eaton, M.A., Balmer, D., Burton, N., Grice, P.V., Musgrove, A.J., Hearn, R., Hilton, G., Leech, D., Noble, D.G., Ratcliffe, N., Rehfisch, M.M., Whitehead, S. & Wotton, S. (2008), *The State of the UK's Birds 2007* (RSPB, BTO, WWT, CCW, EHS, NE & SNH, Sandy); http://www.bto.org/research/pop_trends/state_uk_birds.htm.

Field, R.H. & Gregory, R.D. (1999), *Measuring population changes from the Breeding Bird Survey*. BTO Research Report 217 (British Trust for Ornithology, Thetford); http://www.bto.org/bbs/results/bbsreport.htm.

Flegg, J. (2004), *Time to Fly: Exploring Bird Migration* (British Trust for Ornithology, Thetford).

Gibbons, D.W., Reid, J.B. & Chapman, R.A. (1993), *The New Atlas of Breeding Birds in Britain and Ireland 1988–1991* (T. & A.D. Poyser, London).

Gregory, R.D., Wilkinson, N.I., Noble, D.G., Robinson, J.A., Brown, A.F., Hughes, J., Procter, D.A., Gibbons, D.W. & Galbraith, C.A. (2002), The population status of birds in the United Kingdom, Channel Islands & the Isle of Man: an analysis of conservation concern 2002–2007, *British Birds* 95, 410–50.

Hagenmeijer, W.J.M. & Blair, M.J. (eds.) (1997), *The EBCC Atlas of European Breeding Birds: Their Distribution and Abundance* (T. & A.D. Poyser, London).

Lack, P. (compiler) (1986), *The Atlas of Wintering Birds in Britain and Ireland* (T. & A.D. Poyser, Calton).

Marchant, J.H., Hudson, R., Carter, S.P. & Whittington, P.A. (1990), *Population Trends in British Breeding Birds* (British Trust for Ornithology, Tring).

Marchant, J.H., Freeman, S.N., Crick, H.Q.P. & Beaven, L.P. (2004), The BTO Heronries Census of England and Wales 1928–2000: new indices and a comparison of analytical methods, *Ibis* 146, 323–34; http://www.bto.org/birdtrends2007/heronries.htm.

Mead, C. (2000), *The state of the nations' birds* (Whittet Books, Stowmarket).

Miller, G. (2000), *Gloucestershire Biodiversity Action Plan*; www.gloucestershirebap.org.uk.

Mullarney, K., Svensson, L., Zetterström, D., & Grant, P.J. (1999), *Collins Bird Guide* (HarperCollins, London).

Robertson, C.J.R., Hyvőnen, P., Fraser, M.J. & Pickard, C.R. (2007), *Atlas of Bird Distribution in New Zealand 1999–2004* (The Ornithological Society of New Zealand, Wellington).

Robinson, R.A. (2005), *BirdFacts: profiles of birds occurring in Britain & Ireland* (v1.21, June 2008). BTO Research Report 407 (British Trust for Ornithology, Thetford); http://www.bto.org/birdfacts.

Sharpe, C. (principal editor) (2007), *Manx Bird Atlas: an Atlas of Breeding and Wintering Birds on the Isle of Man* (Liverpool University Press, Liverpool).

Sharrock, J.T.R. (1976), *The Atlas of Breeding Birds in Britain and Ireland* (T. & A.D. Poyser, Calton).

Swaine, C.M. (1982), *Birds of Gloucestershire* (Alan Sutton Publishing, Gloucester).

Wernham, C.V., Toms, M.P., Marchant, J.H., Clark, J.A., Siriwardena, G.M. & Baillie, S.R. (eds.) (2002), *The Migration Atlas: movements of the birds of Britain and Ireland* (T. & A.D. Poyser, London); http://www.bto.org/research/projects/atlas.htm.

Wright, M.R., Williams, M.F., Alexander, K.N.A., Dymott, P.H., Hookway, R.C., Owen, S.M., Robinson, M.F., Ralphs, I.L. & Stainton, M.S. (1990), *An Atlas of Cotswold Breeding Birds* (Drinkwater, Shipston-on-Stour).

Index of species

Page numbers in bold type refer to the relevant species account.